BIOCHEMICAL SOCIETY SYMPOSIA

No. 73

TRANSCRIPTION

BIOCHEMICAL SOCIETY SYMPOSIUM No. 73
held at Imperial College, London, April 2005

Transcription

ORGANIZED AND EDITED BY
S.G.E. ROBERTS, R.O.J. WEINZIERL
AND R.J. WHITE

PORTLAND PRESS

Published by Portland Press,
Third Floor, Eagle House, 16 Procter Street,
London WC1V 6NX, U.K.
on behalf of The Biochemical Society
Tel: (+44) 20 7280 4110;
e-mail: editorial@portlandpress.com
http://www.portlandpress.com

ISBN-13: 978 1 85578 163 4
ISBN-10: 1 85578 163 8
ISSN (print): 0067-8964
ISSN (online): 1744-1439

British Library Cataloguing in Publication Data
A catalogue record for this book is available from the British Library

Typeset by Portland Press Ltd
Printed in Great Britain by Cambridge Printing

Contents

Preface

The Biochemical Society's Annual Symposium, Transcription UK, took place at Imperial College London in April 2005. This book brings together the written contributions from many of the presenters and provides a fascinating snapshot of the current state of a wide range of transcription-related topics, ranging from molecules to cells.

Although the major focus of the conference was transcription in eukaryotic systems, the book starts with a contribution from Nora Miroslavova and Steve Busby, who discuss the role of various bacterial promoter elements in directing specific transcription initiation. The availability of high-resolution structures of bacterial RNA polymerases, combined with the wealth of biochemical and genetic data available, allows a functional dissection of the bacterial initiation complex with unprecedented resolution and has led to a detailed understanding of the design features that natural promoters exploit to yield physiologically relevant transcription levels.

The next contribution by Douglas Higgs and colleagues reminds us that everything is much more complicated in eukaryotic systems, where gene regulatory information is only partially encoded by DNA elements. The authors describe the events underlying the transcriptional programme resulting in globin gene activation. These events include the binding of transcription factors and a number of epigenetic modifications, ranging from DNA methylation to repositioning of particular genes within the nucleus. It is likely that such mechanisms are common to many genes involved in lineage commitment, differentiation and development.

In the next chapter, David Heery and colleagues also deal with the regulation of gene activity during haematopoiesis, but highlight the pathological roles of various fusion proteins resulting from chromosomal translocations. A fusion of MOZ (monocytic leukaemia zinc finger protein) to nuclear receptor co-activators interferes with the activities of several other transcription factors that are required for the differentiation of haematopoietic tissue. The knowledge of such mechanisms will provide many opportunities and targets for future therapeutic interventions.

Chapters 4 (by Patrick Cramer) and 5 (by Robert Weinzierl and colleagues), return to some of the most basic aspects of the transcriptional machinery — the structure and function of RNA polymerases and their associated basal factors. Cramer provides an overview of recent progress with X-ray crystallographic studies on yeast RNA polymerase II. New structure-based insights have been obtained from investigating an RNA polymerase II-elongation factor complex, leading to the concept of a tuneable active site whose catalytic properties can be modulated through external factors. This idea is taken up by Weinzierl and colleagues who present an analysis of a basal factor domain that is present in close vicinity to the active site of RNA polymerase and may determine start site selection and other catalytic parameters during transcription initiation.

We then move on to the packaging of genes in higher order chromatin and structure and various aspects of nuclear architecture that impinge on transcriptional regulation. First, Nick Gilbert and Wendy Bickmore describe results of a study based on a novel experimental approach that sheds a new light on the packaging of actively transcribed and repressed genes into chromatin fibres with various degrees of compaction. In contrast to the established view, they argue that gene density, rather than transcriptional status, are the key determinants of the degree of chromatin compaction.

The arrangement of genes within nuclei and their transcriptional activity is also addressed by Peter Cook and colleagues. Their transcription factory model has many interesting functional consequences. They argue that the location of genes relative to transcription factories has kinetic ramifications in keeping genes transcriptionally active or repressed. Some of these factories may even become enriched with particular transcription factors, so they become dedicated to the transcription of those units that require those factors.

Brian McStay and colleagues discuss the role of upstream binding factor (UBF) in setting up what is arguably the ultimate transcription factory: the nucleolus. Large arrays of heterologous UBF-binding sequences, when integrated at ectopic sites on human chromosomes, are morphologically similar to nucleolus organizer regions, demonstrating that UBF plays a major role in organizing the chromatin structure of ribosomal RNA-encoding genes.

Richard Reece and colleagues turn our attention to another important area of differential gene expression, which involves the ability of cells to alter their transcription programme in response to changes in metabolite concentrations. In the classic model system, the yeast GAL genetic switch, direct interactions between individual metabolites and regulators of RNA polymerase II-mediated transcription are involved in signal transduction processes. A similar process also applies to at least several amino acid and nucleotide metabolites. This mode of regulation probably applies to many unicellular eukaryotes that lack the metabolic homoeostasis systems present in higher, multicellular, eukaryotes.

Sofia Aligianni and Patrick Varga-Weisz highlight the key role of DNA replication in the reorganization of gene expression through changes in chromatin structure and other epigenetic markers. This is a particularly significant process during early development when the functional interactions between the replication and transcriptional machineries are likely to be most decisive in maintaining an elegant balance between epigenetic maintenance and change.

Tom Owen-Hughes deals with the process of nucleosome rearrangement, which underlies many of the changes in gene expression patterns. Recent advances in the understanding of conformational transitions of nucleosomes, and their consequences in altering chromatin structures and dynamics are highlighted. Nucleosome repositioning and exchange mechanisms are affected by various combinations of the post-translational modification state of histone tails, the presence of histone variants and the activity of chromatin remodeling complexes.

Shen-Hsi Yang and Andrew Sharrocks take us further into the regulatory network that links cytoplasmic and nuclear signal transduction pathways. The MAPK (mitogen-activated protein kinase) pathways are some of the best-studied signalling routes to the nucleus, resulting in differential phosphorylation and sumoylation of transcriptional regulators. The transcriptional activation potential of some factors, such as Elk-1, is controlled by the functional interaction between the MAPK and SUMO (small ubiquitin-related modifier) pathways.

Alexandre Akoulitchev and colleagues discuss how non-coding RNAs, including snRNAs (small nuclear RNAs), affect transcription initiation and elongation processes. TFIIH (transcription factor IIH), a key basal factor, appears to associate specifically with U1 snRNA (a component of the RNA splicing machinery). This finding provides more evidence for a strong regulatory link between the transcriptional and post-transcriptional machineries. Its also provides an interesting theoretical framework for investigating the molecular mechanisms that are responsible for determining the coupling between *in vitro* promoter strength and the promoter proximal position of the first intron.

Bob White and colleagues address another disease-related aspect of gene regulation, the function of the proto-oncogene product c-Myc. The inappropriate expression of c-Myc is estimated to contribute to one-seventh of all cancer deaths in the U.S.A., but its mechanism of action is still unclear. Studies of c-Myc had, until recently, concentrated exclusively on its ability to regulate transcription of protein-encoding genes, but most of the observed effects were relatively weak. Here, the authors show that c-Myc strongly stimulates RNA polymerase I and RNA polymerase III transcription systems. The production of the components of the protein synthetic machinery in the cell can be rate-limiting, thus the potent effect of c-Myc on rRNA synthesis may contribute substantially to its growth-promoting activity.

Làszlò Tora and colleagues focus on the molecular basis of polyglutamine diseases. They describe their discovery that ataxin-7, a protein containing polyglutamine stretches that can be expanded under pathological conditions, is an integral component of the chromatin-modifying SAGA complex. SAGA acts as a transcriptional co-activator/adapter that can acetylate nucleosomes and create a chromatin environment suitable for transcription of specific genes. The presence of polyglutamine expanses in ataxin-7 does not interfere with its incorporation into SAGA and possibly result in misregulated acetylation activities.

Kirsteen Campbell and Neil Perkins provide us with an overview of another medically important transcription factor, NF-κB, with an amazingly diverse functional range. NF-κB is involved in immune, inflammatory, stress, proliferative and apoptotic responses in most cell types and its activity can be induced by a variety of mechanisms. The complexity of the NF-κB response is due to the specificity and timing of genes regulated by NF-κB under particular circumstances and its tight integration with other signaling pathways in the cell.

The interaction of p53 with an array of cofactors is the theme of the contribution by Amanda Coutts and Nicholas La Thangue. Similar to c-Myc, mutations affecting p53 are frequently found in human cancer cells. p53 is involved in the detection of DNA damage and the orchestration of the cellular response to such damage. The cofactors exerting effects on p53 include other transcription factors and chromatin-modulating enzymes. Some of these cofactors are expressed in a tissues-specific manner which may contribute to the differing response of p53 to damaging agents.

In the next chapter, Stefan Roberts describes yet another disease state, Wilms' tumour, which is caused by mutation in the transcription factor WT1. Just like p53, the functional properties of WT1 are highly dependent on a series of cofactors that turn the protein into either an activator or repressor. WT1 enhances the DNA binding and transcriptional activation potential of p53, affects the transcription of genes involved in cholesterol biosynthesis and plays a tissue-specific role in the development of genito-urinary organs. There is clearly still a lot to learn about WT1 before a common theme about its mode of action emerges.

Next, Jackie Russell and Joost Zomerdijk summarize the current knowledge of the molecular composition and functional characterization of the components of the RNA polymerase I transcriptional machinery. The high levels of rRNA production by this system is tightly linked to the rate of cell growth. A detailed understanding of all the regulatory features of the RNA polymerase I system is thus required.

In the penultimate chapter, Michael Green and colleagues, describe a major technical breakthrough aimed at the identification of the interactions occurring between gene-specific transcription factors and components of the basal transcriptional machinery. This is an area that has created a lot of controversy because many *in vitro* approaches to detect protein–protein interactions are potentially artifact-prone. The method described here depends on FRET (fluorescence resonance energy transfer) analysis in living yeast cells. FRET can be used to measure the spatial proximity between biological macromolecules and can thus be used to identify interaction targets, but also the spatial relationship between subunits within or between multiprotein complexes.

In the final contribution, we return once more to promoter core motifs, but this time in eukaryotes. Petra Gross and Thomas Oelgeschläger emphasize the lack of uniform consensus elements surrounding the RNA polymerase II start sites; in fact, many of the best known elements, such as TATA boxes, are only found in a small subset of promoters. The authors review the functional properties and contribution to specificity of other motifs, such as the initiator DPE (downstream promoter element), and relate these elements to the components of the basal machinery that recognize them.

We thank all the speakers and participants who made the Symposium such a success, all the authors for sending us their contributions and Mike Cunningham and his colleagues at Portland Press for their efforts in the editing and compiling this book.

Stefan Roberts, Robert Weinzierl
and Bob White

Abbreviations

AD1	activation domain 1
AdML	adenovirus major late
AML	acute myeloid leukaemia
ARF	alternative reading frame
ARPP	acidic ribosomal phosphoprotein
ASE	antisense for ERCC
Asf	anti-silencing function
ATM	ataxia telangiectasia mutated
ATR	ataxia telangiectasia-related
BAC	bacterial artifical chromosome
BASP	brain acid-soluble protein
BMZF	bone-marrow zinc finger
BRE	TFIIB recognition element; B-recognition element
3C	chromosome conformation capture
CAF	chromatin assembly factor
CAK	CDK-activating kinase
CARM	co-activator-associated arginine methyltransferase
CAST	CD3ε-associated signal transducer
CBP	CREB-binding protein
CDK	cyclin-dependent kinase
ChIP	chromatin immunoprecipitation
Chk	checkpoint kinase
CID	CTD-interacting domain
CREB	cAMP-response-element-binding protein
CRP	cAMP receptor protein
CTD	C-terminal domain
DAPI	4′,6-diamidino-2-phenylindole
DBTSS	Database of Transcriptional Start Sites
DDS	Denys Drash syndrome
DHS	DNase I-hypersensitive site
DMR	differentially methylated region
DPE	downstream promoter element
DR	death receptor
ECFP	enhanced cyan fluorescent protein
EPD	Eukaryotic Promoter Database
ER	oestrogen receptor
ERK	extracellular-signal-regulated kinase
ERRC	excision repair cross-complementary enzyme gene
EYFP	enhanced yellow fluorescent protein

FHL2	four-and-a-half LIM domain protein
FISH	fluorescence *in situ* hybridization
FRET	fluorescence resonance energy transfer
GADD	growth arrest and DNA-damage-inducible protein
GAPDH	glyceraldehyde-3-phosphate dehydrogenase
GFP	green fluorescent protein
GST	glutathione S-transferase
GTF	general transcription factor
H3ac	acetylated histone H3
HAT	histone acetyltransferase
HDAC	histone deacetylase
HMG	high-mobility group
HMGB	high-mobility-group box
HP1	heterochromatin protein 1
HSC	haematopoietic stem cell
Hsp	heat-shock protein
IAP	inhibitor of apoptosis
IGF	insulin-like growth factor
IGS	intergenic spacer
IκB	inhibitor of NF-κB
IκBSR	super-repressor IκB
IKK	IκB kinase
IL-1	interleukin-1
INR	initiator
IRS-1	insulin receptor substrate-1
ISWI	imitation switch
JMY	junction-mediating and regulatory protein
JNK	c-Jun N-terminal kinase
LCR	locus control region
LINE	long interspersed nuclear element
LMP	latent membrane protein
LPS	lipopolysaccharide
MAPK	mitogen-activated protein kinase
MBII	Myc box II
MCM	mini-chromosome maintenance
Mdm2	murine double minute 2
MEKK	mitogen-activated protein kinase kinase kinase
MEL	mouse erythroleukaemia
MIS	Mullerian inhibiting substance
MORF	MOZ-related factor
MOZ	monocytic leukaemia zinc finger protein
MSE	muscle-specific enhancer
MTE	motif ten element
mTOR	mammalian target of rapamycin
MYST	MOZ, Ybf2/Sas3, Sas2, and TIP60
NAP-1	nucleosome assembly protein-1
NF-E2	nuclear factor-E2

NF-κB	nuclear factor-κB
NMD	nonsense-mediated decay
NOR	nucleolar organizer region
NPC	nuclear pore complex
NR	nuclear receptor
ODC	ornithine decarboxylase
OPT domain	Oct1/PTF/transcription domain
ORC	origin recognition complex
PAF49	polymerase I-associated factor of 49 kDa
Par-4	prostate apoptosis response-4
P/CAF	p300/CBP-associated factor
PCNA	proliferating-cell nuclear antigen
PHD	plant homeodomain
PIAS	protein inhibitor of activated STAT
PIC	pre-initiation complex
PML	promyelocytic leukaemia
p[NH]ppA	adenosine 5′-[β,γ-imido]triphosphate
Pol	RNA polymerase
PPAR	peroxisome-proliferator-activated receptor
PRMT	protein-arginine N-methyltransferase
P-TEF	positive transcription elongation factor
PTF	PSE (proximal sequence element)-binding transcription factor
RANTES	regulated upon activation, normal T-cell expressed and secreted
RAR	retinoic acid receptor
RHD	Rel homlogy domain
RNAP	RNA polymerase (*also* Pol)
SAGA	Spt/Ada/Gcn5 acetyltransferase
SCA	spinocerebellar ataxia
SCL	stem-cell leukaemia
STAGA	SPT3/TAF9/GCN5 acetyltransferase
STAT	signal transduction and activators of transcription
Strap	serine/threonine-kinase-receptor-associated protien
SUMO	small ubiquitin-related modifier
SV40	simian virus 40
TAD	transcriptional activation domain
TAF	TBP-associated factor
TBP	TATA-binding protein
TCF	ternary complex factor
TFB	transcription factor B
TFII	transcription factor II
TFTC	TBP-free TAF complex
TIC	TAF- and INR-dependent cofactor
TIF-IB	transcription initiation factor-IB
TIF	transcription intermediary factor
TNF	tumour necrosis factor

TPR	tetratricopeptide
TRAF	TNFα-receptor-associated factor
TRRAP	transformation/transactivation domain-associated protein; *also* transformation/transcription domain-associated protein
UAS	upstream activating sequence
UBF	upstream binding factor
UCE	upstream control element
WICH	WSTF/ISWI/chromatin-remodelling factor
WSTF	Williams-syndrome transcription factor
WT1	Wilms' tumour 1
WTIP	WT1-interacting protein

Biochem. Soc. Symp. **73**, 1–10
(Printed in Great Britain)

1

Investigations of the modular structure of bacterial promoters

Nora S. Miroslavova and Stephen J.W. Busby[1]

School of Biosciences, The University of Birmingham, Birmingham B15 2TT, U.K.

Abstract

Bacterial RNA polymerase holoenzyme carries different determinants that contact different promoter DNA sequence elements. These contacts are essential for the recognition of promoters prior to transcript initiation. Here, we have investigated how active promoters can be built from different combinations of elements. Our results show that the contribution of different contacts to promoter activity is critically dependent on the overall promoter context, and that certain combinations of contacts can hinder transcription initiation.

Overview

All bacterial genes are transcribed by a multisubunit RNA polymerase that contains two large subunits, β and β′, and two copies of the smaller α subunit. Biochemical, genetic and structural studies have shown that the β and β′ subunits form a cleft that contains the enzyme's active site for RNA polymerization, and that the N-terminal domains of the two α subunits are responsible for holding the assembly together [1]. Although the RNA polymerase $\alpha_2\beta\beta'$ assembly (known as the core enzyme) is competent to make RNA, it is unable to initiate transcription. In order to recognize promoters and to begin transcription, core RNA polymerase must acquire a σ subunit (resulting in the formation of RNA polymerase holoenzyme) [2]. All bacteria contain σ subunits, which function by binding to the surface of RNA polymerase core enzyme, thus providing determinants for the recognition of specific promoter elements. Following promoter recognition, the bound σ subunit choreographs a series of isomerizations that lead to promoter DNA unwinding and entry of the DNA template into the enzyme active site, where transcript formation is initiated. Most σ subunits contain four domains that are connected by flexible linkers [3]. Domains 2, 3 and 4 carry determinants for recognition of different promoter

[1]To whom correspondence should be addressed (email s.j.w.busby@bham.ac.uk).

elements, namely the −10, extended −10 and −35 elements respectively (Figure 1A). Structural studies with *Thermus aquaticus* RNA polymerase holoenzyme in complexes with DNA substrates [4,5] have resulted in a clear view of how the interactions between σ and different promoter elements are organized during promoter recognition (Figure 1B).

Most bacteria contain a few thousand genes that are expressed from 1000–2000 promoters. The best-studied example is *E. coli* K-12, which appears typical of many bacteria, in that the expression of its genes is very tightly regulated, with changes controlled by alterations in the environment. In most conditions, a single *E. coli* cell contains 1000–2000 molecules of RNA polymerase, and thus effective regulation requires controlled distribution of the polymerase between the promoters in its genome [6]. This distribution is mainly a result of the different base sequence determinants found at the promoters, although an important role is also played by supercoiling and the proteins that interact with the bacterial chromosome to fold and organize it (these are known as nucleoid-associated proteins). Changes in the distribution of RNA polymerase between promoters are controlled by the production of different σ factors, and by transcription factors that bind at or near specific promoters and either activate or repress those promoters in response to specific metabolic signals [7]. Although most studies of the regulation of transcription initiation in *E. coli* have focused on promoter sequence determinants, σ factors and transcription factors, it is now clear that other factors, such as small ligands, small effector proteins that contact RNA polymerase without contacting promoter DNA, and changes in the bacterial folded chromosome structure, also play a key role [8,9].

Elements of promoter recognition

In this article, we focus on the primary determinants of promoter activity in *E. coli*, which are the initial contacts between RNA polymerase and different promoter elements. Our study is concerned with RNA polymerase containing the major σ^{70} subunit, though the take-home messages apply to most σ factors. It has long been known [2,10] that promoter −10 elements are recognized by a helix in Region 2.4 of σ^{70} Domain 2, and that a helix–turn–helix in Domain 4 is responsible for recognition of promoter −35 elements (Figure 1). The consensus −10 and −35 elements are established as the hexamer sequences TATAAT and TTGACA respectively. More recent work has shown that a helix in Domain 3 of σ^{70} contacts the extended −10 element (consensus TRTGN), located immediately upstream of the −10 element [3,11]. Finally, the RNA polymerase α subunits play a key role [12,13]. Each α subunit carries an independently folded C-terminal domain (residues 250–329), joined to the α subunit N-terminal domain by a flexible linker. The two C-terminal domains of the two RNA polymerase α subunits can bind to a 20 bp DNA element, known as the UP element, located at promoters immediately upstream of the −35 element. The UP element consists of two half-elements, each recognized by one α subunit C-terminal domain (Figure 1A). A consensus UP element has been determined, and it has been shown that, at certain promoters, the interactions of RNA polymerase with UP

A)

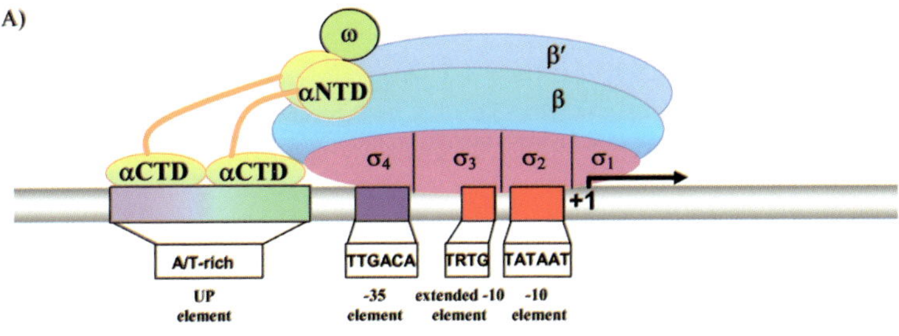

B)

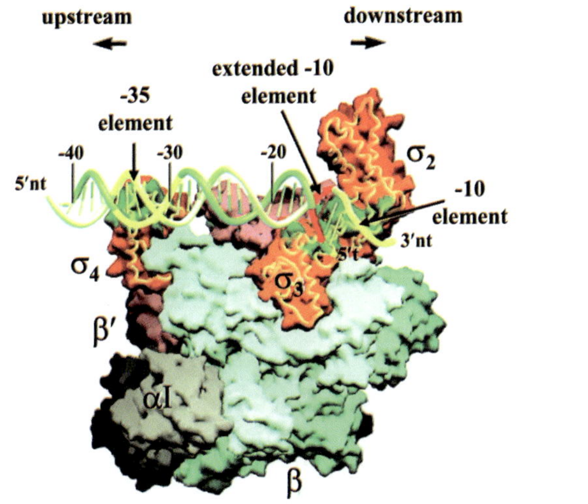

Figure 1 RNA polymerase–promoter interactions. (**A**) The diagram illustrates the subunit structure of RNA polymerase holoenzyme, showing the disposition of the different subunits as the enzyme makes contact with a bacterial promoter. Each α subunit consists of an N-terminal domain (αNTD) and a C-terminal domain (αCTD), joined by a linker. The σ subunit is divided into four segments to illustrate the four domains. The locations of the promoter UP, −35, extended −10 and −10 elements are indicated, together with consensus sequences. The diagram illustrates the different interactions between RNA polymerase and promoter elements. The small 91-amino-acid ω subunit acts as a chaperone for the β′ subunit [24]. (**B**) Model of RNA polymerase holoenzyme docking to promoter DNA, based on structural studies with *Thermus aquaticus* RNA polymerase from Seth Darst and co-workers [4,5]. [Reproduced with permission from Murakami, K.S., Masuda, S., Campbell, E.A., Muzzin, O. and Darst, S.A. (2002) Science **296**, 1285–1290. Copyright 2002 AAAS.] The locations of domains 2, 3 and 4 of the σ subunit (coloured orange) and their interactions with the −10, extended −10 and −35 elements are shown. Note that αCTD and σ domain 1 are not seen in the *T. aquaticus* RNA polymerase structure.

A)

NM513

-97 ↓ anti-UP no -35 element

GAATTCGCATAAAAAACGGCTAAATTCTTGTGTAAACGGACTGCAGTGGTACCTAGGAGATCTACTTTTCGCATC

+1 ↓ -10 element +53 ↓

TTTTTGTACC**TATAAT**TATTTC**A**TACCATAAGCCTAATGGAGCGAATTATGAGAGTTCTGGTTACCGCCAAGCTT

no extended -10 element

B)

Promoter Name	Promoter Sequence	Promoter Activity
	-55 ↓ -35 element -10 element +1 ↓	
NM513	GCAGTGGTACCTAGGAGATCTACTTTTCGCATCTTTTTGTACC**TATAAT**TATTTC**A**	20
NM535	GCAGTGGTACCTAGGAGATC**TTGACA**TCGCATCTTTTTGTACC**TATAAT**TATTTC**A**	6251
NM511	GCAGTGGTACCTAGGAGATCTACTTTTCGCATCTTTTT**TATG**C**TATAAT**TATTTC**A**	1959
NM501	CAC**TAATTTATTCCATGTCACACTT**TTCGCATCTTTTTGTACC**TATAAT**TATTTC**A**	1185
	DNA site for CRP extended -10 element	

C)

Promoter Name	Promoter Sequence	Promoter Activity
	-55 ↓ -10 element +1 ↓	
NM500	CAC**TAATTTATTCCATGTCACACTT**TTCGCATCTTTTT**TATG**C**TATAAT**TATTTC**A**	2645
	DNA site for CRP extended -10 element	

Figure 2 Building a promoter from different elements. (**A**) The complete base sequence of the starting 144 bp NM513 EcoRI–HindIII fragment, which carries a consensus −10 element (bold type) but non-consensus extended −10, −35 and UP elements. The locations of the different elements are indicated by horizontal lines. The transcription start site, +1, is in bold and underlined. (**B**) Partial base sequences of the starting NM513 promoter and three derivatives, carrying different promoter elements, are shown from positions −55 to +1. NM535 carries consensus −35 and −10 elements, shown in bold type. NM511 carries consensus extended −10 and −10 elements (bold). NM501 carries a consensus −10 element and a DNA site for CRP (bold). All the promoters carry identical sequences downstream of the transcription start site. Each promoter was fused to the *lacZ* gene (in the expression vector pRW50 [18]), and its ability to initiate transcription was deduced from measurements of β-galactosidase activities in the Δ*lac E. coli* strain MC4100 growing exponentially. Cells were grown at 37°C in Luria–Bertani medium, supplemented with tetracycline (35 μg/ml). Activities are expressed as nmol of *o*-nitrophenyl β-D-galactoside hydrolysed·min^{-1}·mg dry cell mass^{-1} (Miller units) and are listed in the final column. (**C**) Partial base sequence from positions −55 to +1 of the NM500 promoter, which carries a DNA site for CRP, and consensus extended −10 and −10 elements (shown in bold). Expression from this promoter was deduced from the β-galactosidase activity measurement shown in the final column.

A) Promoters with a consensus -10 hexamer

Promoter Name	Promoter Sequence	Promoter Activity
	-50 ↓ -35 element -10 element +1 ↓	
NM501	ATTTATTCCATGTCACACTTTTCGCATCTTTTTGTACC**TATAAT**TATTTCA	7
NM531	ATTTATTCCATGTCA**TTGACA**TCGCATCTTTTTGTACC**TATAAT**TATTTCA	9594
NM500	ATTTATTCCATGTCACACTTTTCGCATCTTTTT**TATG**C**TATAAT**TATTTCA	1623
NM530	ATTTATTCCATGTCA**TTGACA**TCGCATCTTTTT**TATG**C**TATAAT**TATTTCA	6438
	extended -10 element	

B) Promoters with 4/6 match to the consensus -10 hexamer

Promoter Name	Promoter Sequence	Promoter Activity
	-50 ↓ -35 element -10 element +1 ↓	
NM101	ATTTATTCCATGTCACACTTTTCGCATCTTTTTGTACC**TATGGT**TATTTCA	0
NM131	ATTTATTCCATGTCA**TTGACA**TCGCATCTTTTTGTACC**TATGGT**TATTTCA	4367
NM100	ATTTATTCCATGTCACACTTTTCGCATCTTTTT**TATG**C**TATGGT**TATTTCA	1077
NM130	ATTTATTCCATGTCA**TTGACA**TCGCATCTTTTT**TATG**C**TATGGT**TATTTCA	9164
	extended -10 element	

Figure 3 Reduction of promoter activity due to too many consensus promoter elements. (**A**) Partial base sequences from positions −50 to +1 of four promoters containing a consensus −10 element, with or without a consensus −35 hexamer, and with or without a consensus extended −10 element. Consensus sequence elements are indicated by bold type. Each promoter was fused to the *lacZ* gene, and its ability to initiate transcription was deduced from measurements of β-galactosidase activities in a Δ*crp* derivative of *E. coli* strain MC4100, growing exponentially in Luria–Bertani medium. Activities are expressed as nmol of *o*-nitrophenyl β-D-galactoside hydrolysed·min^{-1}·mg dry cell mass^{-1} (Miller units) and are listed in the final column. (**B**) Derivatives of the four promoters in (**A**) in which the consensus −10 hexamer, TATAAT, was replaced with TATGGT. Activities were assayed exactly as in (**A**).

element sequences can improve promoter activity by a factor of 10-fold or much more [14,15].

Our strategy was to investigate how different combinations of promoter elements could make functional promoters. The starting point was the EcoRI–HindIII fragment illustrated in Figure 2(A) that carries the NM513 promoter. This fragment was derived from the *E. coli* galactose operon promoter region [16] which had been engineered to carry a consensus −10 hexamer element, TATAAT, but no consensus extended −10 or −35 element and no UP element, so that it possessed little or no promoter activity. Our aim was to introduce different combinations of promoter elements and to measure the resulting promoter activities. Figure 2(B) illustrates the NM535 promoter, with an added consensus −35 element, and the NM511 promoter, with a consensus extended −10 element. We also constructed the NM501 promoter, which carries a DNA site for the transcription activator CRP (cAMP receptor protein). The reason for this was to show that promoter activity could be created by the intervention of *trans*-acting factors as well as by the addition of *cis*-acting promoter elements. CRP is a well understood bacterial transcription activator [17], and thus we placed a DNA site for CRP at a location where bound CRP would be competent to activate transcription.

To measure the activity of the new promoter constructions, the different EcoRI–HindIII fragments were transferred to plasmid pRW50, a promoter-probe vector that places the *lac* genes under the control of cloned promoters [18]. The resulting pRW50 derivatives were then transformed into the Δ*lac E. coli* K-12 strain MC4100, and β-galactosidase activities were measured as an index of the activity of the cloned promoter (as described in [18]). Data in Figure 2(B) show that, as expected, the starting NM513 promoter had very low activity (expression was close to background levels), whereas the presence of a consensus −35 element (in NM535), an extended −10 element (in NM511) or bound CRP (in NM501) created promoter activity.

Promoter activity: more can mean less

Figure 2(C) illustrates the NM500 derivative of the NM513 promoter, which carries both an extended −10 element and a DNA site for CRP. The activity of the NM500 promoter is greater than the individual activities of the NM511 and NM501 promoters, showing that an extended −10 element can work together with CRP.

Figure 3(A) illustrates an experiment to investigate whether, in the context of a promoter with a consensus −10 element, an extended −10 element can function positively with a consensus −35 element. The data show that the addition of an extended −10 element to a −35 element in the NM530 promoter caused a decrease in promoter activity. Thus we conclude that too many strong contacts between a promoter and RNA polymerase may hinder rather than help a promoter, presumably because the bound RNA polymerase becomes anchored at the promoter and is less able to escape. Support for this point of view comes from the experiment illustrated in Figure 3(B), in which the −10 element was altered from the consensus, TATAAT, to TATGGT. Thus the activity of the NM130 promoter carrying both a consensus −35 element and an extended −10 element was higher than the activities of the NM131 or NM100 promoters that carry either one element or the other.

UP element contributions to promoter activity are contingent on the −35 element

In the promoters illustrated in Figure 2, the DNA sequence just upstream of the −35 region was chosen so as not to resemble any known UP element.

Indeed, in the starting NM513 promoter, we used a DNA sequence that had been characterized previously as an 'anti-UP' sequence, i.e. the sequence least likely to enhance promoter activity due to binding of the C-terminal domain of the RNA polymerase α subunit [15]. In order to measure the effects of introducing a UP element, a 20 bp sequence, selected previously for its optimal UP element function [15], was cloned upstream of the –35 region at the NM511 or NM535 promoters (Figure 4). This yielded the NM510 promoter (carrying a UP element, and consensus –10 and extended –10 elements) and the NM534 promoter (carrying a UP element, and consensus –10 and –35 elements). To measure the effects of these UP elements, DNA fragments carrying the NM510 and NM534 promoters were cloned into the pRW50 *lac* expression vector as before. The data in Figure 4 show that the UP elements had small effects on promoter activity at these promoters.

We reasoned that the effects of the UP elements at the NM510 and NM534 promoters might be increased if the other promoter elements were weakened. Thus we made a series of derivatives of NM534 in which the consensus –35 element was systematically mutated. These changes are illustrated in Figure 5; positions 2–5 of the consensus –35 hexamer, TTGACA, were changed to each of the three alternative bases (note that position 1 was not altered, as we wished to retain the BglII restriction site immediately upstream of the –35 element). The same set of 15 mutations was also created in the context of the NM535 promoter, which lacked a UP element. Figure 5 shows a comparison of the effects of each base change on promoter activity in either NM534 or NM535. Two clear results emerged. First, at the NM535 promoter without a UP element, the effects of the different –35 hexamer changes were in accordance with those reported previously in similar studies with the *E. coli lacUV5* promoter [19] and

Promoter Name	Promoter Sequence	Promoter Activity
	-64 ↓ UP element -35 element -10 element +1 ↓	
NM511	AAACGGACTGCAGTGGTACCTAGGAGATCTACTTTTCGCATCTTTTT**TATG**C**TATAAT**TATTTC**A**	1959
NM510	AAACG**GGAAAATTTTTTTAAAAAAAAGA**TCTACTTTTCGCATCTTTTT**TATG**C**TATAAT**TATTTC**A**	4415
NM535	AAACGGACTGCAGTGGTACCTAGGAGATC**TTGACA**TCGCATCTTTTTGTACC**TATAAT**TATTTC**A**	6251
NM534	AAACG**GGAAAATTTTTTTAAAAAAAAGA**TC**TTGACA**TCGCATCTTTTTGTACC**TATAAT**TATTTC**A** extended -10 element	7823

Figure 4 Effects of a UP element at different promoters. Partial base sequences of promoters containing or lacking UP element sequences in combination with either consensus extended –10 and –10 elements (NM511 and NM510) or consensus –35 and –10 elements (NM535 and NM534) are shown. Consensus sequence elements are indicated by bold type. All of the promoters carry identical sequences downstream of the transcription start site. Each promoter was fused to the *lacZ* gene, and its ability to initiate transcription was deduced from measurements of β-galactosidase activities in the Δ*lac E. coli* strain MC4100 growing exponentially in Luria–Bertani medium. Activities are expressed as nmol of o-nitrophenyl β-D-galactoside hydrolysed·min^{-1}·mg dry cell mass^{-1} (Miller units), and are listed in the final column.

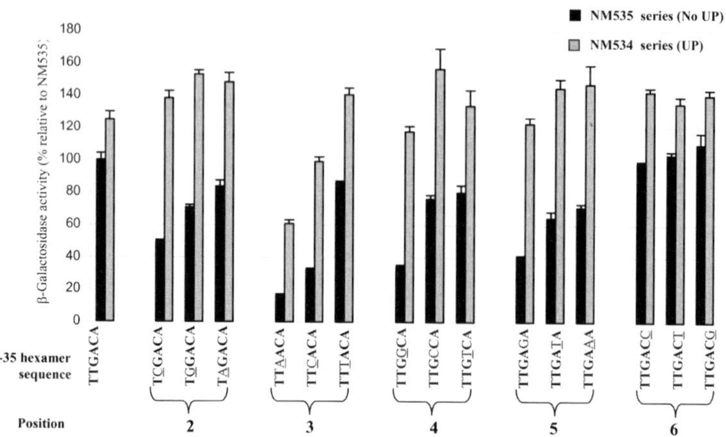

Figure 5 Base changes in the promoter −35 element affect UP element function. The bar chart shows measured β-galactosidase activities during exponential growth of *E. coli* MC4100 Δlac cells containing pRW50 plasmid carrying lac fusions to the NM535 promoter and its derivatives (no UP element; black bars) or to the NM534 promoter and its derivatives (with a UP element; grey bars). The derivatives of both promoters carry single base changes at positions 2–6 of the consensus −35 element, TTGACA, as indicated under the bar chart; the location of each substitution is underlined. Assays were performed as in Figure 2. Activities are expressed as a percentage of the activity with the starting NM535 promoter (100%; equivalent to 625 Miller units). Values are means+S.E.M. for three independent experiments.

the *Salmonella* P22 *ant* promoter [20]. Thus different changes reduce promoter activities by different amounts. Secondly, the presence of the UP element resulted in larger increases in promoter activity at the promoters where the −35 elements caused the biggest decreases. Hence it is clear that the UP element can compensate for defects in the −35 element.

To explore further the idea that the effects of the UP element are greater at weaker promoters, mutations that altered the consensus −10 element from TATAAT to CATAAT or TGTAAT were made in the NM534 and NM535 promoters carrying a consensus −35 element, TTGACA, and derivatives in which the −35 element was changed to either TCGACA or TTGGCA. The results illustrated in Figure 6 show that UP-element-dependent stimulation of promoter activity is greater at promoters that have a defective −35 element than at promoters with a defective −10 element.

Conclusions

The recognition of promoters by the multisubunit bacterial RNA polymerase involves four different protein determinants that recognize four different promoter elements. Our results show that an efficient promoter can be built by different

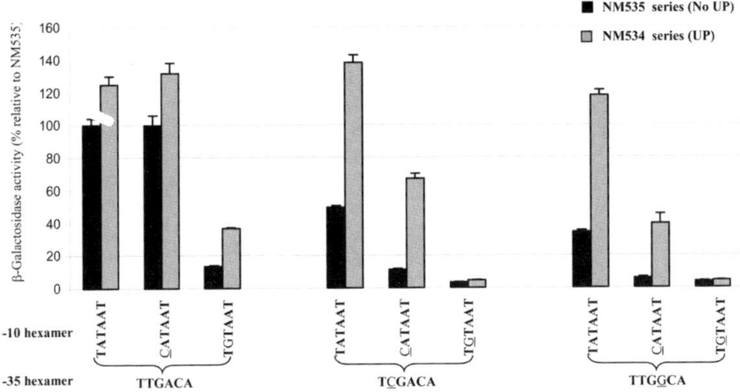

Figure 6 Effects of −10 and −35 element sequences on UP element function. The bar chart shows measured β-galactosidase activities during exponential growth of *E. coli* MC4100 Δ*lac* cells containing pRW50 plasmid carrying *lac* fusions to the NM535 promoter and its derivatives (no UP element; black bars) or to the NM534 promoter and its derivatives (with a UP element; grey bars). The derivatives of both promoters carry combinations of the −35 hexamer elements TTGACA, TCGACA and TTGGCA and the −10 hexamer elements TATAAT, CATAAT and TGTAAT, as indicated. The location of each substitution is underlined. Assays were performed as described in Figure 2. Activities are expressed as a percentage of the activity with the starting NM535 promoter (100%; equivalent to 6251 Miller units). Values are means+S.E.M. for three independent experiments.

combinations of these contacts. It is also clear that promoter activity reaches a ceiling if too many contacts are created (see also [21]). Presumably there is a trade-off between contacts being sufficiently strong for promoter binding, but not too strong so that promoter escape is hindered. Our results suggest that, to some extent, the −35 element and the UP element are redundant. Presumably this is because they are both involved in the very initial contacts between a target promoter and RNA polymerase [22]. Hence there is greater synergy between the UP element and the promoter −10 element. Interestingly, all naturally occurring promoters use sub-optimal combinations of the different contacts [23]. There appear to be two reasons for this. First, because expression levels of genes need to be controlled, the 'tuning' of a promoter's activity is more important than the optimization of its activity. Secondly, sub-optimally active promoters can, if needed, be activated by transcription factors. Hence, in *E. coli* (and probably also in other organisms), the government of the cell favours modest promoters and rejects greedy ones!

We thank the Wellcome Trust for funding this work with a project grant, and the Darwin Trust of Edinburgh for supporting N.M.'s Ph.D. programme. Automated DNA sequencing was performed by the University of Birmingham Functional Genomics Laboratory, supported by grant 6/JIF13209 from the U.K. Joint Infrastructure Fund.

References

1. Ebright, R.H. (2000) J. Mol. Biol. **304**, 687–698
2. Gross, C.A., Chan, C., Dombroski, A., Gruber, T., Sharp, M., Tupy, J. and Young, B. (1998) Cold Spring Harbor Symp. Quant. Biol. **63**, 141–155
3. Campbell, E.A., Muzzin, O., Chlenov, M., Sun, J.L., Olson, C.A., Weinman, O., Trester-Zedlitz, M.L. and Darst, S.A. (2002) Mol. Cell **9**, 527–539
4. Murakami, K.S., Masuda, S., Campbell, E.A., Muzzin, O. and Darst, S.A. (2002) Science **296**, 1285–1290
5. Murakami, K.S. and Darst, S.A. (2003) Curr. Opin. Struct. Biol. **13**, 31–39
6. Ishihama, A. (2000) Annu. Rev. Microbiol. **54**, 499–518
7. Browning, D.F. and Busby, S.J.W. (2004) Nat. Rev. Microbiol. **2**, 1–9
8. Gralla, J.D. (2005) Mol. Microbiol. **55**, 973–977
9. Travers, A. and Muskhelishvili, G. (2005) Nat. Rev. Microbiol. **3**, 157–69
10. Busby, S. and Ebright, R.H. (1994) Cell **79**, 743–746
11. Bown, J.A., Barne, K.A., Minchin, S.D. and Busby, S.J.W. (1997) Nucleic Acids Mol. Biol. **11**, 41–52
12. Ebright, R.H. and Busby, S. (1995) Curr. Opin. Genet. Dev. **5**, 197–203
13. Gourse, R.L., Ross, W. and Gaal, T. (2000) Mol. Microbiol. **37**, 687–695
14. Estrem, S.T., Gaal, T., Ross, W. and Gourse, R.L. (1998) Proc. Natl. Acad. Sci. U.S.A. **95**, 9761–9766
15. Estrem, S.T., Ross, W., Gaal, T., Chen, Z.W.S., Niu, W., Ebright, R.H. and Gourse, R.L. (1999) Genes Dev. **13**, 2134–2147
16. Burns, H.D., Belyaeva, T.A., Busby, S.J.W. and Minchin, S.D. (1996) Biochem. J. **317**, 305–311
17. Busby, S. and Ebright, R.H. (1999) J. Mol. Biol. **293**, 199–213
18. Lodge, J., Fear, J., Busby, S., Gunasekaran, P. and Kamini, N.R. (1992) FEMS Microbiol. Lett. **95**, 271–276
19. Kobayashi, M., Nagata, K. and Ishihama, A. (1990) Nucleic Acids Res. **18**, 7367–7372
20. Moyle, H., Waldburger, C. and Susskind, M.M. (1991) J. Bacteriol. **173**, 1944–1950
21. Grana, D., Gardella, T. and Susskind, M.M. (1988) Genetics **120**, 319–327
22. Sclavi, B., Zaychikov, E., Rogozina, A., Walther, F., Buckle, M. and Heumann, H. (2005) Proc. Natl. Acad. Sci. U.S.A. **102**, 4706–4711
23. Lisser, S. and Margalit, H. (1993) Nucleic Acids Res. **21**, 1507–1516
24. Ghosh, P., Ramakrishnan, C. and Chatterji, D. (2003) Biophys. Chem. **103**, 223–237

Biochem. Soc. Symp. **73**, 11–22
(Printed in Great Britain)

2

How transcriptional and epigenetic programmes are played out on an individual mammalian gene cluster during lineage commitment and differentiation

Douglas R. Higgs[1], Douglas Vernimmen, Marco De Gobbi, Eduardo Anguita, Jim Hughes, Veronica Buckle, Francisco Iborra, David Garrick and William G. Wood

MRC Molecular Haematology Unit, Weatherall Institute of Molecular Medicine, John Radcliffe Hospital, Headington, Oxford OX3 9DS, U.K.

Abstract

In the post-genomic era, a great deal of work has focused on understanding how DNA sequence is used to programme complex nuclear, cellular and tissue functions throughout differentiation and development. There are many approaches to these issues, but we have concentrated on understanding how a single mammalian gene cluster is activated or silenced as stem cells undergo lineage commitment, differentiation and maturation. In particular we have analysed the α globin cluster, which is expressed in a cell-type- and developmental stage-specific manner in the haemopoietic system. Our studies include analysis of the transcriptional programme that accompanies globin gene activation, focusing on the expression of relevant transcription factors and cofactors. Binding of these factors to the chromosomal domain containing the α globin cluster has been characterized by ChIP (chromatin immunoprecipitation). In addition, we have monitored the epigenetic modifications (e.g. nuclear position, timing of replication, chromatin modification, DNA methylation) that occur as the genes are activated (in erythroid cells) or silenced (e.g. in granulocytes) as haemopoiesis proceeds. Together, these observations provide a uniquely well-characterized model illustrating the mechanisms that regulate and

[1]To whom correspondence should be addressed (email doug.higgs@imm.ox.ac.uk).

memorize patterns of mammalian gene expression as stem cells undergo lineage specification, differentiation and terminal maturation.

Introduction

The process by which multipotent stem cells undergo lineage specification and differentiation to produce their highly specialized progeny is poorly understood. Transcriptional profiling of stem cells and the analysis of selected genes in single cells known to have multiple potential have led to the concept that such cells are 'primed' to express the transcriptional programmes of many different lineages [1,2]. Commitment and differentiation involve both consolidating expression of the chosen subset of lineage-affiliated genes and repressing the other primed programmes that are not part of the lineage in question, so-called 'multilineage priming'. This term describes the global changes in transcription programmes, but the specific mechanisms underlying commitment and differentiation have not yet been fully explored. Therefore, to take this hypothesis forward, it is necessary to determine how an *individual* gene is switched on or off as multipotent cells differentiate into increasingly more specialized cells.

Haemopoiesis provides an accessible mammalian system to address this question. Haemopoietic stem cells are self-renewing cells supporting the production of at least eight lineages of specialized, functionally and morphologically distinct blood cells [3]. The phenomenon of multilineage priming has been most clearly demonstrated in haemopoiesis [1,2]. The specific cellular pathway committing progenitors to form red blood cells (erythropoiesis) is well defined and results ultimately in highly specialized precursors (erythroblasts) which, in the later stages of differentiation, express almost exclusively the α- and β-like globin genes to synthesize haemoglobin [4]. Therefore the globin genes, which are silent in early haemopoietic progenitors and expressed at high levels during terminal erythroid differentiation, provide ideal candidates for understanding the principles by which individual genes are activated during lineage commitment and differentiation.

Most previous studies of mammalian globin gene expression have concentrated on fully committed erythroid cells. However, to further our understanding of how the globin genes are activated during erythropoiesis, we need to establish the hierarchy and order of events as commitment and differentiation proceed. In a series of ongoing experiments, we have used a variety of approaches to analyse the pattern of transcription factor binding and the associated epigenetic changes across the entire α globin domain, in both mouse and human, in cells representing different stages of the pathway from pluripotent cells to mature erythroblasts.

These studies are designed to define in detail the steps involved in the sequential activation of the α globin genes during lineage commitment and differentiation, and thereby provide a paradigm for the mechanism by which other mammalian genes may be similarly regulated.

Defining the α globin chromosomal domain and the key regulatory elements

Our laboratory concentrates on understanding how the α globin genes are activated during erythropoiesis. To understand fully how these genes are regulated in their natural chromosomal environment, we need to delimit the α globin regulatory domain, and to identify all key regulatory elements, the factors and cofactors that bind to them *in vivo*, how the assembled nucleoprotein complexes interact, and the epigenetic events that develop as the genes are activated or repressed.

It has frequently been suggested that mammalian chromosomes are subdivided into discrete structural and functional domains. This concept was originally based on the observation that chromatin within a nucleus is necessarily highly packaged and, at some level of compaction, is thought to be folded into topologically constrained loops tethered at their bases by nuclear attachment sites [5]. These models of structural organization have been extended to incorporate various aspects of function. For example, it has been suggested that structural loops contain independently regulated gene(s) and their *cis*-acting regulatory elements, protected from the influence of similar sequences in neighbouring domains by insulator elements, which may correspond to the physical boundaries of the structural loops. It has also been suggested that these structural/functional domains may be controlled independently in terms of their epigenetic modification (timing of replication, methylation, chromatin assembly and nuclear localization). The potential importance of these models is that, once identified, the genes within such domains should be correctly regulated regardless of their position in the genome.

Despite their attraction, no gene cluster conforms to all components or predictions of these models. Nevertheless, different aspects of the models are consistent with experimental observations on many genes. Therefore the models provide a reasonable starting point to address the question of whether there is a functional unit of a chromosome above the level of a gene.

Following completion of the Human Genome Project, there has been considerable interest in developing bioinformatic routines for annotating chromosomal domains, genes and their regulatory elements [6]. We have used such an approach, comparing the sequences of 22 species spanning 500 million years of evolution [7,8]. Together, these studies have defined a ~135 kb region of conserved synteny (Figure 1) that has been maintained in all 22 species. It seems reasonable to conclude that this region contains all of the *cis*-acting elements required to fully regulate expression of the α globin genes.

To identify *cis*-acting regulatory elements, we have used a variety of approaches. First, we have mapped all erythroid-specific and constitutive DNase 1-hypersensitive sites across this 135 kb region in human and mouse (Figures 1 and 2, and [9,10]). In addition, we have used three independent bioinformatics approaches, including the MCS protocol [11] and a newly developed algorithm (called GUMBY; S. Prabhakar, unpublished work) which, together, identified a total of 24 multi-species conserved sequences (denoted MCS in Figure 1) in a region of 238 kb. This defined seven erythroid-specific regulatory elements,

Figure 1

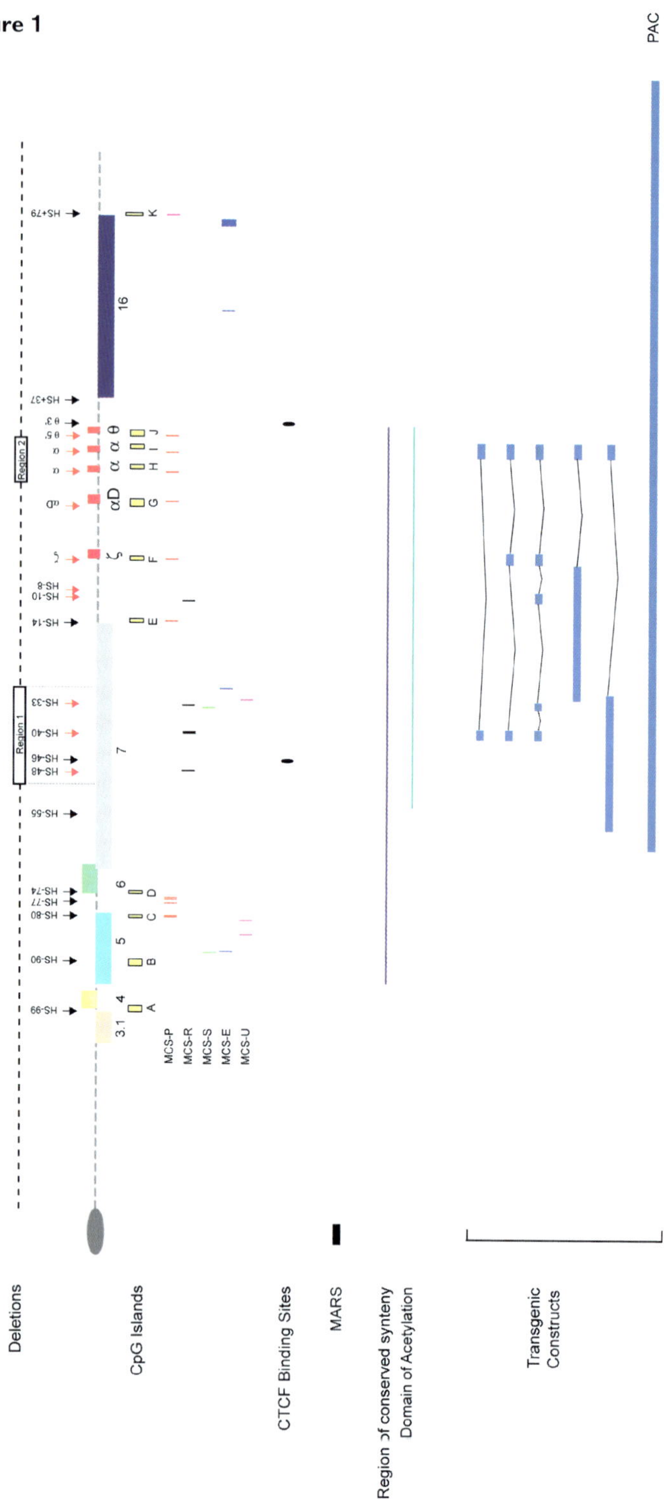
Deletions
CpG Islands
CTCF Binding Sites
MARS
Region of conserved synteny
Domain of Acetylation
Transgenic Constructs
PAC
Region 1
Region 2
MCS-P
MCS-R
MCS-S
MCS-E
MCS-U

including previously uncharacterized conserved non-coding sequences [8]. Close inspection of these small (~100–500 bp) conserved elements has revealed phylogenetically conserved motifs, many of which correspond to known haemopoietic transcription factor binding sites. In particular, we found that most elements contain highly conserved sites with the potential to bind the key erythroid transcription factor GATA-1.

These studies have thus defined a putative regulatory domain and the key *cis*-acting elements located within it.

Functional studies of the regulatory domain

Clearly, normal regulation of α globin expression requires the α genes themselves to be structurally present and normal. The inherited form of anaemia called α thalassaemia, in which α globin synthesis is down-regulated, most frequently results from deletion of the α globin genes (Figure 1, region 2, and [12]). In a series of experiments using stable transfectants [13], interspecific hybrids containing normal and abnormal copies of the human α globin cluster on chromosome 16 [14], transgenic mice [15] and mice in which the α globin cluster has been modified by homologous recombination [16], we have systematically analysed the role of upstream elements in regulating α globin expression *in vivo* (summarized in [17] and Figure 1). These experiments have been complemented by observations on rare patients with α thalassaemia in whom the α globin genes are intact but who harbour deletions of the upstream regulatory elements ([18]; and summarized in Figure 1, region 1). Together, these data show that fully regulated α globin expression depends on these elements (HS-48, HS-40, HS-33 and HS-10 in human; see Figure 1), and it appears (at least in human) that the major regulatory element is HS-40, the most highly conserved of all the elements. The role

Figure 1 The human α globin cluster. The telomere is represented as an oval. The annotated globin genes (red boxes) and flanking genes (coloured boxes) are shown. These genes are annotated as previously [7,31]: 3.1, *POLR3K*; 4, *C16orf33*; 5, *C16orf8*; 6, *MPG*; 7, *C16orf35*; 16, *LUC7L*. Above the line are DNaseI-hypersensitive sites (black arrows for constitutive and red for erythroid-specific sites). The extent of previously described deletions is indicated by the black broken horizontal line, with a small box over the α genes (region 2; SRO2) representing many common deletions [12] that remove these gene(s). The shortest region of overlap (region 1; SRO1) of all upstream deletions [18] is shown. Below are CpG islands (yellow boxes), conserved promoter elements (MCS-P; red lines), conserved regulatory elements (MCS-R; black lines), conserved splicing intronic regulatory elements (MCS-S; green lines), conserved alternative exons (MCS-E; blue lines) and conserved elements of unknown function (MCS-U; pink lines). The region of conserved synteny is shown underneath as a purple line. The domain of acetylation is shown as a green line. Transgenic constructs are as described previously; boxes represent the segments of DNA joined together (lines) in each construct. The PAC construct which gives optimal expression is shown as a blue box at the bottom. CTCF, CCCTC-binding factor; MARS, mamx attachment regions.

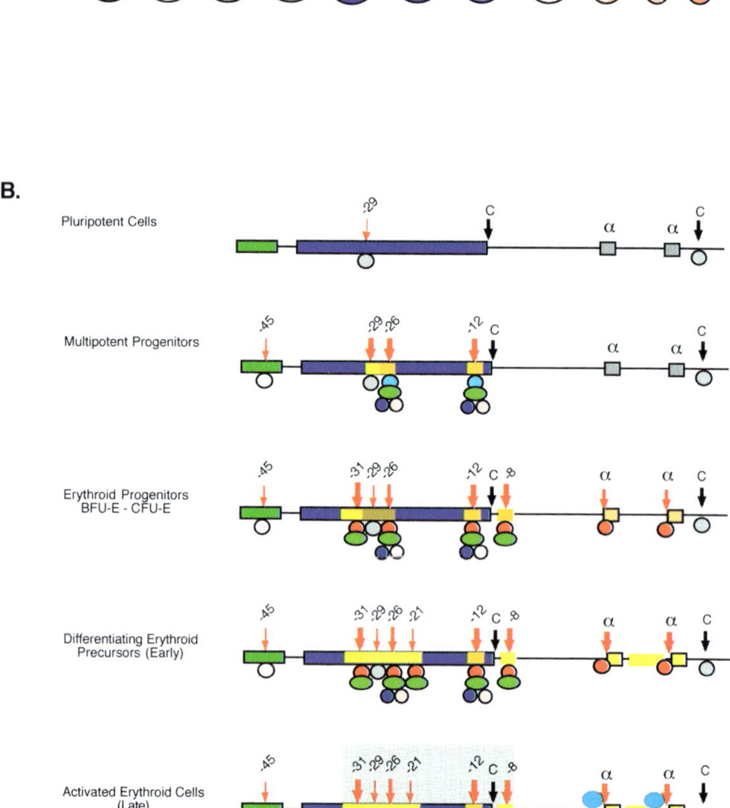

A.
Progenitors
Precursors
HSC
CFU-GEMM
BFU-E Early
BFU-E Late
CFU-E
Pro
Bas
Pol
Ort
Retic
RBC
B.
Pluripotent Cells
Multipotent Progenitors
Erythroid Progenitors BFU-E - CFU-E
Differentiating Erythroid Precursors (Early)
Activated Erythroid Cells (Late)
-45
-31
-29
-26
-21
-12
-8
C
α
Unknown Factors
CTCF
GATA-2
GATA-1/FOG-1
SCL Complex
p18/p45 NF-E2
PIC
RNA Polymerase II

of other regulatory elements and any potential redundancy is still under investigation, although we have noted that a large (120 kb) transgenic construct (Figure 1), including the entire cluster and all regulatory elements, is consistently expressed at higher levels than smaller transgenes [17].

Sequential binding of key haematopoietic transcription factors during lineage commitment

Many of the key factors that regulate globin gene expression have been well characterized [3] and their patterns of expression during erythropoiesis well documented (summarized in [10]). We have monitored when, during erythropoiesis, these factors bind the α globin cluster by characterizing the development of DNase 1-hypersensitive sites and by using ChIP, measured by quantitative PCR, across the entire mouse and human α globin domains ([10]; M. De Gobbi, unpublished work). The α globin gene cluster appears to be inert in pluripotent cells, but priming of expression begins in multipotent haemopoietic progenitors via GATA-2, which is expressed early in haematopoiesis [3]. In committed erythroid progenitors, GATA-2 is replaced by GATA-1 and binding is extended to additional sites, including the α globin promoters. Both GATA-1 and GATA-2 nucleate the binding of various protein complexes, including SCL (stem cell leukaemia)/LMO2/E2A/Ldb-1 and NF-E2 (nuclear factor-E2). At this stage of erythroid commitment (see differentiating erythroid cells in Figure 2A), the entire α globin cluster appears to be poised for 'action', even though little or no globin mRNA is made.

Figure 2 Summary of α globin activation during erythroid differentiation. In pluripotent cells, the α globin cluster appears relatively inert; only HS-29 appears to be bound, and the significance of this is unknown. In multipotent cells the cluster is primed in the upstream region by multiprotein complexes containing SCL and NF-E2 nucleated by GATA-2. HS-29 becomes very prominent and the associated histones are highly acetylated. Otherwise, only limited histone modifications (indicated by yellow boxes) are present. In committed erythroid progenitors, most HSs associated with conserved non-coding sequences are bound by multiprotein complexes containing various combinations of SCL and NF-E2, now nucleated by GATA-1. The α globin promoters are also bound by GATA-1, either alone or as part of a different multiprotein complex. In differentiating erythroid cells, histone modifications extend throughout much of the locus, creating an erythroid-specific domain of histone hyperacetylation. The pattern of binding differs very little between cells expressing globin and those that do not. The exponential increase in globin mRNA synthesis would be consistent with co-operative interactions between proteins bound upstream of the cluster (shaded box) and multiprotein complexes, including Pol II at the α globin promoters. Abbreviations: HSC, haemopoietic stem cell; CFU-GEMM, colony-forming units, granulocyte, erythroid, megakaryocyte and monocyte; BFU-E, burst-forming units, erythroid; CFU-E, colony-forming units, erythroid; Pro, proerythroblast; Bas, basophillic erythroblast; Pol, polychromatic erythroblast; Ort, orthochromatic erythroblast; Retic, reticulocyte; RBC, red blood cell; CTCF, CCCTC-binding factor; FOG, friend of GATA.

Epigenetic changes that occur during haematopoiesis

We have examined nuclear position [19], nuclear attachment [20], timing and pattern of replication [21], location with respect to chromosome territory (V. Buckle, unpublished work) and DNA methylation of CpG islands [9], and found no differences between erythroid and non-erythroid cells. Studies are currently under way to investigate these features during erythropoiesis (V. Buckle, unpublished work).

We have also compared activating and repressing histone modifications in erythroid and non-erythroid cells. Whereas we see a major difference in the activating modifications [H3ac (acetylated histone H3), H4ac and H3K4me2 (methylated histone H3)], as yet we have not identified any consistent repressing modifications in non-erythroid cells. Preliminary studies [22] demonstrated that a domain of acetylation, broadly corresponding to the region containing the conserved regulatory elements and genes, develops in differentiating erythroid cells of human, mouse and chicken. More recently, we examined this at high resolution [10]. As before, we found a low level of acetylation, greater than in heterochromatin, across the entire α globin cluster in non-erythroid cells. We found a similar pattern in pluripotent embryonic stem cells. A broad, erythroid-specific domain of histone acetylation (H3ac and H4ac) first appears in multipotent haematopoietic cells, where it initiates far upstream of the α globin cluster. The domain of acetylation extends during differentiation to become fully developed in mature erythroblasts (Figure 2). A similar pattern was seen for H3K4me2-modified chromatin. The broad domain of acetylation was punctuated by several distinct peaks of acetylation and H3K4me2 modification, which closely reflect the patterns of DNase 1-hypersensitive sites seen in corresponding cell types. At the resolution of these observations, it appears that histone modifications develop in parallel with the binding of transcription factors.

Recruitment of the PIC (pre-initiation complex) and Pol II (RNA polymerase II) during erythropoiesis

Over the past year, we have continued the description of globin gene activation by establishing when during erythropoiesis the PIC [including TFIIA (transcription factor IIA), TFIIB, TFIID, TFIIE, TFIIF and TFIIH, many of which are themselves multiprotein complexes] and Pol II associate with the α globin promoters (D. Vernimmen et al., unpublished work). Whether this occurs in a stepwise manner or involves interaction with a pre-formed holocomplex containing Pol II and at least some components of the PIC is not clear, although recent evidence suggests that active genes associate with pre-formed transcription factories [23].

Initially, we analysed the point during erythropoiesis at which Pol II first binds the complex. This appears to occur quite late in both mouse and human ([10]; M. De Gobbi, unpublished work), co incident with the onset of globin gene transcription, as determined by *in situ* RNA analysis ([24]; V. Buckle, unpublished work). To look at this more specifically, we examined binding in uninduced MEL (mouse erythroleukaemia) cells (representing pro-erythroblasts) and induced MEL cells (representing intermediate and late erythroblasts). In uninduced cells there is little or no Pol II or PIC bound at the α globin promoters, even though, at this stage of erythropoiesis, all of the upstream elements appear fully occupied with the haemopoietic transcription factors (Figure 2). By contrast, in induced MEL cells there is a considerable enrichment of both Pol II and all tested components of the PIC (D. Vernimmen, unpublished work).

These observations suggest that the onset of globin gene transcription coincides with a *de novo* interaction between the α globin cluster, the PIC and Pol II, rather than enhanced initiation or elongation of a pre-bound PIC–Pol II complex.

Role of upstream elements in activating α globin expression

These findings suggest that upstream elements might play a role in facilitating or enhancing the interaction between the α globin complex and the PIC–Pol II complex. To examine this, we have analysed the binding of the PIC and Pol II in a single copy of human chromosome 16 in which the upstream elements (HS-48, HS-40 and HS-33) had been deleted. This chromosome was derived from a previously reported patient with α thalassaemia [25] and was transferred into an erythroid background (MEL cell) by cell fusion. When a normal copy of human chromosome 16 was transferred in this way, induction of the interspecific MEL hybrid greatly up-regulated both mouse and human α globin expression [14]. In the hybrid containing the abnormal copy of chromosome 16, although the endogenous mouse α globin was induced normally, the human α globin gene was not. This could be either because the upstream elements are normally required for the interaction between the α globin complex, the PIC and Pol II or that they are required to enhance initiation or elongation of a complex that forms independently of the upstream elements.

Preliminary data show that the association between the PIC, Pol II and the α globin promoters is significantly reduced in the absence of the upstream regulatory elements (D. Vernimmen, unpublished work), favouring a model in which the conserved, upstream sequences are required to initiate or maintain an interaction between the α globin promoters and the PIC–Pol II complex.

Physical interaction between key regulatory elements and transcription factories?

This raises the question of how the upstream elements, the α globin promoters and the PIC–Pol II complex interact with each other. There is increasing evidence from analysis of the β globin cluster [26–29] that regulatory elements interact physically by a looping mechanism mediated by protein–protein interactions involving complexes nucleated at the key sites. Furthermore, it has been suggested that active gene clusters are recruited to fixed transcription factories containing Pol II, whereas inactive clusters remain outside such factories (Figure 3 and [23]). Is there any evidence that the upstream elements and the α globin promoters interact simultaneously with a transcription factory?

We have noted that when the entire α globin cluster is profiled for binding of the PIC components and Pol II (D. Vernimmen, unpublished work), there are minor peaks of enrichment corresponding precisely to each of the upstream regulatory elements (HS-48, HS-40 and HS-33). It is possible that these sites, which lie within the introns of a widely expressed gene upstream of the α globin cluster, are cryptic promoters. However, it is equally likely that protein–DNA complexes assembled at these elements are physically in contact with each other, the α globin promoter and the PIC–Pol II complex within a transcription factory. In this way, ChIP assays using antibodies against either Pol II or components of the PIC may enrich both the promoter elements (directly) and the upstream regulatory elements (indirectly).

In this case, it should be possible to demonstrate a physical interaction between the upstream elements and the α globin promoters in erythroid cells (when the α globin genes are active), but not in non-erythroid cells (where the α genes are silenced). To address this, we and others [30] have used chromosome conformation capture. In preliminary experiments using oligonucleotide primers normally located close to the murine α globin regulatory element (HS-26) and the α globin promoter, which are normally ~35–55 kb apart, we have shown that PCR products are readily detectable in erythroid, but not in non-erythroid, cells (F. Iborra, unpublished work). This suggests that within the nucleus of erythroid cells there is frequent juxtaposition of the major α globin regulatory

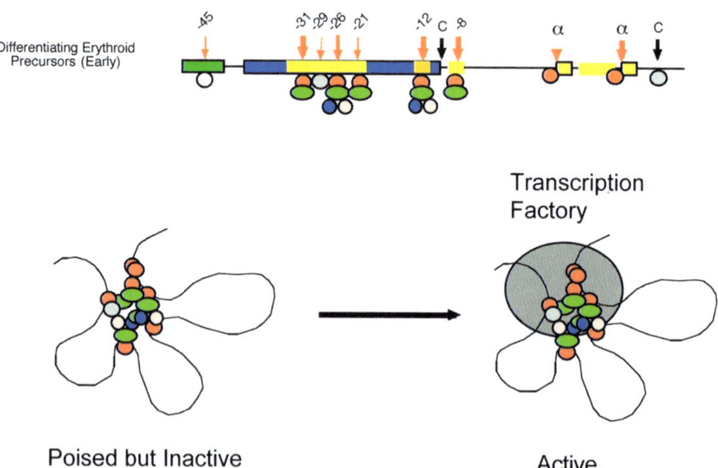

Figure 3 Model for α globin activation. This is based on the model proposed in [23]. Here the α globin cluster is recruited to a transcription factory when active, but lies outwith the factory when inactive, even though it is primed with the full repertoire of transcription factors. Annotation of transcription factors is as for Figure 2.

element and α globin promoters. Further studies are under way to determine whether other regulatory elements also participate in this interaction.

Summary

Our studies on the α globin gene cluster, together with complementary data on the β globin cluster, have confirmed that the globin genes continue to provide one of the best characterized examples of how mammalian genes are regulated *in vivo*. The studies provide insight into the order of events and mechanisms by which mammalian genes are activated by key transcriptional regulators during lineage commitment, differentiation and development.

References

1. Jimenez, G., Griffiths, S.D., Ford, A.M., Greaves, M.F. and Enver, T. (1992) Proc. Natl. Acad. Sci. U.S.A. **89**, 10618–10622
2. Hu, M., Krause, D., Greaves, M., Sharkis, S., Dexter, M., Heyworth, C. and Enver, T. (1997) Genes Dev. **11**, 774–785
3. Cantor, A.B. and Orkin, S.H. (2002) Oncogene **21**, 3368–3376
4. Steinberg, M.H., Forget, B.G., Higgs, D.R. and Nagel, R.L. (2001) Disorders of Hemoglobin, Cambridge University Press, Cambridge
5. Udvardy, A. (1999) EMBO J. **18**, 1–8
6. Boffelli, D., Nobrega, M.A. and Rubin, E.M. (2004) Nat. Rev. Genet. **5**, 456–465
7. Flint, J., Tufarelli, C., Peden, J., Clark, K., Daniels, R.J., Hardison, R., Miller, W., Philipsen, S., Tan-Un, K.C., McMorrow, T. et al. (2001) Hum. Mol. Genet. **10**, 371–382
8. Hughes, J.R., Cheng, J.-F., Ventress, N., Prabhakar, S., Clark, K., Anguita, E., De Gobbi, M., De Jong, P., Rubin, E. and Higgs, D.R. (2005) Proc. Natl. Acad. Sci. U.S.A. **102**, 9830–9835
9. Vyas, P., Vickers, M.A., Simmons, D.L., Ayyub, H., Craddock, C.F. and Higgs, D.R. (1992) Cell **69**, 781–793
10. Anguita, E., Hughes, J., Heyworth, C., Blobel, G.A., Wood, W.G. and Higgs, D.R. (2004) EMBO J. **23**, 2841–2852
11. Margulies, E.H., Blanchette, M., Haussler, D. and Green, E.D. (2003) Genome Res. **13**, 2507–2518
12. Higgs, D.R. (2001) in Disorders of Hemoglobin (Steinberg, M.H., Forget, B.G., Higgs, D.R. and Nagel, R.L., eds), pp. 405–430, Cambridge University Press, Cambridge
13. Sharpe, J.A., Summerhill, R.J., Vyas, P., Gourdon, G., Higgs, D.R. and Wood, W.G. (1993) Blood **82**, 1666–1671
14. Higgs, D.R., Wood, W.G., Jarman, A.P., Sharpe, J., Lida, J., Pretorius, I.M. and Ayyub, H. (1990) Genes Dev. **4**, 1588–1601
15. Sharpe, J.A., Chan-Thomas, P.S., Lida, J., Ayyub, H., Wood, W.G. and Higgs, D.R. (1992) EMBO J. **11**, 4565–4572
16. Anguita, E., Sharpe, J.A., Sloane-Stanley, J.A., Tufarelli, C., Higgs, D.R. and Wood, W.G. (2002) Blood **100**, 3450–3456
17. Higgs, D.R., Sharpe, J.A. and Wood, W.G. (1998) Semin. Hematol. **35**, 93–104
18. Viprakasit, V., Kidd, A.M., Ayyub, H., Horsley, S., Hughes, J. and Higgs, D.R. (2003) Br. J. Haematol. **120**, 867–875
19. Brown, K.E., Amoils, S., Horn, J.M., Buckle, V.J., Higgs, D.R., Merkenschlager, M. and Fisher, A.G. (2001) Nat. Cell Biol. **3**, 602–606
20. Jarman, A.P. and Higgs, D.R. (1988) EMBO J. **7**, 3337–3344
21. Smith, Z.E. and Higgs, D.R. (1999) Hum. Mol. Genet. **8**, 1373–1386
22. Anguita, E., Johnson, C.A., Wood, W.G., Turner, B.M. and Higgs, D.R. (2001) Proc. Natl. Acad. Sci. U.S.A. **98**, 12114–12119

23. Osborne, C.S., Chakalova, L., Brown, K.E., Carter, D., Horton, A., Debrand, E., Goyenechea, B., Mitchell, J.A., Lopes, S., Reik, W. and Fraser, P. (2004) Nat. Genet. **36**, 1065–1071
24. de Krom, M., van de Corput, M., von Lindern, M., Grosveld, F. and Strouboulis, J. (2002) Mol. Cell **9**, 1319–1326
25. Craddock, C.F., Vyas, P., Sharpe, J.A., Ayyub, H., Wood, W.G. and Higgs, D.R. (1995) EMBO J. **14**, 1718–1726
26. Palstra, R.J., Tolhuis, B., Splinter, E., Nijmeijer, R., Grosveld, F. and de Laat, W. (2003) Nat. Genet. **35**, 190–194
27. Tolhuis, B., Palstra, R.J., Splinter, E., Grosveld, F. and de Laat, W. (2002) Mol. Cell **10**, 1453–1465
28. Patrinos, G.P., de Krom, M., de Boer, E., Langeveld, A., Imam, A.M., Strouboulis, J., de Laat, W. and Grosveld, F.G. (2004) Genes Dev. **18**, 1495–1509
29. Carter, D., Chakalova, L., Osborne, C.S., Dai, Y.F. and Fraser, P. (2002) Nat. Genet. **32**, 623–626
30. Drissen, R., Palstra, R.J., Gillemans, N., Splinter, E., Grosveld, F., Philipsen, S. and de Laat, W. (2004) Genes Dev. **18**, 2485–2490
31. Daniels, R.J., Peden, J.F., Lloyd, C., Horsley, S.W., Clark, K., Tufarelli, C., Kearney, L., Buckle, V.J., Doggett, N.A., Flint, J. and Higgs, D.R. (2001) Hum. Mol. Genet. **10**, 339–352

Biochem. Soc. Symp. **73**, 23–39
(Printed in Great Britain)

3

MOZ fusion proteins in acute myeloid leukaemia

Philip J.F. Troke, Karin B. Kindle, Hilary M. Collins and David M. Heery[1]

School of Pharmacy, University of Nottingham, University Park, Nottingham NG7 2RD, U.K.

Abstract

MOZ (monocytic leukaemia zinc finger protein; also known as ZNF220 or MYST3) is a member of the MYST family of protein acetyltransferases. Chromosomal translocations involving the *MOZ* gene are associated with AML (acute myeloid leukaemia), suggesting that it has a role in haematopoiesis. Recurrent reciprocal translocations fuse the *MOZ* gene [or the gene encoding MORF (MOZ-related factor); also known as MYST4] to genes encoding the nuclear receptor co-activators CBP [CREB (cAMP response element-binding protein)-binding protein], p300 or the p160 protein TIF2 (transcription intermediary factor 2). The resulting fusion proteins can transform haematopoietic progenitors *in vitro*, and induce myeloproliferative disease in mice. Recent insights into the molecular mechanisms underlying these effects indicate that MOZ fusion proteins interfere with the activities of transcription factors such as nuclear receptors, p53 and Runx proteins. Our studies suggest that subverting the function of cellular CBP and p300 proteins may play a key role in this process. Here we review the recent progress in understanding the role of MOZ fusion proteins in the aetiology of AML.

MOZ (monocytic leukaemia zinc finger protein) gene fusions in AML (acute myeloid leukaemia)

The chromosomal translocation t(8;16)(p11;p13) defines a distinct AML subtype (M4/M5) with characteristic morphological, cytochemical and clinical features [1–3]. Blasts associated with this translocation are of myelomonocytic origin and express both monocytic and granulocytic enzymes, indicating a block at the granulo-monocytic stage. The blasts exhibit (atypical) erythrophagocytosis,

[1]To whom correspondence should be addressed (email david.heery@nottingham.ac.uk).

and often contain prominent cytoplasmic granules that are not normally present in monocytes or most myeloblasts. Given the poor prognosis of patients presenting with translocation t(8;16) characteristics, the early identification of this atypical subtype is critical for effective treatment [3].

The genes at the breakpoints of t(8;16)(p11;p13) were identified as that encoding the acetyltransferase CBP [CREB (cAMP response element-binding protein)-binding protein] on chromosome 16, and a novel gene termed *MOZ* (also known as *ZNF220* or *MYST3*) on chromosome 8 [2]. Hybridization studies confirmed that the *MOZ* and *CBP* genes were at the breakpoints of t(8;16)(p11;p13) translocations in four additional patients presenting with this subtype of AML [2]. Sequence analysis of the expressed mRNAs by reverse transcriptase–PCR indicated that an in-frame MOZ–CBP mRNA was generated, whereas CBP–MOZ mRNA was out of frame, suggesting that the expressed fusion product was the MOZ–CBP protein. To date, at least 50 patients have been identified with the t(8;16) translocation [4]. It appears likely that expression of a MOZ–CBP fusion protein is critical to the development of AML in these patients [5].

The M4/M5 subtype of AML is also associated with a number of other chromosomal translocations (Table 1). The chromosome 8 inversion, inv8(p11;q13), results in the fusion of the *MOZ* gene to the gene encoding TIF2 (transcription intermediary factor 2; also known as GRIP1 and NCoA2) [6,7], and at least six cases have been reported [8]. In addition, MOZ–p300 [9] and MORF (MOZ-related factor)–CBP [10] translocations have been reported in AML patients with similar clinical features to those with the t(8;16)(p11;p13) translocation. MOZ was also identified at the breakpoint of a t(2;8)(p23;p11.2) translocation, although the identity of the gene on chromosome 2 was not determined [11]. Interestingly, the gene encoding SRC1 (steroid receptor co-activator 1), a homologue of TIF2, maps to the 2p23 locus [12], although it remains to be determined whether the chromosome 2 break occurs within this gene.

Functional domain organization of MOZ

The *MOZ* gene comprises 17 exons that span approx. 80 kb of DNA, with exons 2–17 encoding a 2004-amino-acid (225 kDa) nuclear protein [13]. MOZ contains six identifiable domains based on homology searches (shown schematically in Figure 1) that are conserved in MORF (total sequence identity of 60%) and an additional proline/glutamine-rich repeat. The domain that defines this family is the MYST domain, named after its founding members MOZ, YBF2 (SAS3), SAS2 and TIP60 [2]. This domain consists of a protein HAT (histone acetyltransferase) domain immediately C-terminal to a C2HC zinc finger (amino acids 522–703), which is conserved in a wide range of eukaryotes, including yeast (reviewed in [14]). There are five known human MYST proteins (MOZ, MORF, hMOF, TIP60 and HBO1), which have diverse functions *in vivo* [14]. *In vitro* studies of the MYST domains in MORF and MOZ have confirmed their ability to acetylate histones, with a preference for acetylation of histone H4 [15–18]. Interestingly, the C2HC

Table 1 MOZ and MORF chromosomal translocations in AML. nc, not characterized. MOZ–*** denotes that the gene located at the 2p23 breakpoint has not been identified, although we note that SRC1 maps to this location. [12].

Chromosomal translocation	Resulting fusion protein (codon breakpoints)	References
t(8;16)(p11;p13)	MOZ–CBP (1546–267)	[2]
	MOZ–CBP (1117–29)	[4,13,56–58]
	MOZ–CBP (1386–267)	[5]
	MOZ–CBP (nc)	[59,60]
t(8;22)(p11;q13)	MOZ–p300 (1117–31)	[9]
	MOZ–p300 (1013–244)	[61]
inv(8)(p11;q13)	MOZ–TIF2 (1117–869)	[6,8,62–64]
	MOZ–TIF2 (1117–939)	[7]
t(10;16)(q22;p13)	MORF–CBP (929–29)	[10,65]
	MORF–CBP (929–444)	[66]
	MORF–CBP (832–326)	[67]
t(2;8)(p23;p11.2)	MOZ–*** (nc)	[11]

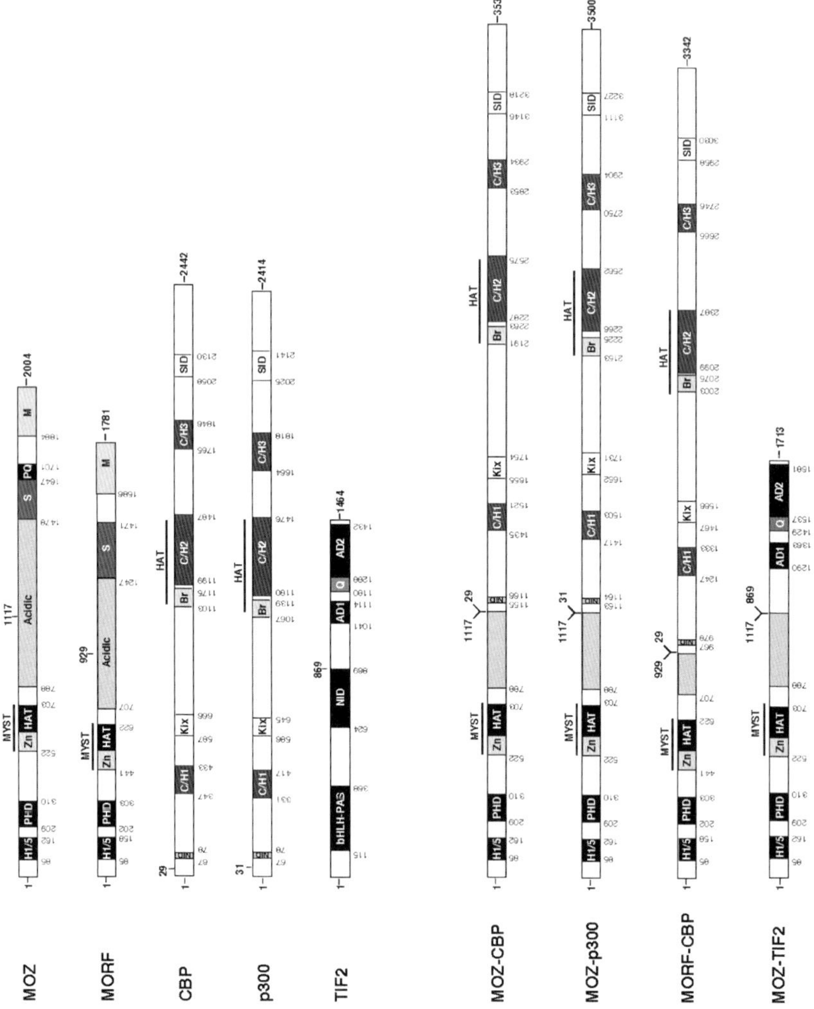

Figure 1 Domain structure of MOZ and the MOZ–CBP and MOZ–TIF2 fusion proteins. Schematic representation of the domain structures of the wild-type MOZ, MORF, CBP, p300 and TIF2 proteins (with the predominant breakpoint indicated above each protein), and the resulting fusion proteins (MOZ–CBP [13], MOZ–p300 [9], MORF–CBP [65] and MOZ–TIF2 [8]). Domain abbreviations: H1/5, histone H1/5-like; Zn, atypical C2HC zinc finger; acidic, acidic-rich; S, serine-rich; PQ, proline/glutamine-rich; M, methionine-rich; NID, NR-interacting domain; C/H 1–3, cysteine/histidine-rich domains 1–3; Kix, CREB-binding domain; Br, bromodomain; bHLH, basic helix–loop–helix domain; PAS, Per–Arndt–Sim zinc finger motif; Q, glutamine-rich domain.

finger at the N-terminus of this domain mediates substrate specificity in other MYST proteins [19]. Analysis of a bacterially expressed MOZ MYST domain (expressed as a fusion protein with maltose-binding protein) indicated that substitution of the first cysteine residue of the C2HC finger (Cys^{543}) with glycine resulted in abrogated HAT activity [20]. Thus this zinc finger contributes to the HAT activity either directly or by influencing the folding of the domain. Consistent with this, we have found that histone acetylation by the MOZ MYST domain (amino acids 500–810) *in vitro* is enhanced by supplementation of the assay buffer with zinc chloride, which may facilitate proper folding of the MYST domain (D.M. Heery, unpublished work).

The N-terminus of MOZ (amino acids 85–162) contains a sequence (termed the H1/5-like domain) that shows weak identity with the linker histone H1 and its variant H5. A similar sequence is found in the *Drosophila* protein Enok, a MYST family protein required for olfactory learning and memory [21]. While the function of this domain is unknown, it has been suggested that it may be involved in nucleosome recognition and/or self-association [14,19]. Deletion of this domain was shown to alter the subcellular localization of a MOZ polypeptide comprising residues 1–311 from the nucleus to the cytoplasm [17], although the mechanism remains unclear.

Adjacent to the H1/5-like domain are two C4HC3 zinc fingers (amino acids 209–310) that together form a region known as a PHD (plant homeodomain) or LAP (leukaemia-associated-protein) domain. Similar zinc-finger motifs are found in a number of chromatin regulators, and some have been shown to bind phosphoinositides, suggesting their possible regulation through nuclear lipid signalling [22]. Interestingly, the MOZ PHD zinc fingers were reported to bind phosphinositides weakly [22]. We have recently identified proteins that interact with the N-terminus of MOZ via this domain (T. Sirisriro and D.M. Heery, unpublished work).

The C-terminal half of the MOZ protein is dominated by a domain rich in acidic residues (amino acids 788–1478), followed by serine-rich (amino acids 1479–1647), proline/glutamine-rich (amino acids 1648–1701) and methionine-rich (amino acids 1884–2004) domains. The slightly shorter (1781 amino acids; 200 kDa) MORF sequence lacks the proline/glutamine-rich sequence, although the functional consequence of this, if any, is unknown. As discussed below, the leukaemogenic breakpoints occur predominantly within intron 16 of the *MOZ* gene [13]. Thus residues C-terminal to amino acid 1117, including the acidic, serine-rich, proline/glutamine-rich and methionine-rich domains, are absent from the MOZ fusion proteins described to date.

MOZ as a transcriptional co-activator

Kitabayashi and colleagues [17] identified MOZ as a component of a large multi-subunit complex that was immunoprecipitated from mouse myeloid L-G cell extracts stably expressing FLAG–AML1b with antibodies recognizing the FLAG epitope. Co-expression of MOZ in P19 cells enhanced AML1b-mediated transcription in reporter assays, and N- and C-terminal sequences of MOZ were found to bind AML1b *in vitro*. Consistent with this, MOZ was reported to co-operate with AML1b to enhance activation of a reporter driven by the MIP1α (macrophage inflammatory protein 1α) promoter [23]. We have also found that MOZ enhances AML1b-mediated reporter gene activation in similar experiments (H.M. Collins, K.B. Kindle, S. Matsuda, P.J.F. Troke, E. Kalkhoven and D.M. Heery, unpublished work). Interestingly, the MOZ MYST domain was not required for the stimulation of reporter activity, as shown by amino acid substitution of an essential glycine residue in the acetyl-CoA-binding domain (G657E) ([17]; H.M. Collins, K.B. Kindle, S. Matsuda, P.J.F. Troke, E. Kalkhoven and D.M. Heery, unpublished work). These results indicate that the acetyltransferase activity of MOZ is not required for its effects in reporter assays, suggesting that other HATs in the complex, such as CBP and p300, may perform any necessary chromatin modifications.

In attempts to identify transcriptional regulatory domains in MOZ, it was shown that fusion of the GAL4 DNA-binding domain to full-length MOZ, the MOZ N-terminus (residues 1–787) or the acidic domain (788–1518) resulted in proteins that repressed transcription of a GAL4-driven reporter, whereas the C-terminus (1517–2004) enhanced the reporter activity [16,17]. However, the role of these sequences in transcriptional regulation by MOZ remains unclear. Genetic studies have shown that MOZ regulates the maintenance of *Hox2* gene expression in zebrafish (*Dani rerio*)embryos, and that the MOZ HAT domain is required for this function [24]. Clearly, transcription profiling and RNAi (RNA interference) approaches will be required to identify genes that are regulated by MOZ.

Structure and function of CBP and TIF2

Human CBP is a 2442-amino-acid nuclear protein (shown schematically in Figure 1) containing a HAT domain, as well as a series of modular domains that allow it to interact with a large number of transcription factors and other proteins. CBP, and its homologue p300, are global co-activators that enhance the transcriptional activity of many transcription factors [25]. The protein interaction domains include the C/H1 and C/H3 domains (also termed TAZ1 and TAZ2 respectively), the KIX domain and the SID (SRC1 interaction domain) [26]. Histone acetylation mediated by CBP/p300 has been demonstrated on a number of promoters, including the oestrogen-regulated pS2 gene [27,28], and this activity is critical for ligand-dependent transcription by the oestrogen receptor. Acetylation of non-histone proteins, including transcription factors and co-activators, is thought to be another important function of CBP/p300, as is their ability to co-ordinate the assembly of multi-protein co-regulator complexes. Thus CBP and p300 perform multiple functions in gene expression.

CBP and p300 interact with haematopoietic transcription factors, and have been shown to play roles in normal haematopoiesis (reviewed in [29]). Interestingly, while loss of either CBP or p300 in mice is associated with haematological tumours, these proteins appear to have distinct functions in blood-cell development [30,31]. Normal CBP content is necessary for the maintenance of the HSC (haematopoietic stem cell) population, whereas p300 is required for correct differentiation of these cells [31]. While $p300^{-/-:+/+}$ chimaeric mice (derived by injecting strain-matched $p300^{-/-}$ cells into wild-type blastocysts) showed an absence of $p300^{-/-}$ mature haematopoietic cells in their peripheral blood, transplantation of $p300^{-/-}$ HSCs into irradiated mice gave rise to abundant, well-differentiated haematopoietic cells [31]. This has been suggested to indicate that normal haematopoiesis requires p300 function in bone-marrow stromal cells, rather than in HSCs. Analysis of mice heterozygous for CBP indicated that mono-allelic loss is sufficient to cause a defect in HSC self-renewal, and a subsequent haematopoietic failure, as mice age [31].

TIF2 (represented schematically in Figure 1) and its homologues SRC1 and ACTR (activator of thyroid hormone and retinoic acid receptors) are NR (nuclear receptor)-specific co-activator proteins, collectively termed the p160 co-activators [32,33]. The ligand-dependent interaction of p160s with NRs is mediated by three conserved LXXLL amino acid motifs within the central NR-interacting domain, although some isoforms of SRC1 and TIF2 contain additional LXXLL motifs towards the C-terminus [34]. One function of the p160s is to facilitate the assembly of NR cofactor complexes on target gene promoters. CBP and p300 are recruited via AD1 (activation domain 1), while the arginine methyltransferase proteins CARM1 (co-activator-associated arginine methyltransferase 1)/PRMT1 (protein arginine N-methyltransferase 1) are recruited via AD2. This facilitates chromatin modifications and further protein–protein interactions required to initiate transcription. Previous experiments from our group have shown that a deletion mutant containing only the minimal NR- and CBP-binding domains is sufficient to function as a potent co-activator of NR-mediated transcription [35]. Other domains in p160s mediate interactions with AP-1 (activator protein-1) [36], NF-κB (nuclear factor-κB) [37], p53 [38], TEFs (transcriptional enhancer factors) [39], MEF2C (myocyte enhancer factor 2C) [40], STATs (signal transducers and activators of transcription) [41] and E2F [42], and may be important in cross-talk with NR signalling pathways. While TIF2 is required for postnatal survival and male reproduction in mice [43], it is not clear whether TIF2 or other p160s play a direct role in haematopoiesis.

Domain structure of MOZ fusion proteins

The domain structures of MOZ–TIF2 and the predominant form of MOZ–CBP are shown in Figure 1. In the majority of the translocations reported to date, the MOZ sequence C-terminal to residue 1117 is replaced by sequence from the fusion partner [5,6,8,13]. This is due to recurrent breakpoints within intron 16 of the *MOZ* gene [13]. Thus both MOZ–CBP and MOZ–TIF2 contain the H1/5, PHD and MYST domains and the N-terminal part of the acidic domain of MOZ. In MOZ–CBP translocations, the majority of the breakpoints occur within intron 2 of the CBP gene [13]. Thus the predominant MOZ–CBP protein expressed contains almost the entire CBP sequence (amino acids 29–2442) (Figure 1). Similarly, the MOZ–p300 and MORF–CBP fusion proteins retain most of the p300 and CBP sequences (amino acids 31–2414 and 29–2442 respectively).

Most MOZ–TIF2 fusion proteins reported contain amino acids 869–1464 of TIF2, comprising AD1 and AD2 required for interaction with CBP/p300 and CARM1/PRMT respectively. A key function of the p160s is to recruit histone-modifying enzymes such as CBP/p300 and CARM1/PRMT1 to target gene promoters [35,44–48]. Thus it has been proposed that MOZ–TIF2 may mimic MOZ–CBP via its ability to recruit the CBP/p300 proteins [6]. In summary, MOZ fusion proteins retain the acetyltransferase activity of the MYST domain and also that of its partner (or the fusion partner's associated factors). Thus the expression of MOZ fusion proteins in haematopoietic progenitors is likely to interfere with normal chromatin modification and gene expression patterns, thereby disturbing the differentiation programmes of these cells.

Transformation of myeloid progenitors and AML induction in mice by MOZ-fusion proteins

In vitro analysis of the MOZ–CBP fusion protein provided the first direct evidence that MOZ fusions could block haematopoietic cell differentiation [17]. Upon stimulation with interleukin-6, the murine myeloid M1 cell line undergoes growth arrest and the cells subsequently differentiate into monocytes/macrophages. However, M1 cells stably expressing MOZ–CBP showed an enhanced growth rate and absence of differentiation upon interleukin-6 stimulation, as determined by the analysis of cell morphology and cell-surface-marker expression [17]. Thus the MOZ–CBP protein inhibits myeloid cell differentiation *in vitro*, although its effects *in vivo* have yet to be characterized.

Two studies have investigated the ability of the MOZ–TIF2 protein to inhibit haematopoietic differentiation [20,49]. Deguchi et al. [20] investigated the effects of MOZ–TIF2 expression in mouse myeloid progenitor cells both *in vitro* and *in vivo*. MOZ–TIF2, but not full-length MOZ, TIF2 or truncated derivatives, induced the immortalization of myeloid progenitors in serial replating experiments *in vitro*. In bone marrow transplantation assays, lineage-deficient progenitors expressing MOZ–TIF2 induced an AML-like disease in mice with 100% penetrance. Secondary transplantation of bone marrow from these mice resulted in AML in the recipients, indicating that the disease arises in cells with self-renewal properties. Interestingly, the latency in secondary recipients was shorter than seen in the initially injected mice, which, combined with the fact that the myeloid blasts generated showed mono- or oligo-clonality, suggested that secondary mutations may be required for the disease to progress to a fully malignant form of AML.

A more recent study compared the leukaemogenic properties of MOZ–TIF2 and BCR–ABL *in vivo* [49]. BCR–ABL is generated by the translocation t(9;22)(q34;q11) and leads to the constitutive activation of the ABL tyrosine kinase, resulting in chronic myeloid leukaemia [50]. This results in the abnormal expansion of haematopoietic cells expressing the fusion protein, although they are still able to achieve terminal differentiation. Expression of MOZ–TIF2, but not BCR–ABL, in normal bone-marrow mononuclear cells, purified HSCs or committed progenitor cells extended their proliferation in serial replating assays

[49]. The finding that common myeloid and granulocyte/monocyte progenitors transduced with MOZ–TIF2 could be serially replated indicated that the fusion protein causes the re-establishment of self-renewal in cells that are normally committed irrevocably to terminal differentiation. Following multiple rounds of replating, these cells formed small blast-like colonies and showed morphological features indistinguishable from those of blasts obtained from patients with MOZ rearrangements. In addition, when transplanted into irradiated mice, these MOZ–TIF2-expressing cells induced an AML-like syndrome that could be transplanted further to secondary recipients. Thus MOZ–TIF2 expression appears to mediate a block in myeloid differentiation that is downstream of the normally committed granulocyte/monocyte progenitors, and confers a specific leukaemogenic phenotype to the cells that includes the ability to self-renew. In contrast, the BCR–ABL fusion protein confers short-term proliferative advantages to transduced cells, but not self-renewal properties, and is unable to inhibit terminal differentiation. These findings indicate that there are qualitative differences between fusion proteins generated by different leukaemogenic translocations. It also confirms that the expression of the MOZ–TIF2 protein is sufficient to induce a leukaemic or pre-leukaemic state through the inhibition of myeloid differentiation and re-establishment of self-renewal in progenitor cells committed to terminal differentiation.

MOZ fusion protein-mediated inhibition of transcription by NRs, p53 and AML1

In order to understand the molecular mechanisms underlying the action of MOZ fusion proteins, several studies have compared the function of wild-type and fusion proteins in biochemical and cell-based assays. A study of the effect of MOZ–CBP on the transcriptional activity of AML1b showed that, in contrast with the enhancement activity observed with wild-type MOZ, MOZ–CBP strongly inhibited AML1b-mediated transcription in a dose-dependent manner [17]. Deletion mapping experiments indicated that the HAT domain and bromodomain of CBP were required for inhibition of AML1b by MOZ–CBP. In contrast, deletion of the PHD finger and MYST domains of MOZ did not hamper the transcription-inhibitory properties of MOZ–CBP [17]. CBP and p300 also function as co-activators for AML1 [51], and thus the inhibition of transcription by MOZ–CBP may be mediated through dominant-negative effects on the normal function of either MOZ or CBP, or of both proteins. To address this, reporter assays using proteins consisting of the GAL4 DNA-binding domain fused to full-length MOZ or CBP were performed. These revealed that MOZ–CBP inhibited transactivation by GAL4–CBP, but not GAL4–MOZ. This finding was supported by the observation that CBP, but not MOZ, rescued AML1b activity in the presence of MOZ–CBP. Thus the inhibition of AML1b activity by MOZ–CBP appears to involve interference with the function of CBP, rather than MOZ. Further analysis will be required to understand the molecular mechanism underlying this effect.

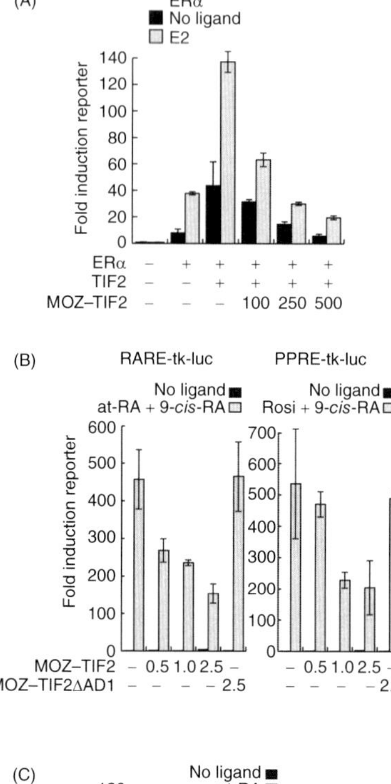

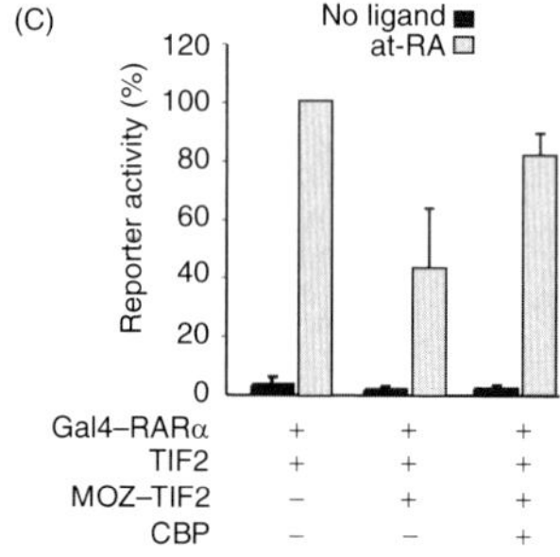

Figure 2 Inhibition of NR-mediated transcription by the MOZ–TIF2 fusion protein. (**A**) Expression of MOZ–TIF2 in COS-1 cells results in a dose-dependent inhibition of luciferase reporter gene activation by exogenous oestrogen receptor α (ERα) in combination with TIF2. Black bars indicate reporter activity in the presence of vehicle only (no ligand). Shaded bars indicate reporter activity in the presence of 17β-oestradiol (E2). Cells were transfected with 100 ng of ERα and/or TIF2 expression vectors (+). The amount (ng) of transfected MOZ–TIF2 vector is indicated. Reporter activation is represented as fold induction over basal activity (in the absence of ERα or ligand). Error bars indicate S.E.M. (**B**) Effects of MOZ–TIF2 or MOZ–TIF2ΔAD1 on ligand-induced activation of RAR- and PPAR-regulated reporter genes in U937 myeloid leukaemia cells. The reporter plasmid RARE-tk-Luc consists of a minimal thymidine kinase

As both TIF2 and CBP play important roles in ligand-dependent transcription by NRs, we investigated the effects of MOZ–TIF2 on NR-mediated transcription [52]. The MOZ–TIF2 fusion protein sequence lacks the TIF2 NR-interaction domain, but retains its CBP-binding domain (AD1; see Figure 1). Protein interaction experiments confirmed that MOZ–TIF2 interacts with CBP both *in vitro* and *in vivo* through its AD1 [52]. Moreover, MOZ–TIF2 strongly inhibited ligand-dependent reporter activation by oestrogen receptors in COS-1 cells in a dose-dependent manner (Figure 2A). Similarly, the ability of endogenous RARs (retinoic acid receptors) and PPARs (peroxisome-proliferator-activated receptors) to activate reporter genes was severely compromised in U937 myeloid leukaemia cells expressing MOZ–TIF2 (Figure 2). In contrast, MOZ–TIF2ΔAD1, which lacks the CBP-binding domain, failed to inhibit transcription by RARs and PPARs (Figure 2B) or other NRs [52] under similar conditions. This strongly suggests that transcriptional inhibition by MOZ–TIF2 is dependent on its interaction with CBP. Consistent with this, and as observed for the effect of MOZ–CBP on AML1b activity, exogenous CBP, but not MOZ, rescued the inhibitory effects of MOZ–TIF2 (Figure 2C). Thus MOZ–TIF2 acts as a dominant-negative inhibitor of NR-mediated transcription through mechanisms reliant on sequestration of CBP. Given that NRs and their ligands are important regulators of haematopoietic cell differentiation [53,54], the inhibition of specific NR activities may be a contributory factor in the transformation of haematopoietic progenitors expressing MOZ fusion proteins.

As MOZ fusion proteins appear to impair the function of CBP, we investigated whether CBP-dependent transcription factors other than NRs are also susceptible to their effects. The tumour suppressor p53 is a CBP-dependent transcription factor that functions in cell proliferation and in response to DNA damage, and is a frequent target of mutations in cancers. Our results showed that MOZ–TIF2 also inhibited p53-mediated transcription of reporter genes via a number of different p53-regulated promoters [52]. Once again, a mutant lacking AD1 failed to inhibit p53 transcription, and exogenous CBP rescued p53 activity [52]. Other leukaemogenic fusion proteins such as MLL (mixed lineage leukaemia protein)–ELL (eleven-nineteen lysine-rich leukaemia protein) have been shown to interfere with the interaction of CBP and p53, resulting in the inhibition of p53 transcriptional activity [55]. Thus, while the p53 gene is normally intact in leukaemias, the function of the p53 protein may be impaired by leukaemogenic proteins such as the MOZ fusion proteins.

promoter containing a DR5 response element upstream of the luciferase gene. PPRE-tk-Luc contains a PPARα response element within the minimal thymidine kinase promoter driving luciferase expression. Black bars indicate reporter activity in the presence of vehicle only (no ligand). Shaded bars indicate reporter activity in the presence of all-*trans*-retinoic acid (at-RA) plus 9-*cis*-retinoic acid (9-*cis*-RA) (for RARE-tk-Luc), or rosiglitazone (Rosi) plus 9-*cis*-retinoic acid for the PPRE-tk-Luc reporter. Amounts of transfected DNA are indicated (μg). (C) CBP rescues the inhibition of RAR activity by MOZ–TIF2. Activation of a GAL4 reporter gene by GAL4–RARα fusion protein in the absence (black bars) or presence (shaded bars) of at-RA in COS-1 cells. Cells were transfected with 100 ng of expression vector as indicated. Reproduced in part (with permission from the American Society for Microbiology) from Kindle et al. [52].

Reduced levels of CBP protein in cells expressing MOZ–TIF2

We have investigated the subcellular distribution of MOZ proteins in a variety of cell lines and shown that MOZ localizes to discrete nuclear speckles [52]. In contrast, MOZ–TIF2 has a predominantly non-speckled, mesh-like nuclear distribution. CBP is also associated with nuclear speckles in many cells [the PML (promyelocytic leukaemia) bodies]. The CBP-binding domain AD1 is critical for the ability of MOZ–TIF2 to perturb transcription factor activity, and also for the induction of AML in mice. As FRET (fluorescence resonance energy transfer) experiments had confirmed that MOZ–TIF2 and CBP proteins interact *in vivo*, we investigated the effects of various MOZ fusion proteins on the subcellular localization of CBP. Cells expressing MOZ–TIF2 showed a marked reduction of CBP staining in PML bodies, although the PML bodies were intact, suggesting that the interaction results in mislocalization of CBP. Indeed, the specificity of this effect was confirmed by the observation that association of CBP with PML bodies was normal in cells expressing the MOZ–TIF2ΔAD1 deletion mutant [52]. However, cells expressing MOZ–TIF2 displayed a dramatic reduction in levels of cellular CBP, as determined by Western blots. CBP depletion was also observed in murine lineage-deficient HSCs transduced with MOZ–TIF2 vector, but not controls or the MOZ–TIF2ΔAD1 expression vector. These results are consistent with the observed decreases in the transcriptional activities of NRs, p53 and other CBP-dependent transcription factors that regulate haematopoietic cell differentiation. Treatment of MOZ–TIF2-transduced cells with the proteasome inhibitor lactacystine failed to rescue CBP levels, indicating that the degradation is proteasome-independent [52]. The mechanism of CBP depletion is currently under investigation. In summary, the subversion of the function of CBP/p300, and potentially other chromatin modification enzymes, is likely to be important in the leukaemogenicity of MOZ fusion proteins.

In addition to AD1, the C2HC zinc finger present at the N-terminus of the MYST domain is another motif that is essential for the induction of AML in mice [20]. In the MYST family protein MOF, the C2HC zinc finger was reported to mediate an interaction with nucleosomes [19]. The subcellular distribution of MOZ fusion proteins is distinct from patterns observed for either parental protein ([52]; H.M. Collins, K.B. Kindle, S. Matsuda, P.J.F. Troke, E. Kalkhoven and D.M. Heery, unpublished work), suggesting that MOZ fusion proteins are recruited to sites other than those occupied by the parental proteins. Thus, in addition to mislocalization/degradation of CBP, gene expression may also be perturbed by aberrant recruitment of MOZ and CBP HAT domains to chromatin.

Concluding remarks

Chromosomal translocations associated with AML subtype M4/M5 involve genes encoding the protein acetyltransferases MOZ, MORF, CBP and p300, or the p160 protein TIF2. In addition to haplo-insufficiency effects, these events generate novel fusion proteins containing the MYST domain and N-terminal

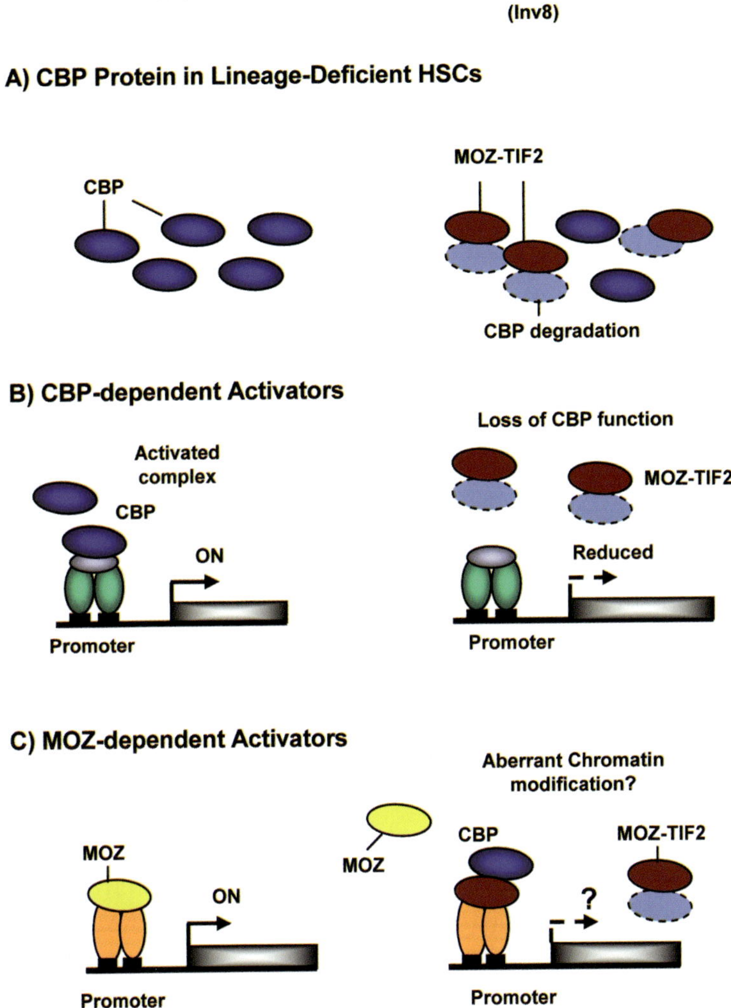

Figure 3 Molecular action of MOZ–TIF2 *in vivo.* (**A**) Schematic representation of the effects of MOZ–TIF2 *in vivo*. Proteins are represented by colour-coded ellipses. (**A**) MOZ–TIF2 proteins (red) in inv8(p11;q13) leukaemic blasts interact with cellular CBP (dark blue), leading to its degradation (light blue). (**B**) Transcription by CBP-dependent activators (green) is compromised by MOZ–TIF2 due to the sequestration and degradation of CBP. (**C**) Transcription by MOZ-dependent activators (orange) may be perturbed by competition between MOZ (yellow) and MOZ–TIF2 for protein targets, resulting in aberrant chromatin modification due to recruitment of MOZ–TIF2/CBP complex to MOZ target promoters.

sequences of MOZ or MORF fused to almost full-length HATs CBP and p300, or the C-terminal sequence of TIF2. Similarities in the leukaemogenic phenotypes of the MOZ–TIF2 and MOZ–CBP fusions may be explained by the ability of MOZ–TIF2 to bind CBP and thus form a complex that mimics the MOZ–CBP and MOZ–p300 fusion proteins. A number of studies have established that the expression of MOZ fusion proteins is sufficient to transform haematopoietic progenitors and induce AML-like disease in mice [17,20,49,52]. Here we have reviewed recent investigations into the molecular events underlying the leukaemogenicity of MOZ fusion proteins. These studies indicate that multiple molecular processes are likely to be perturbed by MOZ fusion proteins. The interaction of MOZ–TIF2 with CBP results in mislocalization of CBP from PML bodies, which are a nuclear sump for this protein. This is dependent on AD1, which forms a stable complex with the SID of CBP [35]. Similarly, MOZ–CBP is mislocalized from PML bodies, and both fusion proteins show a distinct subcellular localization to MOZ (H.M. Collins, K.B. Kindle, S. Matsuda, P.J.F. Troke, E. Kalkhoven and D.M. Heery, unpublished work). MOZ–TIF2 expression in HSCs and other cell lines results in a marked decrease in cellular levels of the CBP and p300 proteins (Figure 3A) [52]. The degradation of CBP is achieved by a non-proteasomal pathway, which remains to be characterized. In the t8;16 translocations, loss of one allele of the CBP gene may also result in a decrease in wild-type CBP protein levels through haplo-insufficiency, although this remains to be determined.

The mislocalization of CBP and the decrease in CBP levels is likely to be a key factor in the observed loss of transcriptional activity of CBP-dependent activators such as NRs and p53, and other activators that are required for the differentiation of haematopoietic tissue. This hypothesis is supported by the observation that exogenous CBP can rescue transcription of affected reporter genes. Moreover, disabling AD1 of MOZ–TIF2, so that it can no longer interact with CBP, results in proteins that are unable to transform haematopoietic progenitors *in vitro* [49,52] or cause AML in a mouse model [20]. In addition to its effects on CBP/p300 function, MOZ–TIF2 may perturb other chromatin-modifying enzymes such as CARM1 and PRMT1, which can associate with AD2 of TIF2.

The effects of MOZ fusion proteins on MOZ-dependent activators are less clear. Although MOZ–CBP was reported to inhibit AML1-dependent reporter activity [17], we have observed that MOZ–TIF2 enhances AML1 reporter activation (H.C. Collins, K.B. Kindle, S. Matsuda, P.J.F. Troke, E. Kalkhoven and D.M. Heery, unpublished work). Consistent with this, MOZ–TIF2 enhanced or repressed the expression of different subsets of genes in comparative microarray studies (H.M. Collins and D.M. Heery, unpublished work), and thus is likely to have differential effects on gene expression *in vivo*. MOZ–TIF2 and MOZ–CBP display aberrant nuclear localization that may indicate that they are recruited to novel chromatin sites by a mechanism involving various nucleosome-binding motifs in the N-terminus of MOZ (Figure 3C). This may result in aberrant chromatin modifications that affect gene expression and thus differentiation pathways. Consistent with the above observations, the CBP- and nucleosome-binding domains were found to be critical for the leukaemogenicity of MOZ–TIF2 [20]. Further understanding of the molecular mechanisms by which MOZ fusion proteins subvert haematopoietic differentiation may enable the design of therapeutic approaches for the treatment of these malignancies.

References

1. Lai, J.L., Zandecki, M., Jouet, J.P., Savary, J.B., Lambiliotte, A., Bauters, F., Cosson, A. and Deminatti, M. (1987) Cancer Genet. Cytogenet. **27**, 101–109
2. Borrow, J., Stanton, Jr, V.P., Andresen, J.M., Becher, R., Behm, F.G., Chaganti, R.S., Civin, C.I., Disteche, C., Dube, I., Frischauf, A.M. et al. (1996) Nat. Genet. **14**, 33–41
3. Sun, T. and Wu, E. (2001) Am. J. Hematol. **66**, 207–212
4. Schmidt, H.H., Strehl, S., Thaler, D., Strunk, D., Sill, H., Linkesch, W., Jager, U., Sperr, W., Greinix, H.T., Konig, M. et al. (2004) Leukemia **18**, 1115–1121
5. Panagopoulos, I., Fioretos, T., Isaksson, M., Mitelman, F., Johansson, B., Theorin, N. and Juliusson, G. (2002) Genes Chromosomes Cancer **35**, 372–374
6. Carapeti, M., Aguiar, R.C., Goldman, J.M. and Cross, N.C. (1998) Blood **91**, 3127–3133
7. Liang, J., Prouty, L., Williams, B.J., Dayton, M.A. and Blanchard, K.L. (1998) Blood **92**, 2118–2122
8. Murati, A., Adelaide, J., Popovici, C., Mozziconacci, M.J., Arnoulet, C., Lafage-Pochitaloff, M., Sainty, D., Birnbaum, D. and Chaffanet, M. (2003) Int. J. Mol. Med. **12**, 423–428
9. Chaffanet, M., Gressin, L., Preudhomme, C., Soenen-Cornu, V., Birnbaum, D. and Pebusque, M.J. (2000) Genes Chromosomes Cancer **28**, 138–144
10. Panagopoulos, I., Fioretos, T., Isaksson, M., Samuelsson, U., Billstrom, R., Strombeck, B., Mitelman, F. and Johansson, B. (2001) Hum. Mol. Genet **10**, 395–404
11. Imamura, T., Kakazu, N., Hibi, S., Morimoto, A., Fukushima, Y., Ijuin, I., Hada, S., Kitabayashi, I., Abe, T. and Imashuku, S. (2003) Genes Chromosomes Cancer **36**, 413–419
12. Carapeti, M., Aguiar, R.C.T., Chase, A., Goldman, J.M. and Cross, N.C.P. (1998) Genomics **52**, 242–244
13. Panagopoulos, L., Isaksson, M., Lindvall, C., Hagemeijer, A., Mitelman, F. and Johansson, B. (2003) Genes Chromosomes Cancer **36**, 90–98
14. Yang, X.J. (2004) Nucleic Acids Res. **32**, 959–976
15. Champagne, N., Bertos, N.R., Pelletier, N., Wang, A.H., Vezmar, M., Yang, Y., Heng, H.H. and Yang, X.J. (1999) J. Biol. Chem. **274**, 28528–28536
16. Champagne, N., Pelletier, N. and Yang, X.J. (2001) Oncogene **20**, 404–409
17. Kitabayashi, I., Aikawa, Y., Nguyen, L.A., Yokoyama, A. and Ohki, M. (2001) EMBO J. **20**, 7184–7196
18. Pelletier, N., Champagne, N., Lim, H. and Yang, X.J. (2003) Methods **31**, 24–32
19. Akhtar, A. and Becker, P.B. (2001) EMBO Rep. **2**, 113–118
20. Deguchi, K., Ayton, P.M., Carapeti, M., Kutok, J.L., Snyder, C.S., Williams, I.R., Cross, N.C.P., Glass, C.K., Cleary, M.L. and Gilliland, D.G. (2003) Cancer Cell **3**, 259–271
21. Scott, E.K., Lee, T. and Luo, L.Q. (2001) Curr. Biol. **11**, 99–104
22. Gozani, O., Karuman, P., Jones, D.R., Ivanov, D., Cha, J., Lugovskoy, A.A., Baird, C.L., Zhu, H., Field, S.J., Lessnick, S.L. et al. (2003) Cell **114**, 99–111
23. Bristow, C.A.P. and Shore, P. (2003) Nucleic Acids Res. **31**, 2735–2744
24. Miller, C.T., Maves, L. and Kimmel, C.B. (2004) Development **131**, 2443–2461
25. Chan, H.M. and La Thangue, N.B. (2001) J. Cell Sci. **114**, 2363–2373
26. Dyson, H.J. and Wright, P.E. (2005) Nat. Rev. Mol. Cell Biol. **6**, 197–208
27. Daujat, S., Bauer, U.M., Shah, V., Turner, B., Berger, S. and Kouzarides, T. (2002) Curr. Biol. **12**, 2090–2097
28. Metivier, R., Penot, G., Hubner, M.R., Reid, G., Brand, H., Kos, M. and Gannon, F. (2003) Cell **115**, 751–763
29. Blobel, G.A. (2002) J. Leukocyte Biol. **71**, 545–556
30. Kung, A.L., Rebel, V.I., Bronson, R.T., Ch'ng, L.E., Sieff, C.A., Livingston, D.M. and Yao, T.P. (2000) Genes Dev. **14**, 272–277

31. Rebel, V.I., Kung, A.L., Tanner, E.A., Yang, H., Bronson, R.T. and Livingston, D.M. (2002) Proc. Natl. Acad. Sci. U.S.A. **99**, 14789–14794
32. Leo, C. and Chen, J.D. (2000) Gene **245**, 1–11
33. Xu, J.M. and Li, Q.T. (2003) Mol. Endocrinol. **17**, 1681–1692
34. Heery, D.M., Kalkhoven, E., Hoare, S. and Parker, M.G. (1997) Nature (London) **387**, 733–736
35. Sheppard, H.M., Harries, J.C., Hussain, S., Bevan, C. and Heery, D.M. (2001) Mol. Cell. Biol. **21**, 39–50
36. Lee, S.K., Kim, H.J., Na, S.Y., Kim, T.S., Choi, H.S., Im, S.Y. and Lee, J.W. (1998) J. Biol. Chem. **273**, 16651–16654
37. Na, S.Y., Lee, S.K., Han, S.J., Choi, H.S., Im, S.Y. and Lee, J.W. (1998) J. Biol. Chem. **273**, 10831–10834
38. Lee, S.K., Kim, H.J., Kim, J.W. and Lee, J.W. (1999) Mol. Endocrinol. **13**, 1924–1933
39. Belandia, B. and Parker, M.G. (2000) J. Biol. Chem. **275**, 30801–30805
40. Chen, S.L., Wang, S.C.M., Hosking, B. and Muscat, G.E.O. (2001) Mol. Endocrinol. **15**, 783–796
41. Giraud, S., Bienvenu, F., Avril, S., Gascan, H., Heery, D.M. and Coqueret, O. (2002) J. Biol. Chem. **277**, 8004–8011
42. Louie, M.C., Zou, J.X., Rabinovich, A. and Chen, H.W. (2004) Mol. Cell. Biol. **24**, 5157–5171
43. Mark, M., Yoshida-Komiya, H., Gehin, M., Liao, L., Tsai, M.J., O'Malley, B.W., Chambon, P. and Xu, J.M. (2004) Proc. Natl. Acad. Sci. U.S.A. **101**, 4453–4458
44. Chen, H., Lin, R.J., Schiltz, R.L., Chakravarti, D., Nash, A., Nagy, L., Privalsky, M.L., Nakatani, Y. and Evans, R.M. (1997) Cell **90**, 569–580
45. Kalkhoven, E., Valentine, J.E., Heery, D.M. and Parker, M.G. (1998) EMBO J. **17**, 232–243
46. Voegel, J.J., Heine, M.J., Tini, M., Vivat, V., Chambon, P. and Gronemeyer, H. (1998) EMBO J. **17**, 507–519
47. Chen, D., Ma, H., Hong, H., Koh, S.S., Huang, S.M., Schurter, B.T., Aswad, D.W. and Stallcup, M.R. (1999) Science **284**, 2174–2177
48. Stallcup, M.R., Chen, D., Koh, S.S., Ma, H., Lee, Y.H., Li, H., Schurter, B.T. and Aswad, D.W. (2000) Biochem. Soc. Trans. **28**, 415–418
49. Huntly, B.J.P., Shigematsu, H., Deguchi, K., Lee, B.H., Mizuno, S., Duclos, N., Rowan, R., Amaral, S., Curley, D., Williams, I.R. et al. (2004) Cancer Cell **6**, 587–596
50. Saglio, G. and Cilloni, D. (2004) Cell. Mol. Life Sci. **61**, 2897–2911
51. Kitabayashi, I., Yokoyama, A., Shimizu, K. and Ohki, M. (1998) EMBO J. **17**, 2994–3004
52. Kindle, K.B., Troke, P.J.F., Collins, H.M., Matsuda, S., Bossi, D., Bellodi, C., Kalkhoven, E., Salomoni, P., Pelicci, P.G., Minucci, S. and Heery, D.M. (2005) Mol. Cell. Biol. **25**, 988–1002
53. Tenen, D.G., Hromas, R., Licht, J.D. and Zhang, D.E. (1997) Blood **90**, 489–519
54. Zelent, A., Guidez, F., Melnick, A., Waxman, S. and Licht, J.D. (2001) Oncogene **20**, 7186–7203
55. Wiederschain, D., Kawai, H., Gu, J.J., Shilatifard, A. and Yuan, Z.M. (2003) Mol. Cell. Biol. **23**, 4230–4246
56. Panagopoulos, I., Isaksson, M., Lindvall, C., Bjorkholm, M., Ahlgren, T., Fioretos, T., Heim, S., Mitelman, F. and Johansson, B. (2000) Genes Chromosomes Cancer **28**, 415–424
57. Rozman, M., Camos, M., Colomer, D., Villamor, N., Esteve, J., Costa, D., Carrio, A., Aymerich, M., Aguilar, J.L., Domingo, A. et al. (2004) Genes Chromosomes Cancer **40**, 140–145
58. Tasaka, T., Matsuhashi, Y., Uehara, E., Tamura, T., Kakazu, N., Abe, T. and Nagai, M. (2004) Leukemia Lymphoma **45**, 621–625
59. Giles, R.H., Dauwerse, J.G., Higgins, C., Petrij, F., Wessels, J.W., Beverstock, G.C., Dohner, H., Jotterand-Bellomo, M., Falkenburg, J.H., Slater, R.M. et al. (1997) Leukemia **11**, 2087–2096
60. Bernasconi, P., Orlandi, E., Cavigliano, P., Calatroni, S., Boni, M., Astori, C., Pagnucco, G., Giglio, S., Caresana, M., Lazzarino, M. and Bernasconi, C. (2000) Haematologica **85**, 1087–1091
61. Kitabayashi, I., Aikawa, Y., Yokoyama, A., Hosoda, F., Nagai, M., Kakazu, N., Abe, T. and Ohki, M. (2001) Leukemia **15**, 89–94
62. Carapeti, M., Aguiar, R.C., Watmore, A.E., Goldman, J.M. and Cross, N.C. (1999) Cancer

Genet. Cytogenet. **113**, 70–72

63. Panagopoulos, I., Teixeira, M.R., Micci, F., Hammerstrom, J., Isaksson, M., Johansson, B., Mitelman, F. and Heim, S. (2000) Leukemia Lymphoma **39**, 651–656
64. Billio, A., Steer, E.J., Pianezze, G., Svaldi, M., Casin, M., Amato, B., Coser, P. and Cross, N.C. (2002) Haematologica **87**, ECR15
65. Vizmanos, J.L., Larrayoz, M.J., Lahortiga, I., Floristan, F., Alvarez, C., Odero, M.D., Novo, F.J. and Calasanz, M.J. (2003) Genes Chromosomes Cancer **36**, 402–405
66. Murati, A., Adelaide, J., Mozziconacci, M.J., Popovici, C., Carbuccia, N., Letessier, A., Birg, F., Birnbaum, D. and Chaffanet, M. (2004) Br. J. Haematol. **125**, 601–604
67. Kojima, K., Kaneda, K., Yoshida, C., Dansako, H., Fujii, N., Yano, T., Shinagawa, K., Yasukawa, M., Fujita, S. and Tanimoto, M. (2003) Br. J. Haematol. **120**, 271–273

Biochem. Soc. Symp. **73**, 41–47
(Printed in Great Britain)

4

Mechanistic studies of the mRNA transcription cycle

Patrick Cramer[1]

Gene Center, Ludwig-Maximilians-University of Munich, Department of Chemistry and Biochemistry, Feodor-Lynen-Str. 25, 81377 Munich, Germany

Abstract

We have now completed an atomic crystallographic model of the 12-subunit yeast RNA polymerase II in elongation mode, with DNA and RNA in the active-centre cleft, and the NTP substrate at the growing end of the RNA. From these studies has emerged a detailed three-dimensional view of mRNA elongation. We have extended this structural analysis to a polymerase elongation complex bound by the transcript cleavage factor TFIIS (transcription factor IIS), which is required for polymerase escape from DNA arrest sites. A detailed model of this complex reveals a single tuneable active site for RNA polymerization and cleavage, and changes in the position of the RNA and polymerase domains, reflecting the dynamic nature of the elongation complex. An additional structure of a polymerase CTD (C-terminal domain) phosphopeptide bound by the 3′-RNA processing factor Pcf11 provides insights into the coupling of transcription elongation to mRNA processing. The structure of the CTD phosphatase Scp1 trapped in an intermediary enzymatic state explains CTD dephosphorylation during recycling of the polymerase. We also recently reported the first crystal structure of a Mediator subcomplex, which reveals an extended helical fold with a conserved hinge.

Introduction

Following the structure determination of the ten-subunit core of Pol II (RNA polymerase II) from yeast in free form and bound to a DNA–RNA hybrid in the Kornberg laboratory [1–4], our group has concentrated over the last 4 years on working towards the complete 12-subunit Pol II structure, and on structures of functional complexes of the complete Pol II with nucleic acids and with additional protein factors. In addition, several structures of

[1]email cramer@lmb.uni-muenchen.de

Pol II-associated factors have been resolved. The main technique used in our laboratory is X-ray crystallography, but a wide range of genetic, biochemical and biophysical methods have also been used in the course of these studies. The group also develops structural biology methods for solving transient multicomponent crystal structures that can help to elucidate the transcription mechanism. Several previous reviews have summarized parts of our work and have compared our results with those from other laboratories [5–7]. Here we review our main results from the last 4 years.

Reconstitution, crystallization and backbone model of the complete 12-subunit Pol II [8]

In order to prepare the complete Pol II, including the heterodimeric subcomplex Rpb4/7 that was lacking in previous studies, for crystallization, we reconstituted the enzyme from the endogenous 10-subunit Pol II core, purified from yeast culture by immunoaffinity chromatography, and from the recombinant Rpb4/7 subcomplex, obtained by co-expression of Rpb4 and Rpb7 in *Escherichia coli* [9]. To obtain a stoichiometric polymerase, an excess of Rpb4/7 was used during reconstitution, and the resulting complete enzyme was purified further by gel filtration. The complete Pol II was crystallized under new conditions and adopted a new crystal form with a very large unit cell that contained approx. 80% solvent. Phases were derived from the refined atomic model of the core Pol II. A backbone model of the complete Pol II was obtained by fitting the refined structure of the Pol II core and a structure of the archaeal counterpart of the Rpb4/7 complex [10] to a 4.2 Å crystallographic electron density map. To confirm the fit of the distantly related Rpb4/7 homologue, it was crucial to label recombinant Rpb4/7 with selenomethionine and to detect the labels with the anomalous selenium signal.

Previous structures of the Pol II core revealed a ‘clamp’ that binds the DNA template strand via three switch regions and a flexible linker to the repeat CTD (C-terminal domain). In the complete Pol II structure, Rpb4/7 protrudes from the polymerase upstream face, on which initiation factors assemble for promoter DNA loading, explaining why Rpb4/7 is essential for initiation. Rpb7 forms a wedge between the clamp and the linker, restricting the clamp to a closed position. The allosteric wedge prevents entry of the promoter DNA duplex into the active-centre cleft, and induces in two switch regions a conformation poised for template strand binding. Interaction of Rpb4/7 with the Pol II linker explains Rpb4-mediated recruitment of the CTD phosphatase to the CTD during Pol II recycling. The core–Rpb7 interaction and some functions of Rpb4/7 are apparently conserved in all eukaryotic and archaeal RNA polymerases, but not in the bacterial enzyme.

Structure of complete Pol II and free Rpb4/7 [11]

The crystals of the complete Pol II are very radiation-sensitive and show a limited resolution. Using elaborate data collection strategies and an optimized cryo-cooling procedure, we were able to extend the resolution to 3.8 Å, but could not extend it further. In order to build and refine an atomic model of Pol II at this resolution, we required a high-quality structure for Rpb4/7. We thus crystallized the free Rpb4/7 complex and solved its structure at 2.3 Å resolution. Structure determination of free Rpb4/7 was difficult, and a flexible N-terminal tail and internal loop in Rpb4 had to be removed to obtain diffraction-quality crystals. The structure was solved using selenomethionine-labelled protein and multi-wavelength anomalous diffraction. We combined the atomic Rpb4/7 structure with the previously refined core Pol II structure, and could refine a complete atomic model of Pol II at 3.8 Å resolution (Figure 1). Comparison of the free Pol II core and the free Rpb4/7 structures with the structure of the complete Pol II revealed that the core–Rpb4/7 interaction involves induced folding of a part of the linker domain of the largest subunit and of the 'tip' loop of Rpb7.

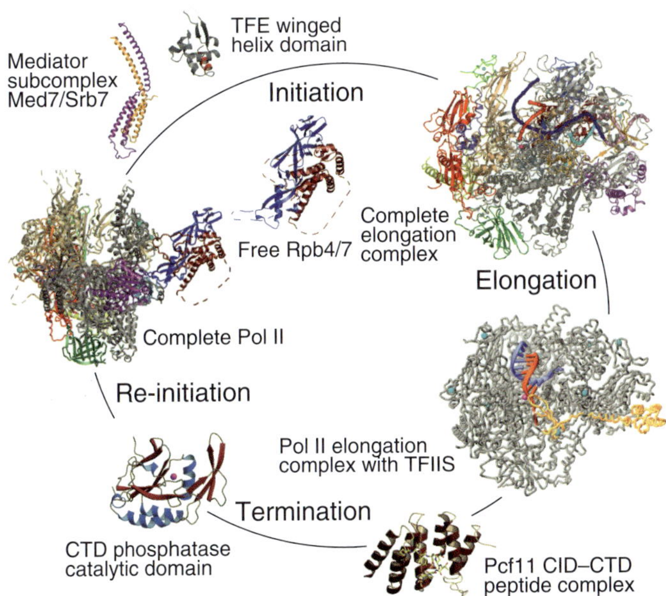

Figure 1 Structures elucidating the mRNA transcription cycle. Crystal structures are arranged in a clockwise circular fashion according to their relevance for understanding the mRNA transcription cycle. The phases of the cycle are indicated (initiation, elongation, termination and re-initiation). For details on the structures and references, please refer to the text.

Mechanisms of elongation: structure of the complete Pol II–transcription bubble–RNA complex with substrate NTP [12]

Based on the refined complete Pol II structure, we could solve the structure of a complex of the complete Pol II bound to a DNA transcription bubble and product RNA (a complete elongation complex; Figure 1). The complex was obtained by incubating reconstituted complete Pol II with a pre-annealed nucleic acid construct consisting of a DNA 41-mer with a central mismatch bubble and an RNA 20-mer. The crystal structure reveals entering template and non-template DNA, a 7 bp DNA/RNA hybrid, and three nucleotides of separating DNA and RNA. The complex adopts the post-translocation state and accommodates a co-crystallized non-reactive analogue of an NTP substrate. The NTP binds in the active-site pore, at a position to interact with a DNA template base. Residues surrounding the NTP are conserved in all cellular RNA polymerases, suggesting a universal mechanism of NTP selection and incorporation. DNA–DNA and DNA–RNA strand separation can be explained by Pol II-induced duplex distortions. Four protein loops partition the active-centre cleft, contribute to embedding the hybrid, prevent strand re-association and create an RNA exit tunnel. This recently described structure reveals all protein subunits and all nucleic acid elements in a natural Pol II transcription elongation complex and thus represents a milestone in our analysis of the mechanism of gene transcription.

Switching from mRNA elongation to cleavage: the Pol II–TFIIS (transcription factor IIS) complex [12,13]

We could further extend structural studies of Pol II to a complex with the transcription elongation factor TFIIS, which binds transiently to the polymerase. This is the first structure of a Pol II–transcription-factor complex. TFIIS was the first Pol II-associated factor to be purified, and induces mRNA cleavage by enhancing a weak intrinsic nuclease activity of Pol II. We were able to allow TFIIS to diffuse into preformed Pol II crystals. This allowed us to use high TFIIS concentrations, and to achieve full occupancy of the factor on the Pol II surface. To monitor TFIIS soaking, we have developed a novel fluoresecence-based assay (H. Kettenberger and P. Cramer, unpublished work). A similar assay can be used to monitor the presence of nucleic acids in potential polymerase–nucleic-acid complex crystals. A backbone model of the Pol II–TFIIS complex from X-ray data to 3.8 Å resolution shows that TFIIS binds to Pol II in a manner that could not have been predicted, since both TFIIS and Pol II undergo several folding transitions and conformational changes. TFIIS extends from the polymerase surface via a pore to the internal active site, spanning a distance of 100 Å. Two essential and invariant acidic residues in a TFIIS hairpin complement the Pol II active site and could position a metal ion and a water molecule for hydrolytic RNA cleavage. TFIIS also induces structural changes in Pol II. Soaking of preformed elongation complex crystals with TFIIS revealed that TFIIS re-aligns

RNA in the active centre. The work provides strong evidence for the idea that Pol II has a single 'tuneable' active site for RNA polymerization and cleavage, in contrast with DNA polymerases, which have two separate sites for DNA polymerization and cleavage. Also, transcription factors and their target surfaces on Pol II can undergo major structural rearrangements, suggesting that even more dramatic structural transitions will be seen in the future [6].

Coupling mRNA elongation and processing: recognition of the RNA polymerase CTD by a CTD-binding domain [14]

Transcription elongation is coupled to RNA processing by the CTD of Pol II, which consists of up to 52 repeats of the sequence Tyr^1-Ser^2-Pro^3-Thr^4-Ser^5-Pro^6-Ser^7. The phosphorylated CTD binds tightly to a conserved CID (CTD-interacting domain) that is present in the proteins Pcf11 and Nrd1, which are essential and evolutionarily conserved factors for polyadenylation-dependent and -independent 3′-RNA processing respectively. We have solved the structure of a Ser^2-phosphorylated CTD peptide bound to the CID of Pcf11. The CTD motif Ser^2-Pro^3-Thr^4-Ser^5 forms a β-turn that binds to a conserved groove in the CID. The Ser^2 phosphate group does not contact the CID, consistent with reports that the free CTD also shows some affinity for the CID. However, since phosphorylation at Ser^2 had been shown to enhance CID affinity, we suggested that the Ser^2 phosphate group is recognized indirectly. The mechanism of indirect recognition, however, remains unclear. Since the Ser^2 phosphate apparently does not form any intra-CTD interactions in a free CTD peptide [15], it is unlikely that this phosphorylation induces a conformation in the free CTD that is bound in preference.

Pol II recycling: structure and mechanism of an RNA polymerase CTD phosphatase [16]

After a round of transcription, Pol II must be recycled, which requires dephosphorylation of the CTD by the phosphatase Fcp1. We have determined the X-ray structure of the small CTD phosphatase Scp1, which is homologous with the Fcp1 catalytic domain. The structure shows similarity to phosphotransferases and phosphohydrolases that solely share a DXDX(V/T) signature motif with Fcp1/Scp1. The first aspartate residue in the signature motif undergoes metal-assisted phosphorylation during catalysis, resulting in a phosphoaspartate intermediate that we could mimic structurally with the inhibitor beryllofluoride. The basis for CTD specificity remains unclear, but may involve an insertion domain that is unique to Fcp1/Scp1. Fcp1 specificity may arise additionally from phosphatase recruitment near the CTD via the Pol II subcomplex Rpb4/7, which is required for Fcp1 binding to the polymerase *in vitro*. Another CTD phosphatase, unrelated to the Fcp1/Scp1 family, has recently been characterized [17–19].

Towards initiation: structure of an essential domain of the initiation factor TFE/TFIIE [20]

In order to understand the mechanism of transcription initiation, we ultimately have to obtain the structure of the initiation complex, which comprises, in addition to Pol II, the general transcription factors TFIIB, TFIID, TFIIE, TFIIF and TFIIH. As a first step towards this remote goal, we have to determine structures of the general transcription factors and their domains. We defined a conserved and functionally essential N-terminal domain in TFE, the archaeal homologue of the large TFIIE α subunit. We showed by X-ray crystallography that this TFE domain adopts a winged helix–turn–helix (winged helix) fold (Figure 1), extended by specific helices at the N- and C-termini. Although the winged helix fold is often found in DNA-binding proteins, TFE is not a typical DNA-binding winged helix protein, since its putative DNA-binding face shows a negatively charged groove and an unusually long wing, and since the domain lacks DNA-binding activity *in vitro*. The groove and a conserved hydrophobic surface patch on the additional N-terminal helix may, however, allow for interactions with other general transcription factors and RNA polymerase. Homology modelling shows that the TFE domain is conserved in TFIIE, including the potential functional surfaces.

Towards regulation: the first Mediator subunit structures [21]

Understanding the mechanism of transcription initiation will be a prerequisite for further studies that could reveal mechanisms of transcription activation and repression. The Mediator of transcriptional regulation is a central co-activator multiprotein complex that regulates Pol II initiation and enables a response of Pol II to activators and repressors. We solved the crystal structure of a highly conserved part of the Mediator, the MED7/MED21 (Med7/Srb7) heterodimer. The structure consists of a four-helix bundle domain and a coiled-coil protrusion connected by a flexible hinge (Figure 1). The heterodimer is very extended, spanning one-third of the Mediator length, and almost the diameter of Pol II. It contains putative protein-binding sites on the surface that allow for assembly of the Mediator middle module. The MED7/MED21 hinge could account for changes in overall Mediator structure upon binding to Pol II or activators. The structure serves as a nucleus for structure determination of larger Mediator subcomplexes.

References

1. Cramer, P. (2002) Curr. Opin. Struct. Biol. 12, 89–97
2. Cramer, P., Bushnell, D.A., Fu, J., Gnatt, A.L., Maier-Davis, B., Thompson, N.E., Burgess, R.R., Edwards, A.M., David, P.R. and Kornberg, R.D. (2000) Science 288, 640–649
3. Cramer, P., Bushnell, D.A. and Kornberg, R.D. (2001) Science 292, 1863–1876
4. Gnatt, A.L., Cramer, P., Fu, J., Bushnell, D.A. and Kornberg, R.D. (2001) Science 292, 1876–1882
5. Armache, K.-J., Kettenberger, H. and Cramer, P. (2005) Curr. Opin. Struct. Biol. 15, 197–203

6. Cramer, P. (2004) Curr. Opin. Genet. Dev. 14, 218–226
7. Cramer, P. (2004) Adv. Protein Chem. 67, 1–42
8. Armache, K.-J., Kettenberger, H. and Cramer, P. (2003) Proc. Natl. Acad. Sci. U.S.A. 100, 6964–6968
9. Mitsuzawa, H., Kanda, E. and Ishihama, A. (2003) Nucleic Acids Res. 31, 4696–4701
10. Todone, F., Brick, P., Werner, F., Weinzierl, R.O. and Onesti, S. (2001) Mol. Cell 8, 1137–1143
11. Armache, K.-J., Mitterweger, S., Meinhart, A. and Cramer, P. (2005) J. Biol. Chem. 280, 7131–7134
12. Kettenberger, H., Armache, K.-J. and Cramer, P. (2004) Mol. Cell 16, 955–965
13. Kettenberger, H., Armache, K.-J. and Cramer, P. (2003) Cell 114, 347–357
14. Meinhart, A. and Cramer, P. (2004) Nature (London) 430, 223–226
15. Noble, C.G., Hollingworth, D., Martin, S.R., Ennis-Adeniran, V., Smerdon, S.J., Kelly, G., Taylor, I.A. and Ramos, A. (2005) Nat. Struct. Mol. Biol. 12, 144–151
16. Kamenski, T., Heilmeier, S., Meinhart, A. and Cramer, P. (2004) Mol. Cell 15, 399–407
17. Ganem, C., Devaux, F., Torchet, C., Jacq, C., Quevillon-Cheruel, S., Labesse, G., Facca, C. and Faye, G. (2003) EMBO J. 22, 1588–1598
18. Krishnamurthy, S., He, X., Reyes-Reyes, M., Moore, C. and Hampsey, M. (2004) Mol. Cell 14, 387–394
19. Meinhart, A., Silberzahn, T. and Cramer, P. (2003) J. Biol. Chem. 278, 15917–15921
20. Meinhart, A., Blobel, J. and Cramer, P. (2003) J. Biol. Chem. 278, 48267–48274
21. Baumli, S., Hoeppner, S. and Cramer, P. (2005) J. Biol. Chem. 280, 18171–18178

Biochem. Soc. Symp. **73**, 49–58
(Printed in Great Britain)

5

Modulation of RNA polymerase core functions by basal transcription factor TFB/TFIIB

Finn Werner, Simone Wiesler, Sven Nottebaum and Robert O.J. Weinzierl

Imperial College London, Division of Cell and Molecular Biology, Sir Alexander Fleming Building, Exhibition Road, London SW7 2AZ, U.K.

Abstract

The archaeal basal transcriptional machinery consists of TBP (TATA-binding protein), TFB (transcription factor B; a homologue of eukaryotic TFIIB) and an RNA polymerase that is structurally very similar to eukaryotic RNA polymerase II. This constellation of factors is sufficient to assemble specifically on a TATA box-containing promoter and to initiate transcription at a specific start site. We have used this system to study the functional interaction between basal transcription factors and RNA polymerase, with special emphasis on the post-recruitment function of TFB. A bioinformatics analysis of the B-finger of archaeal TFB and eukaryotic TFIIB reveals that this structure undergoes rapid and apparently systematic evolution in archaeal and eukaryotic evolutionary domains. We provide a detailed analysis of these changes and discuss their possible functional implications.

Introduction

Almost every biological process depends critically on the regulated expression of genes through the combined activities of chromatin structures, gene-specific transcription factors and the basal transcriptional machinery. RNA polymerases (RNAPs; also abbreviated as Pols) play a central role in this process, as their activity leads directly to the production of transcripts representing the expressed coding information of the genome. Especially in eukaryotic systems, RNAPs are constantly surrounded by numerous basal factors that assist the enzymes at all stages of the transcription cycle and control their various activities

[1]To whom correspondence should be addressed (email r.weinzierl@imperial.ac.uk).

[1]. The biochemical and structural characterization of the most intensely studied system, the eukaryotic RNAPII apparatus, has proceeded at a rapid pace during the last two decades, but even now much remains to be learned about the structure and function of cellular RNAPs. The large size (~400 kDa) and high degree of complexity (12–18 different subunits) of eukaryotic RNAPs poses a considerable challenge to unravelling on the molecular level the many subfunctions that are known to reside in these enzymes [2]. Recent breakthroughs resulting in high-resolution X-ray structures of bacterial RNAPs and yeast RNAPII have provided valuable insights into the molecular organization of these enzymes [3], and one of the next tasks will be to correlate these structural insights with function. This will require experimental systems where structural information can be directly linked with site-directed mutagenesis approaches and coupled with *in vitro* assays to identify the functional consequences of the introduced changes. This goal is already achievable with bacterial RNAPs, but for eukaryotic RNAPs the situation is less satisfactory – although many

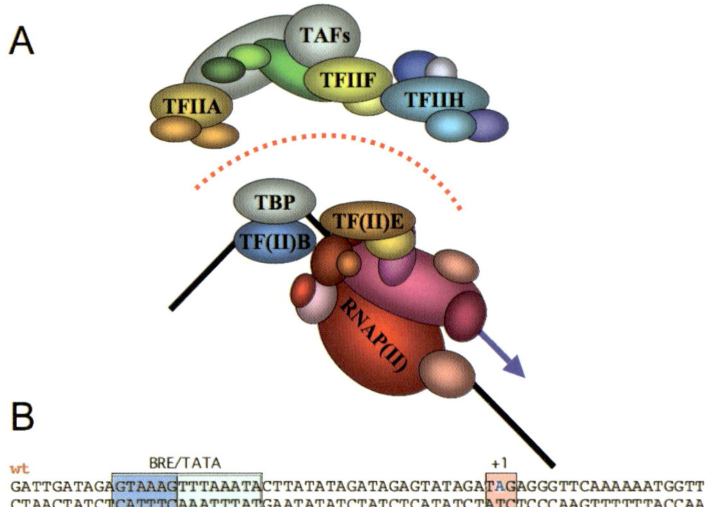

Figure 1 Structural and functional similarity between the archaeal and eukaryotic (RNAPII) transcriptional machineries. (**A**) The basal transcription factors present in the archaeal transcriptional machinery (TBP, TFB, TFE and RNAP) are schematically represented bound to the DNA template (shown as a black line). The eukaryotic RNAPII machinery contains the same factors and the additional factors [TFIIA, TAFs (TBP-associated factors), TFIIF and TFIIH] shown above the red dotted line. (**B**) A schematic diagram of an archaeal promoter. The TATA motif (centred around position −27) is shaded in light blue, and the adjacent BRE (B-recognition element) is shaded in dark blue. The transcription start site is marked as +1, and the surrounding initiator sequence is depicted in red. Note the similarity of these elements with a typical arrangement of core elements in an RNAPII promoter.

attempts have been made, it has not yet been possible to reconstitute an active eukaryotic RNAPII enzyme from recombinant subunits.

In the absence of a reconstitution system for eukaryotic RNAPII, we have turned to an alternative system to bridge the gap. The subunits present in eukaryotic RNAPs, especially in RNAPII, display a high degree of primary sequence homology throughout the entire eukaryotic range. Surprisingly, RNAPII-like subunits are also found in archaea, rather primitive organisms that are thought to resemble some of the earliest forms of life on Earth [4]. Many molecular components of the archaeal transcriptional machinery are structurally similar to the ones found in eukaryotic cells (Figure 1; recently reviewed in [5]). In addition, archaea also contain other features that are strongly reminiscent of the RNAPII transcriptional machinery. Many archaeal genes contain TATA box-like sequence elements in their promoters [6,7] that are specifically recognized by archaeal homologues of TBP (TATA-binding protein) [8]. Furthermore, the binding of archaeal TBP to TATA motifs is stabilized by TFB (transcription factor B), a homologue of eukaryotic TFIIB playing a functionally equivalent role [9]. In fact, TFB and TFIIB are so similar in their functional properties, size and arrangement of various subdomains that we will refer to them as TFB/TFIIB in any subsequent statement that is equally valid for both of them. The three archaeal basal transcription factors TBP, TFB and RNAP constitute a minimal basal *in vitro* transcription system that is both structurally and functionally directly comparable with the core components of the eukaryotic RNAPII transcriptional machinery [10,11]. It is therefore clear that many of the functions carried out by the archaeal and core eukaryotic RNAPII transcriptional machineries are directly comparable, and it therefore makes sense to study the fundamental mechanisms in the 'stripped-down' archaeal model systems.

We have successfully reconstituted the entire archaeal transcriptional machinery from *Methanocaldococcus jannaschii* in an entirely recombinant form under defined *in vitro* conditions. The reconstituted system is capable of both promoter-specific basal and activator-stimulated transcription, and thus faithfully reproduces all the known properties of the *in vitro* transcription systems derived from archaeal cellular extracts [12,13]. We are therefore in an ideal situation to use the reconstituted system to identify essential components, study their molecular interactions, introduce various targeted mutations and investigate the structural changes within the protein/nucleic acid components of the basal transcriptional machinery at any stage of the transcription cycle.

Functions of basal factor TFB/TFIIB

Structural aspects of the interactions of TFB/TFIIB with RNAP

During the last few years, we have focused extensively on the functional interplay between RNAPs and basal transcription factors, with a special emphasis on TFB. We have developed or adopted a range of *in vitro* assays suitable for monitoring diverse aspects of basal factor and RNAP function. One of the most intriguing observations emerging from these studies was the discovery that a subdomain of TFB – the B-finger – functionally participates in transcription initiation events by stimulating an early transcription initiation event ('abortive transcription') in a recruitment-independent function (F. Werner and R.O.J. Weinzierl, unpublished work). This discovery prompted us to investigate the structural features of the B-finger domains of archaeal TFBs and eukaryotic TFIIBs in more detail. This task was greatly facilitated by the recent publication of a yeast TFIIB–RNAPII co-crystal structure [14]. The B-finger domain enters the interior of RNAP by looping in and out via the putative transcript exit

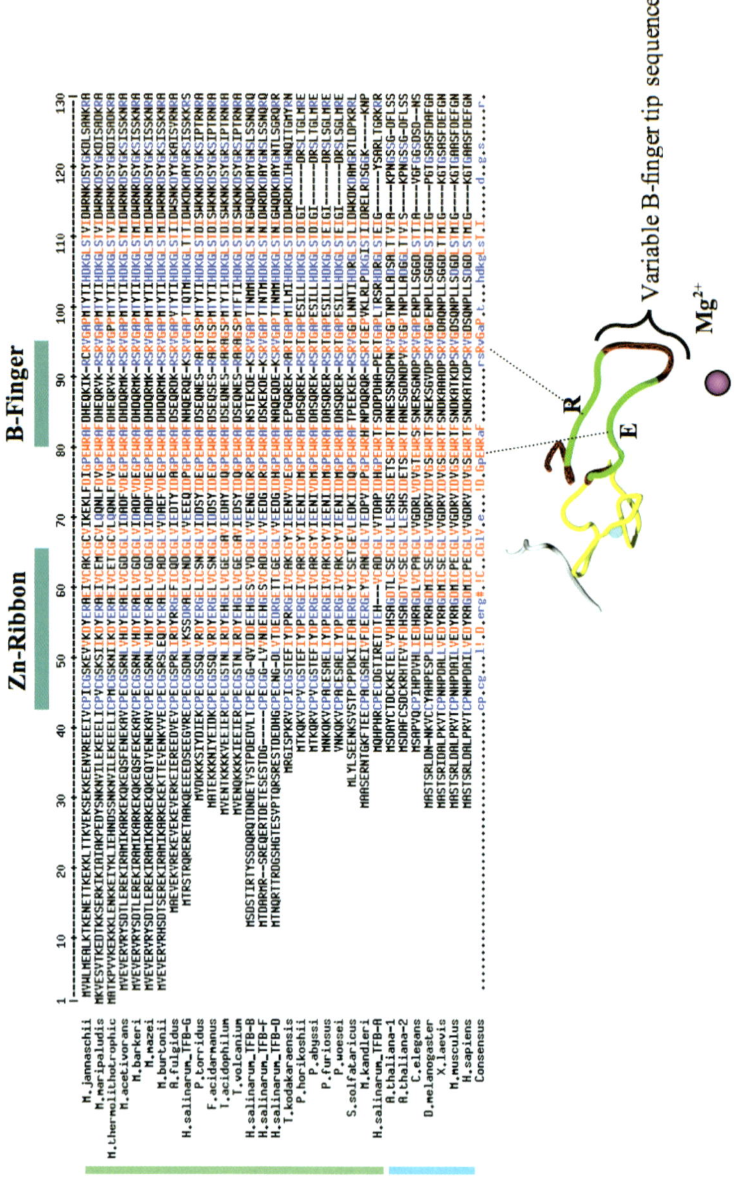

channel of the polymerase (this exit channel is, of course, not occupied by RNA during the assembly of the preinitiation complex). The structure thus clearly shows that the tip of the B-finger loop of TFB/TFIIB becomes a temporary, but integral, part of the active site during the transcription initiation process. Since the structure of the co-crystal did not include any nucleic acid components (e.g. template strand and nascent transcript), the full functional implications of this arrangement are still not completely understood. Bushnell et al. [14] suggested that the location of the B-finger loop implies a possible role in positioning the initiation NTP and/or in defining the length of abortive transcription initiation products. Other functions, such as B-finger-mediated positioning of the DNA-template strand, or even the sequence-specific recognition of short nucleic acid motifs, are also distinct possibilities. Whatever the precise function(s) of the B-finger domain eventually turn out to be, it is already clear that it is strategically placed to govern one or more key kinetic parameters of the transcription initiation process in a quantitative and/or qualitative manner.

The B-finger tip domain sequence is highly variable during evolution

The X-ray structure of the yeast TFIIB/RNAPII co-crystal helped to elucidate the functional contributions of different parts of the TFB/TFIIB N-termini [14]. The availability of several dozen archaeal and eukaryotic TFB/TFIIB sequences in various databases makes it possible to investigate the degree of evolutionary conservation of the various domains. Taking into account the fact that the B-finger domain penetrates into the RNAP towards the active site, we initially assumed that the B-finger domain would display a high degree of structural conservation. In the sequence alignments (Figure 2) the B-finger domain can be located in a highly consistent and reproducible manner in all TFB/TFIIB sequences due to the presence of invariant residues at the N- and C-terminal border: an N-terminal glutamine residue is typically followed by 13–15 amino acid residues (that form the tip of the B-finger) and a C-terminal arginine residue. The invariant glutamine and arginine residues are thought to form a stabilizing salt bridge and are therefore present in all characterized TFB/TFIIB variants. This creates a structurally defined framework that can be easily identified from primary amino acid sequences.

Figure 2 Sequence alignment of a selection of archaeal TFB and eukaryotic TFIIB N-terminal sequences. Two of the main structural motifs (the zinc ribbon and B-finger tip domain) present in the TFB/TFIIB N-terminus are marked above the aligned sequences by vertical bars. The tip of the B-finger domain which is spatially close to the catalytic site of RNAP (Mg^{2+} ion, marked by purple sphere) is flanked by highly conserved sequence motifs (**E**WRnFnnnnn[n]**R**). Invariant residues are marked in red, and residues conserved in the majority of sequences are shown in blue. The X-ray structure of the B-finger domain is shown underneath [14]. The sequence alignment was performed using software described in [21]. On the left-hand side of the aligned sequences, all archaeal TFBs are indicated by a light green vertical line, whereas eukaryotic TFIIBs are marked with a light blue line.

A systematic alignment of all available archaeal and eukaryotic TFB/TFIIB sequences reveals that, contrary to expectation, the B-finger motifs show an astonishing degree of variability, especially near the tip of the loop that is close to the catalytic site of RNAP (Figure 2). The B-finger sequences embody a mosaic of highly diverged sequences, with neither small motifs, nor even individual residues, emerging as being clearly conserved.

Taking into account the presence of numerous highly negatively charged components with the active site (rNTPs, template DNA and nascent RNA), we next investigated whether a more coherent pattern emerged if, instead of primary sequence, only the overall electrostatic charge of the various B-finger tip domains was considered. We therefore determined the predicted isoelectric points (pI) of all currently available archaeal and eukaryotic B-finger sequences by bioinformatics. The results reveal a substantially simpler pattern of charge variation. Among species belonging to the same evolutionary Domains, Kingdoms and phyla, the B-finger charge is usually highly consistent, but there is clearly a substantial degree of highly correlated charge variation across the

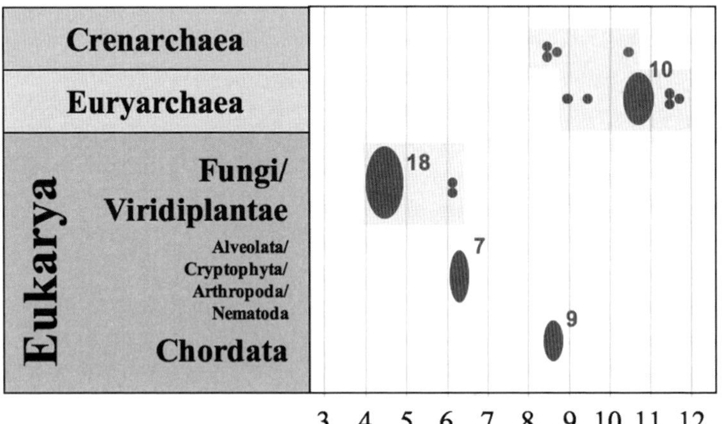

Figure 3 Predicted pI values of B-finger tip domains across the archaeal and eukaryotic evolutionary domains. The amino acid sequences of the B-finger tip domain (located between the putative glutamine–arginine salt bridge residues in various TFB and TFIIB sequences) were used to predict their pI values (http://us.expasy.org/tools/pi_tool.html). The resulting pI values were plotted on the *x*-axis against the evolutionary position of the organisms from which they were derived (*y*-axis). TFB or TFIIB B-finger tip domains with same pI are represented by larger symbols, and the number of values represented by such symbols is indicated in the top right position (e.g. the 'Chordata' symbol represents data from nine different species). The predicted pI values from a small subset of extreme halophilic and acidophilic euryarchaea are not shown here (see Figure 4)

evolutionary range (Figure 3). This variation is most distinct in the eukaryotic lineage, where the progression of B-finger charges from acidic towards basic correlates distinctly with the position of the organisms encoding them in the evolutionary tree. Fungi and green plants contain B-fingers with acidic pI values (~4.5), whereas a number of lower eukaryotes (including *Caenorhabditis elegans* and *Drosophila melanogaster*) have B-fingers with near-neutral pI values (pI ~6.5). On the other hand, the vertebrate B-fingers consistently display high pI values (pI ~8.5).

In archaea, the charge properties of B-fingers revealed that all crenarchaeal and the majority of euryarchaeal TFBs contain positively charged B-fingers. The predicted pI values range from 8.5 to 11.8, and are therefore similar to, or higher than, those observed for vertebrate B-fingers (Figure 3). The only exceptions to this rule are the B-fingers of TFBs encoded by a small group of extreme halophiles and acidophiles (Figure 4 and Discussion).

To exclude the possibility that these systematic charge variations within the B-finger domains reflect a generic property of various TFB/TFIIB protein domains, we determined the charge distributions of the other domains of all TFB/TFIIB sequences in the same manner. The resulting data (based on zinc ribbon and C-terminal domains) revealed no detectable correlation between the values obtained and the phylogenetic origin of the TFB or TFIIB from which they were derived (results not shown). We therefore conclude that the consistent variations in the electrostatic properties of B-fingers is a distinct phenomenon that is likely to reflect genuine structural (and by implication functional) differences within the key B-finger domains from organisms spanning the evolutionary range from archaea to vertebrates.

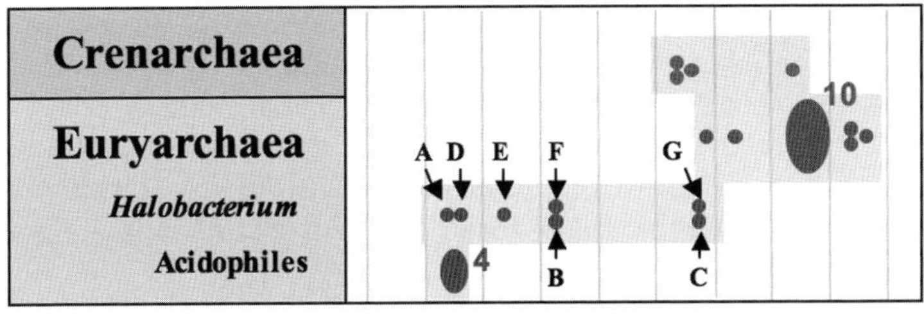

Figure 4 pI values of B-finger tip domains, including those from extreme halophilic and acidophilic euryarchaea. B-finger tip domains with same pI values are represented by larger symbols, and the number of values represented by such symbols is indicated in the top right position. The symbols representing the pI values of B-finger tip domains from the different *Halobacterium* TFB variants (TFB-A, TFB-B, TFB-C, TFB-D, TFB-E, TFB-F and TFB-G) are labelled individually.

Discussion

The bioinformatics analysis presented above suggests that TFB/TFIIBs contain a structurally distinct and constantly evolving B-finger. Within the eukaryotic evolutionary domain, the sequences forming the tip of the B-finger show the greatest degree of charge variation (Figures 2 and 3). This is especially intriguing, because the results predict that yeast, fungal and plant TFIIBs could differ radically in their functional properties from TFIIBs of higher eukaryotes (e.g. *D. melanogaster* or *C. elegans*), especially vertebrates. A clearer understanding of these functional differences will have to await more extensive studies of TFIIB properties (especially post-recruitment functions) in these organisms, and therefore cannot be proposed with any degree of confidence at this stage. It is, however, noteworthy that promoter core elements (reviewed in [15]), and especially initiator sequences ('INR'), may eventually emerge as part of the solution. The precise function of initiators and the protein factors recognizing them are still poorly understood, but the electrostatic charge properties of B-fingers seem to correlate with the presence of these elements in core promoters. Yeast, fungal and plant promoters lack initiators, but they are present in invertebrate, vertebrate and possibly archaeal promoters. It is therefore feasible that B-fingers co-evolved with the DNA sequences surrounding the transcription start sites. If this scenario is true, then the absence of initiator elements in lower eukaryotes and plants is the consequence of acidic B-fingers failing to interact with the template DNA strand during transcription initiation due to electrostatic repulsion [14]. Subsequent evolution of the B-finger tip domain may have led to a gradual adjustment of these charges, resulting eventually in vertebrate B-fingers capable of forming relatively stable electrostatic interactions with the template strand. Such species-specific differences in the interaction of TFIIB with promoter core elements are not unprecedented: the residues in the TFIIB C-terminal domain (core) that are involved in the sequence-specific recognition of the B-recognition element ('BRE') also differ substantially between yeast/plant TFIIBs and their vertebrate homologues, suggesting that BREs are also recognized differentially within the eukaryotic evolutionary domain [16].

The observations outlined above strongly indicate a dynamic change in B-finger function during eukaryotic evolution. Additional conclusions concerning B-finger domains can be drawn from the archaea. The predicted pI values of most archaeal B-fingers consistently cluster in the alkaline range (pI ~8.5–11.8) and are therefore as high, or even higher, than those of vertebrate B-fingers. The only exceptions to this rule are TFBs encoded by a small group of highly unusual archaeal extremophiles. Rather than being difficult to reconcile, these exceptions actually provide substantial support for the importance of electrostatic charge for B-finger function. The genome of the extreme halophilic archaeon *Halobacterium salinarum* (the example used in Figure 4) contains separate genes encoding seven distinct full-length TFB variants (TFB-A–TFB-G). The pI values of the B-finger regions of these variants range from 4.43 via intermediate values (4.87, 5.43, 6.35 and 6.38) all the way into the alkaline range (8.85). The intracellular ionic concentrations of *Halobacterium* are extremely high (4 M K^+ and 0.7 M Na^+; [17]), so that electrostatic interactions cannot play a significant role in these organisms. As a consequence, the differences between the pI values of the B-fingers of the TFB variants are either mostly insignificant or play only a very minor role in creating subtle functional differences between the TFB variants. A similar argument applies to the other exceptions, i.e. the acidophilic archaea. The low intracellular pH of these organisms renders the normally highly negatively charged nucleic acid and rNTP substrates within the active site of RNAP mostly uncharged and not capable of electrostatic interactions [18]. It therefore seems clear that exceptions in the isoelectric points of archaeal B-fingers occur only in the extreme halophiles and acidophiles where, due to their highly unusual intracellular environments, charge properties cannot play any significant role in molecular recognition. Preliminary evidence suggests that the B-finger domains of *Halobacterium* TFB variants may depend on systematic differences in hydrophobicity, rather than electrostatic charge (R.O.J. Weinzierl, unpublished work).

Conclusions

Taken together, these observations suggest that variation in the electrostatic properties of the B-finger domains could be used as a simple and powerful mechanism for influencing the functional properties of archaeal and eukaryotic TFB/TFIIBs. The concept of a 'tuneable' active site in RNAPs, the function of which can be modified by external factors, has already been well documented in other instances. The clearest examples are the elongation factors (*GreA* or *GreB* in bacteria [19] and TFIIS in eukaryotes [20]) which help RNAPs to overcome transcription pause or arrest sites by temporarily inducing nuclease-like behaviour within the catalytic centre. Similar to TFB/TFIIB, these factors insert a finger-like domain containing two negatively charged amino acid residues at the tip via the RNA exit channel into the active site, resulting in a specific repositioning of a Mg^{2+} ion to facilitate transcript cleavage. We believe that the charged amino acid residues located at the tip of the B-finger domain will be found to play a similar role in the modulation of the functional properties of the active site during transcription initiation, especially during the abortive transcription cycle preceding promoter escape.

The differences in B-finger structure raise many interesting questions whose answers can currently only be guessed at and which need to be experimentally addressed. We recently initiated a systematic high-throughput *in vitro* mutagenesis screen of *Methanocaldococcus* TFB, designed to analyse the functional consequences of altering the structure and charge properties of the B-finger domain within the context of the catalytic site of archaeal RNAPs. We expect to see differences in RNAP recruitment, transcription start site selection and the abortive transcription patterns as a consequence of these mutations. It is clear that similar systematic mutagenesis approaches also need to be carried out for other transcription model systems (e.g. yeast, *Drosophila* and humans) to test the various hypotheses outlined above before a clearer picture of B-finger function, and its changing role during evolution, emerges.

This work was supported by a Wellcome Trust Project grant (no. 068764/Z/02/Z) to R.O.J.W. We thank members of Martin Buck's laboratory for many helpful discussions.

References

1. Lemon, B. and Tjian, R. (2000) Genes Dev. **14**, 2551–2569
2. Young, R.A. (1991) Annu. Rev. Biochem. **60**, 689–715
3. Boeger, H., Bushnell, D.A., Davis, R., Griesenbeck, J., Lorch, Y., Strattan, J.S., Westover, K.D. and Kornberg, R.D. (2005) FEBS Lett. **579**, 899–903
4. Woese, C.R., Kandler, O. and Wheelis, M.L. (1990) Proc. Natl. Acad. Sci. U.S.A. **87**, 4576–4579
5. Ouhammouch, M. (2004) Curr. Opin. Genet. Dev. **14**, 133–138
6. Reiter, W.D., Palm, P. and Zillig, W. (1988) Nucleic Acids Res. **16**, 1–19
7. Reiter, W.D., Hydepohl, U. and Zillig, W. (1989) Proc. Natl. Acad. Sci. U.S.A. **87**, 9509–9513
8. Rowlands, T., Baumann, P. and Jackson, S.P. (1994) Science **264**, 1326–1329
9. Wettach, J., Gohl, H.P., Tschochner, H. and Thomm, M. (1995) Proc. Natl. Acad. Sci. U.S.A. **92**, 472–476
10. Hausner, W., Wettach, J., Hethke, C. and Thomm, M. (1996) J. Biol. Chem. **271**, 30144–30148
11. Qureshi, S.A., Bell, S.D. and Jackson, S.P. (1997) EMBO J. **16**, 2927–2936
12. Werner, F. and Weinzierl, R.O.J. (2002) Mol. Cell **10**, 635–646
13. Ouhammouch, M., Werner, F., Weinzierl, R.O.J. and Geiduschek, E.P. (2004) J. Biol. Chem. **279**, 51719–51721
14. Bushnell, D.A., Westover, K.D., Davis, R.E. and Kornberg, R.D. (2004) Science **303**, 983–988
15. Smale, S.T. and Kadonaga, J.T. (2003) Annu. Rev. Biochem. **72**, 449–479
16. Littlefield, O., Korkhin, Y. and Sigler, P.B. (1999) Proc. Natl. Acad. Sci. U.S.A. **96**, 13668–13673
17. Oren, A. (1986) Can. J. Microbiol. **32**, 4–9
18. van de Vossenberg, J.L.C.M., Driessen, A.J., Zillig, W. and Konings, W.N. (1998) Extremophiles **2**, 67–74
19. Opalka, N., Chlenov, M., Chacon, P., Rice, W.J., Wriggers, W. and Darst, S.A. (2003) Cell **114**, 35–45
20. Kettenberger, H., Armache, K.J. and Cramer, P. (2003) Cell **114**, 347–357
21. Corpet, F. (1988) Nucleic Acids Res. **16**, 10881–10890

Biochem. Soc. Symp. **73**, 59–66
(Printed in Great Britain)

6

The relationship between higher-order chromatin structure and transcription

Nick Gilbert and Wendy A. Bickmore[1]

MRC Human Genetics Unit, Western General Hospital, Edinburgh EH4 2XU, U.K.

Abstract

It has generally been assumed that transcriptionally active genes are in an ‘open’ chromatin structure and that silent genes have a ‘closed’ chromatin structure. Here we re-assess this axiom in the light of genome-wide studies of chromatin fibre structure. Using a combination of sucrose gradient sedimentation and genomic microarrays of the human genome, we argue that open chromatin fibres originate from regions of high gene density, whether or not those genes are transcriptionally active.

Introduction

Chromatin structure, and its modification, is central to the regulation of gene expression. This is best understood and studied at the level of covalent histone modifications. For example, acetylation and methylation of lysine residues in histones H3 and H4 has been correlated with either active transcription or gene repression, depending on the nature of the modification [1]. However, beyond the nucleosome itself, there are other structural states of chromatin that will influence how the underlying DNA sequence is read. Here, we define these as ‘higher-order’ chromatin structures. Although we know little about the detailed structure of chromatin beyond that of individual nucleosomes, it has been assumed that higher-order chromatin structures will also impact on transcription. The supposition has been that transcriptionally active regions have an ‘open’ chromatin structure, and that ‘closing’ of chromatin structure brings about transcriptional silencing [2–4] (Figure 1). Here we review the evidence for this supposition, and instead suggest that there is not a simple ‘one size fits all’ relationship between transcription and higher-order chromatin structure.

[1]To whom correspondence should be addressed (email W.Bickmore@hgu.mrc.ac.uk).

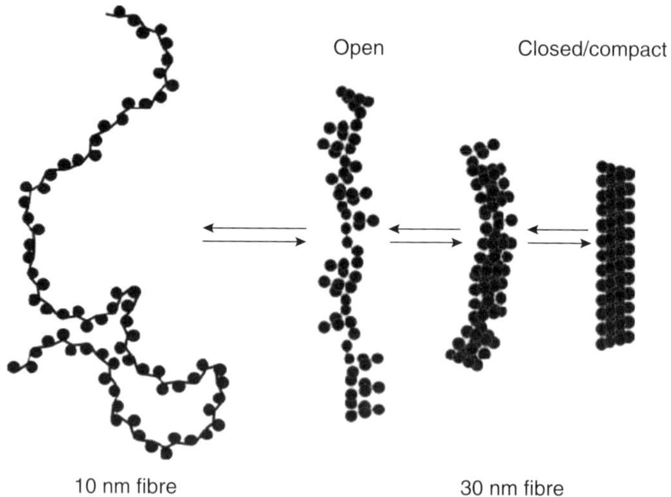

Figure 1 Structures of the 30 nm chromatin fibre. Schematic representation of the 10 nm 'beads on a string' nucleosome array and the folding of this into 30 nm chromatin fibres that are closed/compact rod-like structures, or that are more open in structure, perhaps because of the presence of discontinuities, or distortions in the folding.

What do we know about higher-order chromatin structure?

Even though we have detailed structural information on the nucleosome [5], we still do not understand how nucleosome arrays are arranged into the 30 nm fibres that can be detected in cells by low-angle X-ray diffraction [6]. Recent analysis of nucleosome arrays assembled *in vitro* has been interpreted in favour of the two-start helix model for the 30 nm chromatin fibre [7].

It is highly unlikely that 30 nm chromatin fibres have a uniform structure across the genome. Indeed, sucrose gradient sedimentation of chromatin has shown that chromatin fibres from different parts of the vertebrate genome have different levels of compaction. [8–12]. One compacted region that has been especially well studied biophysically is a 15.5 kb segment of the chicken genome that lies between the β-globin locus and an erythroid-specific folate receptor gene [13]. This region contains repeated sequences, and no gene could be identified in it. It is long enough to accommodate approx. 75 nucleosomes, so should be able to retain higher-order chromatin structures. By sucrose gradient sedimentation this fragment was found to sediment as a 128 S particle, and CsCl buoyant density gradients allowed its shape to be inferred from its frictional coefficient. The authors concluded that this chromatin fibre was a rod ~170 nm long and 40 nm in diameter [13]. Sedimentation analysis suggests a similar compact structure for regions of satellite repeats (constitutive heterochromatin) in the mammalian genome [10].

What types of factor can compact chromatin fibres?

Folding of the chromatin fibre has been shown to be independent of the N-terminal tails of the core histones, except for that of H4 [14]. This histone tail contacts an acidic patch on the surface of a neighbouring nucleosome. This acidic patch is increased in nucleosomes that contain the variant histone H2A.Z [5]. Therefore inter-nucleosome contacts, and hence chromatin fibre structure, might be altered in nucleosome arrays that contain this variant histone. Sedimentation analysis of H2A.Z-containing nucleosome arrays has confirmed that they do indeed form more compact structures than arrays that contain H2A [15]. HP1 (heterochromatin protein 1) may then act further on this compact structure. However, the study was carried out in the absence of linker histones, which clearly have a role in the formation and compaction of the chromatin fibre [16,17].

Recent electron microscopic analysis of nucleosome arrays has suggested that PcG (polycomb group) complexes can compact nucleosome arrays [18]. Since both HP1 and polycomb have been associated with transcriptional silencing, these observations are entirely consistent with the axiom that transcriptionally silent regions have a compact chromatin structure.

So where do open chromatin fibres originate?

Support for the idea that regions of 'open' chromatin fibres correspond to transcriptionally active genes came from sedimentation analysis of the chicken β-globin locus. When active, i.e. in erythrocytes, chromatin from this locus sediments more slowly than bulk chromatin or a non-expressed gene, whereas it sediments with bulk chromatin in non-expressing cells [8,9,11]. However, can the β-globin locus, whose regulation is known to be very complex in mammalian cells, be used as a paradigm for the other ~25 000 genes in the genome? We would argue that it cannot. To understand the global relationships between chromatin fibre structures and gene expression, chromatin fibre structure needs to be investigated at a genomic level.

To do this, we fractionated open chromatin fibres from human cells by sucrose gradient sedimentation and hybridized them, together with differentially labelled total chromatin, to human genomic microarrays [12]. The first array used was assembled from BAC (bacterial artifical chromosome) clones, spaced at ~1 Mb intervals, from the 'golden path' used in the sequencing of the human genome [19]. Domains of the human genome enriched in open chromatin have a $\log_2$ (open/input chromatin) ratio of >0.

The results of this experiment showed that open chromatin fibres originate from the most gene-rich regions (T-bands) of the human genome [12]. For example, the human (HSA) chromosomes with the highest overall open/input chromatin ratio are HSA17, HSA19 and HSA22, which are the most gene-rich chromosomes in the human karyotype, whereas HSA18 is one of the most gene-poor chromosomes and is quite depleted of open chromatin [20] (Figure 2A).

However, at this low (1 Mb) resolution, we could find no correlation between the level, or the probability, of gene expression and the abundance of

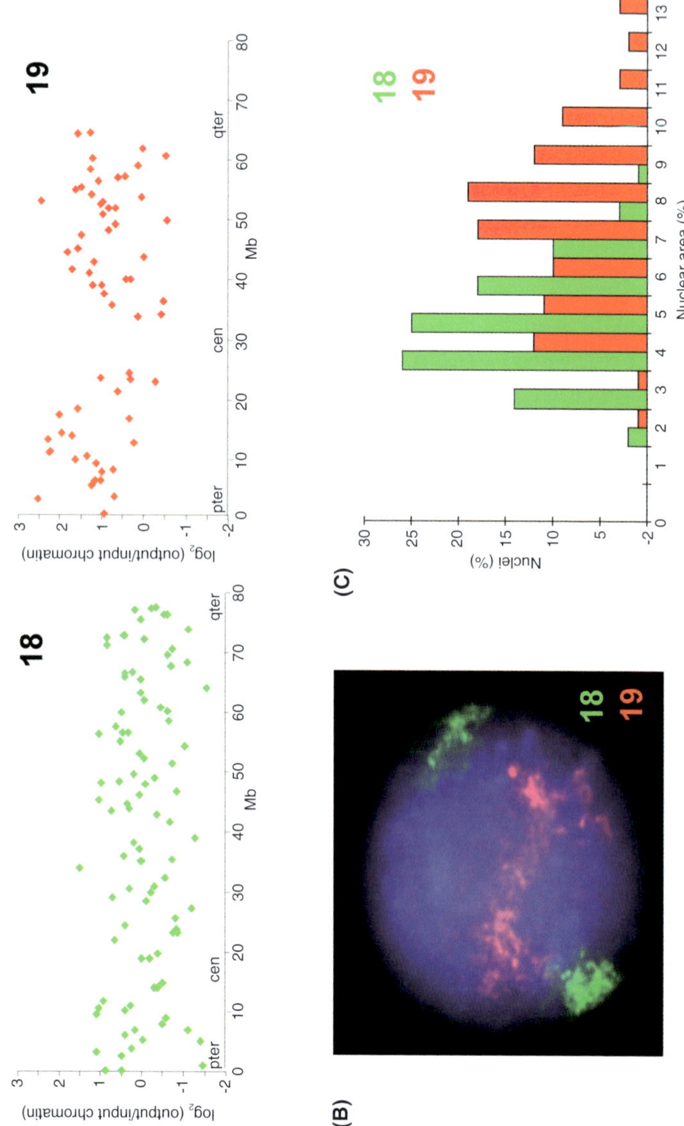

Figure 2 Chromatin compaction of human chromosomes 18 and 19. (**A**) A Sau3AI linker-ligated open chromatin fraction was hybridized to a whole-genome 1 Mb microarray using an input chromatin control. Graphs show the average (n=4) $\log_2$ (open/input chromatin) ratio for BAC along human chromosomes 18 (green) and 19 (red). (**B**) Interphase FISH with chromosome paints for chromosomes 18 (green) and 19 (red) in a human nucleus counterstained with DAPI (4,6-diamidino-2-phenylindole) (blue). (**C**) Histogram of the proportion of nuclear area taken up by the hybridization signal from paints for chromosome 18 (green) or 19 (red) (n=50).

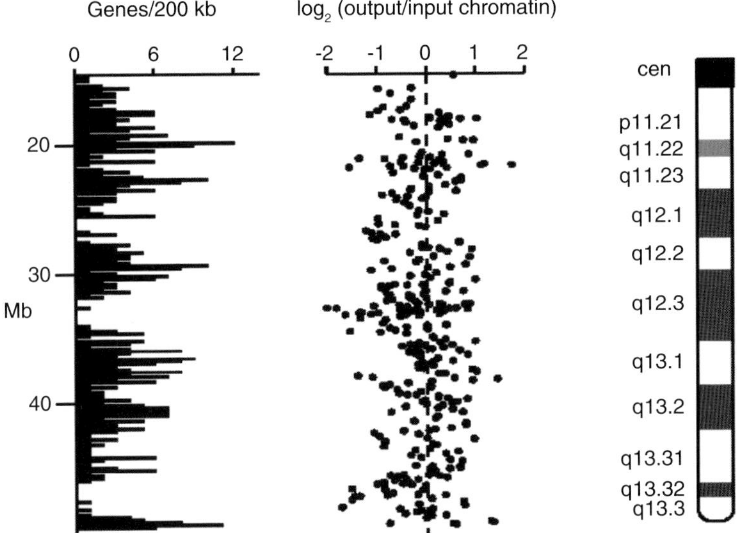

Figure 3 High-resolution analysis of open chromatin on chromosome 22q. A Sau3AI linker-ligated open chromatin fraction was hybridized to a high-resolution contiguous tiling path array of chromosome 22q. The value for the $\log_2$ (open/input chromatin) ratio signal for each clone is aligned to the DNA sequence (Mb) and, to the left, the gene density/200 kb window along chromosome 22q. On the right the array data are aligned to the ideogram for 22q.

open 30 nm chromatin fibres [12]. Analysis of open chromatin fibre distribution at higher resolution is required to verify this. Therefore we hybridized open and input chromatin fractions to a human chromosome 22q genomic array consisting of overlapping sequencing tiling path clones [21]. The average resolution of this array is 78 kb, but it contains regions where the clones are even smaller than the size of the chromatin fibres being examined (~20 kb).

The data from this microarray, like those from the 1 Mb whole genome array, showed a strong correspondence between gene density and the abundance of open chromatin fibres (Figure 3). On this generally gene-rich chromosome arm, domains depleted of open chromatin fibres corresponded to the gene-poor regions, such as 22q12.3 (30.7–32.2 Mb) and 22q13.31–q13.32 (45.5–47.4 Mb).

To ascertain the relationship between the presence of open chromatin fibres on chromosome 22q and gene expression *per se*, data from a gene expression microarray analysis were compared [21]. This used RNA prepared from the same cell type (lymphoblastoid cells) used to prepare the chromatin fibres. Surprisingly, this revealed no simple correlation between gene expression and the presence of open chromatin fibres. We illustrate this here for a 1.2 Mb region of 22q12.1–q12.2 (Figure 4). The most centromeric 500 kb region (in the 22q12.1-G band) contains no known genes and is packaged in compact chromatin ($\log_2$\open/input ≈ −1). Distal to this, there are two domains enriched in open chromatin (27.4–27.6 Mb and 27.9–28.1 Mb) that correspond to closely packed clusters of genes. The expression status, in lymphoblastoid cells, of seven of the

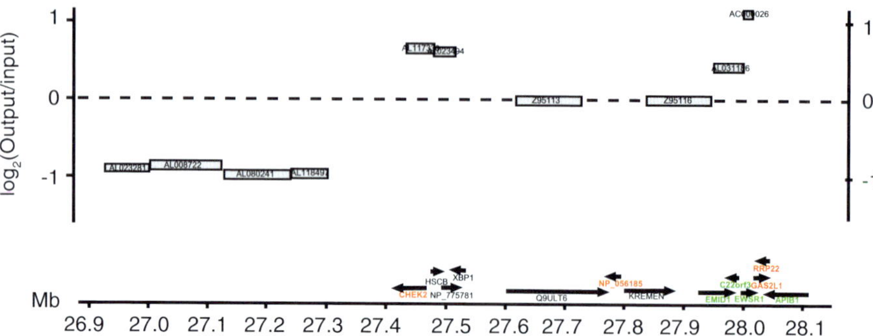

Figure 4 Correlation between gene expression and open chromatin. Values are shown of $\log_2$ (open/input chromatin) ratio for clones from the 22q tile-path genomic microarray, from positions 26.9 to 28.1 Mb according to NCBI build35 (http://www.ensembl.org/Homo_sapiens/mapview?chr=22). Known genes are shown underneath. Their expression status was ascertained in the same cell type (lymphoblastoid) as was used to prepare the chromatin fibres [21]. Green, expressed genes; red, not expressed; black, not determined.

ten genes in these regions could be ascertained. Only four were expressed; the other three genes were silent. Therefore we have demonstrated that silent genes are not necessarily in compact chromatin; rather, open chromatin domains can contain both transcriptionally active and inactive genes.

We have suggested that domains of open chromatin fibres may mark out regions that are competent for transcription, given the right transcription factor environment. This could impose a constraint to maintain clusters of genes together in the genome during evolution, and so partly explain the clustering of functionally unrelated genes together in the human genome [22–24].

What factors might open up the structure of chromatin fibres?

The slowed sedimentation of what we have termed 'open' chromatin fibres may be due to the presence of multiple discontinuities in the 30 nm fibre structure (Figure 1). These might be attributable to DHSs (DNase I-hypersensitive sites). For example, the slowed sedimentation of a 6 kb fragment of the chicken β-globin locus purified from erythrocytes was attributed to a tissue-specific hypersensitive site [9]. Again, the axiom has been that DHSs are associated with transcriptionally active genes, and their regulatory elements. However, two recent large-scale characterizations of DHSs from the human genome found that they were associated with both active and inactive genes [25,26]. This is consistent with the distribution of open chromatin fibres that we found in the human genome.

So far, there is very little biophysical evidence that the modifications of the N-terminal tails of the core histones have much impact on the structure of the 30 nm chromatin fibre [27]. Indeed, histone acetylation and methylation seem to correlate more closely with actual transcriptional activity than with the potential for transcription.

A clue as to the types of enzymes that could mediate the opening of chromatin fibre structure came from an analysis of the sedimentation properties of an inducible plasmid in yeast. Chromatin remodelling events that lead to the formation of slowly sedimenting chromatin fibres were independent of transcription itself, but dependent on the activity of Swi/Snf [28]. *In vitro* evidence to support the notion that Swi/Snf may operate *in vivo* to modulate the folding of nucleosome arrays is the effect of histone SIN (Swi/Snf independent) mutants on the intramolecular folding of nucleosome arrays [29].

Tertiary levels of chromatin structure

So far, we have focused on chromatin structure at the level of the 30 nm chromatin fibre, largely because this is the only level of higher-order chromatin structure that we have reasonable models for. However, it is clear from electron microscopy studies that a lot of mammalian chromatin is packaged into levels beyond this. The few assays that we have for tertiary levels of chromatin structure are, in the main, cytological.

There does seem to be a link between chromatin fibre structure and visible levels of chromatin condensation in the nucleus. At the level of the chromatin fibre, HSA19 is more 'open' than HSA18 (Figure 2A). This is also true at the level of chromatin compaction in the nucleus. Using FISH (fluorescence *in situ* hybridization) with chromosomes paints for these two human chromosomes, we showed previously that HSA19 is more decondensed (i.e. occupies a larger proportion of the nuclear space) than HSA18 (Figure 2B) [30]. At a sub-chromosomal level of resolution, we used the mean squared-interphase separation between FISH probes to show that the regions of the human genome enriched in open 30 nm chromatin fibres are also cytologically condensed [12].

Although we have argued that open chromatin structures confer transcriptional potential to a region, rather than leading to transcription *per se*, we do think that there are some specialized regions of the mammalian genome where opening of the chromatin fibres is directly related to the induction of gene expression. Such regions would include co-ordinately regulated gene clusters. Recently, visible decondensation of the endogenous murine *HoxB* locus was shown to accompany the induction of transcription, both *ex vivo* [31], and *in vivo* during mammalian embryogenesis [32]. In addition, in reporter systems, unfolding and decondensation of chromatin fibres is seen by light microscopy when transcriptional regulators are artificially targeted to the mammalian genome [33–36].

Conclusion

As attention is now being re-focused on the structure of higher-order chromatin fibres, a synergy of biophysical, cytological, genetic and genomic approaches promises to tell us not only how nucleosomes are arranged in chromatin fibres, but also how the process of gene expression occurs within the context of cellular chromatin.

W.A.B is a Centennial Fellow of the James S. McDonnell Foundation. This work was supported by the UK Medical Research Council.

References

1. Fischle, W., Wang, Y. and Allis, C.D. (2003) Curr. Opin. Cell Biol. **15**, 172–183
2. Felsenfeld, G. and Groudine, M. (2003) Nature (London) **421**, 448–453
3. Vermaak, D., Ahmad, K. and Henikoff, S. (2003) Curr. Opin. Cell Biol. **15**, 266–274
4. Mohd-Sarip, A. and Verrijzer, C.P. (2004) Science **306**, 1484–1485
5. Suto, R.K., Clarkson, M.J., Tremethick, D.J. and Luger, K. (2000) Nat. Struct. Biol. **7**, 1121–1124
6. Langmore, J.P. and Paulson, J.R. (1983) J. Cell Biol. **96**, 1120–1131
7. Dorigo, B., Schalch, T., Kulangara, A., Duda, S., Schroeder, R.R. and Richmond, T.J. (2004) Science **306**, 1571–1573
8. Kimura, T., Mills, F.C., Allan, J. and Gould, H. (1983) Nature (London) **306**, 709–712
9. Caplan, A., Kimura, T., Gould, H. and Allan, J. (1987) J. Mol. Biol. **193**, 57–70
10. Gilbert, N. and Allan, J. (2001) Proc. Natl. Acad. Sci. U.S.A. **98**, 11949–11954
11. Fisher, E.A. and Felsenfeld, G. (1986) Biochemistry **25**, 8010–8016
12. Gilbert, N., Boyle, S., Fiegler H., Woodfine K., Carter, N.P. and Bickmore, W.A. (2004) Cell **118**, 555–566
13. Ghirlando, R., Litt, M.D., Prioleau, M.N., Recillas-Targa, F. and Felsenfeld, G. (2004) J. Mol. Biol. **336**, 597–605
14. Dorigo, B., Schalch, T., Bystricky, K. and Richmond, T.J. (2003) J. Mol. Biol. **327**, 85–96
15. Fan, J.Y., Rangasamy, D., Luger, K. and Tremethick, D.J. (2004) Mol. Cell **16**, 655–661
16. Bednar, J., Horowitz, R.A., Grigoryev, S.A., Carruthers, L.M., Hansen, J.C., Koster, A.J. and Woodcock, C.L. (1998) Proc. Natl. Acad. Sci. U.S.A. **95**, 14173–14178
17. Allan, J., Mitchell, T., Harborne, N., Bohm, L. and Crane-Robinson, C. (1986) J. Mol. Biol. **187**, 591–601
18. Francis, N.J., Kingston, R.E. and Woodcock, C.L. (2004) Science **306**, 1574–1577
19. Fiegler, H., Carr, P., Douglas, E.J., Burford, D.C., Hunt, S., Scott, C.E., Smith, J., Vetrie, D., Gorman, P., Tomlinson, I.P. and Carter, N.P. (2003) Genes Chromosomes Cancer **36**, 361–374
20. Craig, J.M. and Bickmore, W.A. (1994) Nat. Genet. **7**, 376–382
21. Woodfine, K., Fiegler, H., Beare, D.M., Collins, J.E., McCann, O.T., Young, B.D., Debernardi, S., Mott, R., Dunham, I. and Carter, N.P. (2004) Hum. Mol. Genet. **13**, 191–202
22. Caron, H., van Schaik, B., van der Mee, M., Baas, F., Riggins, G., van Sluis, P., Hermus, M.C., van Asperen, R., Boon, K., Voute, P.A., Heisterkamp, S., van Kampen, A. and Versteeg, R. (2001) Science **291**, 1289–1292
23. Lercher, M.J., Urrutia, A.O. and Hurst, L.D. (2002) Nat. Genet. **31**, 180–183
24. Versteeg, R., van Schaik, B.D., van Batenburg, M.F., Roos, M., Monajemi, R., Caron, H., Bussemaker, H.J. and van Kampen, A.H. (2003) Genome Res. **13**, 1998–2004
25. Sabo, P.J., Humbert, R., Hawrylycz, M., Wallace, J.C., Dorschner, M.O., McArthur, M. and Stamatoyannopoulos, J.A. (2004) Proc. Natl. Acad. Sci. U.S.A. **101**, 4537–4542
26. Crawford, G.E., Holt, I.E., Mullikin, J.C., Tai, D., Blakesley, R., Bouffard, G., Young, A., Masiello, C., Green, E.D., Wolfsberg, T.G. and Collins, F.S. (2004) Proc. Natl. Acad. Sci. U.S.A. **101**, 992–997
27. Wang, X., He, C., Moore, S.C. and Ausio, J. (2001) J. Biol. Chem. **276**, 12764–12768
28. Kim, Y. and Clark, D.J. (2002) Proc. Natl. Acad. Sci. U.S.A. **99**, 15381–15386
29. Horn, P.J., Crowley, K.A., Carruthers, L.M., Hansen, J.C. and Peterson, C.L. (2002) Nat. Struct. Biol. **9**, 167–171
30. Croft, J.A., Bridger, J.M., Boyle, S., Perry, P., Teague, P. and Bickmore, W.A. (1999) J. Cell Biol. **145**, 1119–1131
31. Chambeyron, S. and Bickmore, W.A. (2004) Genes Dev. **18**, 1119–1130
32. Chambeyron, S., Da Silva, N.R., Lawson, K.A. and Bickmore, W.A. (2005) Development **132**, 2215–2223
33. Tumbar, T., Sudlow, G. and Belmont, A.S. (1999) J. Cell Biol. **145**, 1341–1354
34. Tsukamoto, T., Hashiguchi, N., Janicki, S.M., Tumbar, T., Belmont, A.S. and Spector, D.L. (2000) Nat. Cell Biol. **2**, 871–878
35. Muller, W.G., Walker, D., Hager, G.L. and McNally, J.G. (2001) J. Cell Biol. **154**, 33–48
36. Nye, A.C., Rajendran, R.R., Stenoien, D.L., Mancini, M.A., Katzenellenbogen, B.S. and Belmont, A.S. (2002) Mol. Cell. Biol. **22**, 3437–3449

Biochem. Soc. Symp. **73**, 67–75
(Printed in Great Britain)

7

Specialized transcription factories

Jon Bartlett, Jelena Blagojevic, David Carter, Christopher Eskiw, Maud Fromaget, Christy Job, Monee Shamsher, Inês Faro Trindade, Meng Xu and Peter R. Cook

Sir William Dunn School of Pathology, University of Oxford, South Parks Road, Oxford OX1 3RE, U.K.

Abstract

We have previously suggested a model for the eukaryotic genome based on the structure of the bacterial nucleoid where active RNA polymerases cluster to loop the intervening DNA. This organization of polymerases into clusters – which we call transcription 'factories' – has important consequences. For example, in the nucleus of a HeLa cell the concentration of soluble RNA polymerase II is ~1 mM, but the local concentration in a factory is 1000-fold higher. Because a promoter can diffuse ~100 nm in 15 s, one lying near a factory is likely to initiate; moreover, when released at termination, it will still lie near a factory, and the movement and modifications (e.g. acetylation) accompanying elongation will leave it in an 'open' conformation. Another promoter out in a long loop is less likely to initiate, because the promoter concentration falls off with the cube of the distance from the factory. Moreover, a long tether will buffer it from transcription-induced movement, making it prone to deacetylation, deposition of HP1 (heterochromatin protein 1), and incorporation into heterochromatin. The context around a promoter will then be self-sustaining: productive collisions of an active promoter with the factory will attract factors increasing the frequency of initiation, and the longer an inactive promoter remains inactive, the more it becomes embedded in heterochromatin. We review here the evidence that different factories may specialize in the transcription of different groups of genes.

Introduction

A 'factory' is defined in The Oxford English Dictionary as "a building or range of buildings with plant for the manufacture of goods". We use the term

[1]To whom correspondence should be addressed (email peter.cook@path.ox.ac.uk).

to describe the sites of replication, transcription and repair of damaged DNA in eukaryotic cells, as each site contains many different polymerizing complexes working on many different templates [1–5]. We have also proposed a general model for the organization of all genomes in which the factories involved in transcription play a central role (Figure 1) [6,7]. When genes strung along a template are transcribed, active polymerases and transcription factor–DNA complexes cluster to form the factories, with a consequential looping of intervening DNA. Each factory would then be surrounded by a 'cloud' of loops, and strings of nucleosomes and factories (plus surrounding clouds) would constitute the major architectural motifs responsible for organizing the genome. Active polymerases do not track along their templates; they are bound to a factory, and act both as motors that reel in their templates and as one of the critical structural ties that maintain the loops. DNA within a loop is free to diffuse within the volume set by the length of the tether. As engaged polymerases and transcription factors constitute the molecular ties, loops inevitably appear and disappear as polymerases initiate and terminate, and the factors bind and dissociate.

Clustering ensures that the local concentration of polymerases and promoters in and around a factory is high, enabling them to interact efficiently. For example, HeLa cell nuclei contain a dispersed pool of RNA polymerase II present at a concentration of ~1 μM, but a fraction is found in transcription factories, where the local concentration is ~1000-fold higher [7]; as a result, few – if any – mRNAs are made outside these factories (e.g. [8]). A promoter lying close to a factory is likely to collide with a polymerase and initiate; moreover, when released at termination, it will still lie near a factory, and the movement and modifications (e.g. acetylation) accompanying elongation will leave it in an 'open' conformation. Another promoter out in a long loop is less likely to initiate, because the promoter concentration falls off with the cube of the distance from the factory. Moreover, a long tether will buffer it from transcription-induced movement, making it prone to deacetylation, and incorporation into heterochromatin. The context around a promoter will then be self-sustaining: productive collisions of an active promoter with the factory will attract the appropriate factors, which will inevitably increase the frequency of initiation; on the other hand, the longer an inactive promoter remains inactive, the more it becomes embedded in heterochromatin. Here we review the evidence that different factories might specialize in the transcription of different groups of genes.

Nucleolar transcription factories

Nucleoli provide an important precedent for the kind of specialization that we will consider [9]. Eukaryotic nuclei contain three different kinds of RNA polymerase (i.e. polymerases I, II and III); one of these – RNA polymerase I – is concentrated in nucleoli, where it transcribes a specific group of genes, the repeated ribosomal cistrons. Each nucleolus contains one or more clusters of polymerases at its core. Each is dedicated to both the synthesis of 45 S rRNA and the packaging of the resulting processed transcripts into ribosomes. The human loci encoding 45 S rRNA are carried on chromosomes 13, 14, 15, 21 and

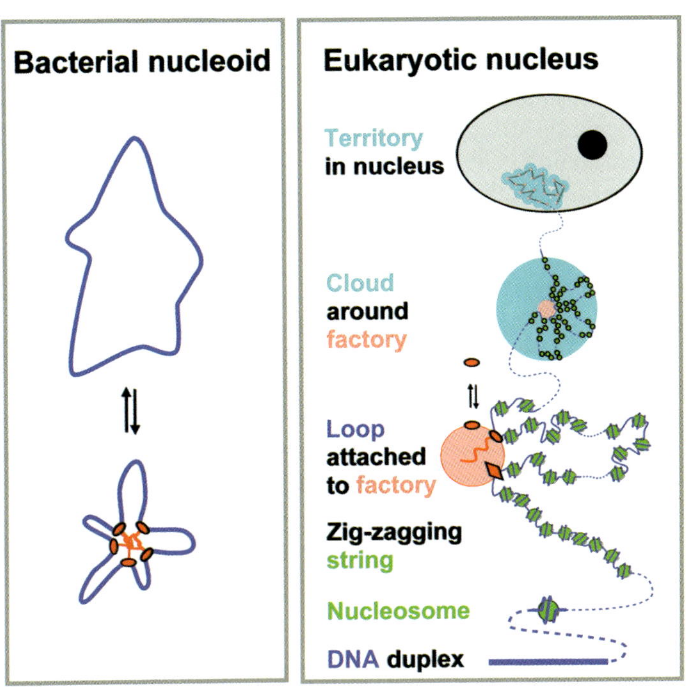

Figure 1 Models for genome structure. Left panel: In bacteria, transcription of the circular chromosome (top), followed by aggregation of polymerases (ovals) and transcripts (red lines), generates a looped structure (bottom) that is self-sustaining (as promoters in active genes now lie close to polymerases). Reproduced from [7]. Right panel: In eukaryotes (and specifically in a HeLa cell), DNA is coiled around the histone octamer, and runs of nucleosomes form a zig-zagging string. At the intermediate level in the hierarchy, this string is organized into loops (average contour length 86 kbp; range 5–200 kbp) by attachment to factories through transcription factors (diamond) and engaged RNA polymerases (ovals). Around 10–20 such loops (only a few are shown) form a 'cloud' around the factory, to give a structure equivalent to that of the bacterial nucleoid. Active polymerases do not track along their templates; they are bound to a factory and act both as motors that reel in their templates and as one of the critical structural ties that maintain the loops. Loops inevitably appear and disappear as polymerases initiate and terminate, and the factors bind and dissociate. Each factory contains one type of RNA polymerase (i.e. I, II or III) to the exclusion of others, and some factories are richer in certain transcription factors than others (and so are involved in the transcription of specific sets of genes). Individual components in the factory exchange continually with others in the soluble pool. A territory is formed by 50–200 successive clouds strung along the chromosome (the general path of DNA between clouds is shown). Modified from Cook [40], with permission of John Wiley & Sons, Inc.

22, with each locus carrying ~80 tandem repeats of 43 kbp containing the gene and an untranscribed spacer. These repeats appear as secondary constrictions in mitotic chromosomes – the NORs (nucleolar organizing regions). Inactive RNA polymerase I and its transcription factor UBF (upstream binding factor) are bound to some NORs, and – on exit from mitosis – these NORs fuse into one or more nucleoli. NORs lacking bound UBF and the polymerase remain inactive and are not initially incorporated into nucleoli, but they may be later if the (heterochromatic) centromeres to which they are linked associate with nucleolar heterochromatin.

Three distinct regions can be seen by electron microscopy: a granular component, in which are embedded one or more fibrillar centres and associated dense fibrillar component(s) [10,11]. The fibrillar centre contains high concentrations of polymerase I and UBF, and transcription takes place on its surface, as nascent Br-RNA is found in the enveloping dense fibrillar component. Nascent transcripts can be labelled by incubating permeabilized cells in Br-UTP. Newly completed transcripts are processed in the granular component to emerge as mature ribosomal subunits into the nucleoplasm. A typical nucleolar factory in a HeLa cell contains ~500 polymerases engaged on around four templates, with zones dedicated to storage, RNA synthesis and processing [10,12].

Nucleoplasmic factories containing either RNA polymerase II or III

Active RNA polymerase I is clearly contained in a specialized compartment, where it transcribes a particular group of genes, but what of the nucleoplasmic polymerases? As quantitative analysis of the numbers of nascent transcripts and sites had revealed that each site must contain many active polymerases, we can envisage two different types of organization. In the first, a cluster of polymerase II molecules might form one factory where all the enzymes work on class II transcription units; an analogous cluster of polymerase III molecules would work only on class III units. In the second, one factory might contain both types of enzyme to generate both class II and III transcripts. Three approaches suggest that the former applies, and that sites contain one type of active enzyme but not the other [8]. In the first, we compared site densities after inhibiting polymerase II with α-amanitin. If the two types of polymerase are intermingled within a site, and if most sites are detected (which was shown to be the case), we would expect inhibition of polymerase II to have little effect on site density. However, site density fell to one-fifth, consistent with one-fifth of the sites being dedicated to transcription by polymerase III. The second approach involved seeing if a polymerase lay next to transcripts made by the other enzyme. Although polymerases II and III often lay immediately next to their own transcripts, they rarely (if at all) lay next to those made by the other enzyme. However, such negative results are indecisive, especially when the approach is bedevilled by steric hindrance between immunolabelling probes. Convincing evidence against co-localization was obtained using a third approach that exploited this steric hindrance; cryosections of appropriately labelled nuclei were preincubated with an antibody to see if it blocked access of a second. We found that an anti-(polymerase II) antibody blocked access to transcripts made by polymerase II, but not to polymerase III protein or its transcripts; conversely, an antibody against polymerase III blocked access to polymerase III transcripts, but not to polymerase II protein or its transcripts. Therefore the results of all three approaches are consistent with polymerases II and III being active in spatially distinct sites.

These results suggest that the two nucleoplasmic polymerases – like the nucleolar enzyme – are concentrated in their own dedicated factories. Quantitative analysis shows that each HeLa cell nucleus typically contains ~90 000 nascent transcripts, with ~15 000, ~65 000 and ~10 000 being made by polymerases I, II and III respectively [12]. Each of the ~8000 polymerase II factories contains around eight polymerases engaged on eight different templates, while each of the ~2000 polymerase III factories contains around five polymerases active on five different templates [8].

Complementary functions in one RNA polymerase II factory

The functions required to generate a 'standard' mRNA made by RNA polymerase II (e.g. capping, splicing and polyadenylation) were originally thought to occur independently, but it is now clear that they are interdependent [13]. Thus the C-terminal domain of the catalytic subunit of polymerase II – which in humans contains 52 heptad repeats with the consensus sequence YSPTSPS – forms a platform that integrates these functions. This enormous complex also proofreads newly made transcripts before going on to destroy faulty ones [14–16]. Proofreading involves the NMD (nonsense-mediated decay) pathway [17], and probably uses ribosomes to scan transcripts for inappropriately placed (i.e. premature) termination codons. Nascent peptides made during ribosomal scanning are found in nucleoplasmic factories, together with components of the translation and NMD machineries. As inhibiting transcription immediately inhibits this nuclear protein synthesis, the two processes must be tightly coupled. Moreover, the transcriptional, translational, NMD and degradative machinery (which destroys faulty transcripts and any misfolded peptides resulting from translational proofreading) all co-localize and co-immunoprecipitate; selected components (translational initiation factor eIF4E, ribosomal subunit S6, NMD factors UPF1 and UPF2) also co-purify with the catalytic subunit of the polymerase, probably by binding to its C-terminal domain. However, it remains unclear whether all the different machineries are bound at any given moment [18].

Further specialization of nucleoplasmic factories

Do nucleoplasmic factories specialize further, to transcribe particular gene sets? It seems that they might, as active sets associate with specific sites (much as active NORs associate with nucleoli). Some examples will now be discussed.

The OPT domain

PTF [PSE (proximal sequence element)-binding transcription factor] activates transcription of snRNA (small nuclear RNA) and related genes. Immunofluorescence shows that it is spread throughout the nucleoplasm of HeLa cells in small foci. In some cells, it is also concentrated in one – or a few – discrete foci (diameter ~1.3 μm) that appear during G_1 phase and disappear in S phase. Oct1 (a transcription factor that interacts with PTF), RNA polymerase II, TBP (TATA-binding protein) and Sp1 are also present in these foci. Each focus typically contains two or three transcription factories that can incorporate Br-UTP into nascent transcripts, so they were christened OPT (Oct1/PTF/transcription) domains. Significantly, a small region on chromosome 6 (i.e. band 6p21) that contains only ~30 Mbp DNA, plus chromosomes 6 and 7, associate with the domain significantly more than other chromosomes. This suggests that these domains may act like nucleoli to bring particular genes on specific chromosomes together to a region where the appropriate transcription and processing factors are concentrated, thereby facilitating the expression of those genes [19].

The β-globin transcription factory

Results from various methods – 3C (chromosome conformation capture), RNA TRAP (RNA tagging and recovery of associated proteins), ChIP (chromatin immunoprecipitation) and FISH (fluorescence *in situ* hybridization) – now provide good evidence for the existence of chromatin loops tethered through active transcription units to a 'hub' or factory (as in Figure 1), as well as providing suggestive evidence that one particular factory specializes in the transcription of a particular set of genes. The first three methods involve fixation, before analysis of which DNA sequences lie next to each other in three-dimensional space – after cutting, dilution and ligation (in 3C), or purifying complexes containing either a specified nascent transcript (in RNA TRAP) or protein bound to DNA (in ChIP).

The mouse *Hbb-b1* (β-globin) gene lies tens of kilobase-pairs away on chromosome 7 from its LCR (locus control region), and ~25 Mbp away from a gene (*Eraf*) encoding the α-globin stabilizing protein. 3C shows that *Hbb-b1* contacts the LCR and *Eraf* in erythroid nuclei (where all three are transcribed, e.g. see [20]), but not in brain nuclei, where all are inactive [21–24]. RNA TRAP confirms that *Hbb1* contacts the LCR [25]. DNA- and RNA-FISH coupled to immunolabelling also show that active *Hbb-b1* and *Eraf* are found together in sites rich in polymerase II; moreover, inactive alleles in the same cell are generally distant from such sites [23]. These factories – in which some of the most highly expressed genes in the erythroid lineage are concentrated – are larger than those in HeLa cells, and have been described as "active chromatin hubs" [21].

Note that we have argued elsewhere that a wide range of other LCRs, including those controlling the expression of α-globin, keratin 18, adenosine deaminase, growth hormone and MHC class II genes, all encode non-genic transcription units, and that transcription of these LCRs probably underlies their function [26]. Note also that where these new methods have been used to demonstrate looping in other cases, the tethering points also turn out to be transcription units, in accordance with the model illustrated in Figure 1. For example, 3C shows that the two barriers flanking the *Drosophila* 87A7 heat-shock locus – *scs* and *scs′* – contact each other [27], and both barriers encode promoters [28]. Moreover, 3C and ChIP show that the mouse *H19* DMR (differentially methylated region) interacts with the *Igf2* DMR1 (but not DMR2) when *H19* is expressed from the maternal allele; conversely, it interacts with the *Igf2* DMR2 (but not DMR1) when *Igf2* is expressed from the paternal allele [29]. Both DMR1 and DMR2 are transcribed – the former encodes both sense and antisense transcription units, and the latter lies within *Igf2* [30]. (Unfortunately, we do not know whether the *H19* DMR is transcribed, but promoters do lie nearby [31].) So in all of these cases, it seems that transcription units contact each other – but only when they are active.

Stress granules

On heat shock, satellite III repeats on human chromosome 9q12 are transcribed and associate with a factory containing heat-shock factor 1 [32]. It becomes tempting to suggest that this factory contains the factors necessary for the transcription of this group of units.

Cajal bodies

Some class II genes encode transcripts with characteristic 3′ stem-loops but no introns or poly(A) tails (e.g. histone mRNAs, U1–U4, U11 and snRNAs). They are often found near Cajal bodies, unlike the class III genes that encode U6 snRNA and 5 S rRNA. This association depends on transcription by RNA polymerase II, as a promoter mutation or transcriptional inhibition prevent it [33–35]. Here, these genes encoding transcripts that are not processed in the usual way associate with 'non-standard' factories that lack the 'normal' processing machinery.

PML (promyelocytic leukaemia) bodies

Gene-dense regions of various human chromosomes, including the major histocompatibility locus, are transcribed at the surface of these structures [36,37]. Again, do these structures contain the appropriate high concentrations of transcription factors required for the transcription of this sub-set of genes?

Peri-nucleolar polymerase III factories

Yeast tRNA genes on different chromosomes associate with the nucleolar surface, and a point mutation in the promoter of one eliminates this association [38]. This suggests that polymerase III factories may associate with the surface of the nucleolus.

In all the described cases, particular groups of genes associate with a particular structure in order to be transcribed, presumably because that structure contains the appropriate machinery.

Conclusions

We have reviewed the evidence that the three different types of nuclear RNA polymerases (i.e. I, II and III) are each concentrated in their own dedicated factories, and that the different types of transcription unit (i.e. classes I, II and III) are only transcribed once they associate with a factory of the appropriate type. We have gone on to suggest that specialization is carried even further, with particular factories becoming enriched in particular factors so they become dedicated to the transcription of those units that require those factors. We imagine these structures to be dynamic, with the factors, polymerases and transcription units all exchanging continuously. Then, loop attachments mediated by transcription factors would persist for seconds, and those mediated by polymerases for the minutes required for the cycle of initiation, elongation and termination [39]. This turnover ensures that factories can evolve as conditions change; the chance transcription of a new unit would inevitably leave a new set of factors bound to the factory, and this would increase the chances that the same (or related) units might be transcribed there.

We thank the BBSRC, Cancer Research UK, the MRC and the Wellcome Trust for support.

References

1. Hozák, P., Hassan, A.B., Jackson, D.A. and Cook, P.R. (1993) Cell **73**, 361–373
2. Jackson, D.A., Hassan, A.B., Errington, R.J. and Cook, P.R. (1993) EMBO J. **12**, 1059–1065
3. Jackson, D.A., Balajee, A.S., Mullenders, L. and Cook, P.R. (1994) J. Cell Sci. **107**, 1745–1752
4. Jackson, D.A., Hassan, A.B., Errington, R.J. and Cook, P.R. (1994) J. Cell Sci. **107**, 1753–1760
5. Cook, P.R. (1999) Science **284**, 1790–1795
6. Cook, P.R. (1995) J. Cell Sci. **108**, 2927–2935
7. Cook, P.R. (2002) Nat. Genet. **32**, 347–352
8. Pombo, A., Jackson, D.A., Hollinshead, M., Wang, Z., Roeder, R.G. and Cook, P.R. (1999) EMBO J. **18**, 2241–2253
9. Shaw, P.J. and Jordan, E.G. (1995) Annu. Rev. Cell Dev. Biol. **11**, 93–121
10. Hozák, P., Cook, P.R., Schöfer, C., Mosgöller, W. and Wachtler, F. (1994) J. Cell Sci. **107**, 639–648
11. Shaw, P.J., Highett, M.I., Bevan, A.F. and Jordan, E.G. (1995) EMBO J. **14**, 2896–2906
12. Jackson, D.A., Iborra, F.J., Manders, E.M.M. and Cook, P.R. (1998) Mol. Biol. Cell **9**, 1523–1536
13. Maniatis, T. and Reed, R. (2002) Nature (London) **416**, 499–506
14. Iborra, F.J., Jackson, D.A. and Cook, P.R. (2001) Science **293**, 1139–1142
15. Iborra, F.J., Escargueil, A.E., Kwek, K.Y., Akoulitchev, A. and Cook, P.R. (2004) J. Cell Sci. **117**, 899–906
16. Iborra, F.J., Jackson, D.A. and Cook, P.R. (2004) J. Cell Sci. **117**, 5713–5720
17. Hilleren, P. and Parker, R. (1999) Annu. Rev. Genet. **33**, 229–260
18. Wetterberg, I., Zhao, J., Masich, S., Wieslander, L. and Skoglund, U. (2001) EMBO J. **20**, 2564–2574
19. Pombo, A., Cuello, P., Schul, W., Yoon, J.-B., Roeder, R.G., Cook, P.R. and Murphy, S. (1998) EMBO J. **17**, 1768–1778
20. Routledge, S.J. and Proudfoot, N.J. (2002) J. Mol. Biol. **323**, 601–611
21. Tolhuis, B., Palstra, R.J., Splinter, E., Grosveld, F. and de Laat, W. (2002) Mol. Cell **10**, 1453–1465
22. Palstra, R.J., Tolhuis, B., Splinter. E., Nijmeijer, R., Grosveld, F. and de Laat, W. (2003) Nat. Genet. **35**, 190–194

23. Osborne, C.S., Chakalova, C., Brown, K.E., Carter, D., Horton, A., Debrand, E., Goyenechea, B., Mitchell, J.A., Lopes, S., Reik, W. and Fraser, P. (2004) Nat. Genet. **36**, 1065–1071
24. Patrinos, G.P., de Krom, M., de Boer, E., Langeveld, A., Imam, A.M., Strouboulis, J., de Laat, W. and Grosveld, F.G. (2004) Genes Dev. **18**, 1495–1509
25. Carter, D., Chakalova, L., Osborne, C.S., Dai, Y.-f. and Fraser, P. (2002) Nat. Genet. **32**, 623–626
26. Cook, P.R. (2003) J. Cell Sci. **116**, 4483–4491
27. Blanton, J., Gaszner, M. and Schedl, P. (2003) Genes Dev. **17**, 664–675
28. Avramova, Z. and Tikhonov, A. (1999) Trends Genet. **15**, 138–139
29. Murrell, A., Heeson, S. and Reik, W. (2004) Nat. Genet. **36**, 889–893
30. Moore, T., Constancia, M., Zubair, M., Bailleul, B., Feil, R., Sasaki, H. and Reik, W. (1997) Proc. Natl. Acad. Sci. U.S.A. **94**, 12509–12514
31. Drewell, R.A., Arney, K.L., Arima, T., Barton, S.C., Brenton, J.D. and Surani, M.A. (2002) Development **129**, 1205–1213
32. Jolly, C., Metz, A., Govin, J., Vigneron, M., Turner, B.M., Khochbin, S. and Vourc'h, C. (2004) J. Cell Biol. **164**, 25–33
33. Callan, H.G., Gall, J.G. and Murphy, C. (1991) Chromosoma **101**, 245–251
34. Frey, M.R., Bailey, A.D., Weiner, A.M. and Matera, A.G. (1999) Curr. Biol. **9**, 126–135
35. Jacobs, E.Y., Frey, M.R., Wu, W., Ingledue, T.C., Gebuhr, T.C., Gao, L., Marzluff, W.F. and Matera, A.G. (1999) Mol. Biol. Cell **10**, 1653–1663
36. Shiels, C., Islam, S.A., Vatcheva, R., Sasieni, P., Sternberg, M.J.E., Freemont, P.S. and Scheer, D. (2001) J. Cell Sci. **114**, 3705–3716
37. Wang, J., Shiels, C., Sasieni, P., Wu, P.J., Islam, S.A., Freemont, P.S. and Sheer, D. (2004). J. Cell Biol. **164**, 515–526
38. Thompson, M., Haeusler, R.A., Good, P.D. and Engelke, D.R. (2003) Science **302**, 1399–1401
39. Kimura, H., Sugaya, K. and Cook, P.R. (2002) J. Cell Biol. **159**, 777–782
40. Cook, P.R. (2001) Principles of Nuclear Structure and Function, J. Wiley and Sons, New York

Biochem. Soc. Symp. **73**, 77–84
(Printed in Great Britain)

8

A role for upstream binding factor in organizing ribosomal gene chromatin

Jane E. Wright, Christine Mais, José-Luis Prieto and Brian McStay[1]

Biomedical Research Centre, Ninewells Hospital and Medical School, University of Dundee, Dundee DD1 9SY, Scotland, UK.

Abstract

Human ribosomal genes are located in NORs (nucleolar organizer regions) on the short arms of acrocentric chromosomes. During metaphase, previously active NORs appear as prominent chromosomal features termed secondary constrictions, which are achromatic in chromosome banding and positive in silver staining. The architectural RNA polymerase I transcription factor UBF (upstream binding factor) binds extensively across the ribosomal gene repeat throughout the cell cycle. Evidence that UBF underpins NOR structure is provided by an examination of cell lines in which large arrays of a heterologous UBF binding sequences are integrated at ectopic sites on human chromosomes. These arrays efficiently recruit UBF even to sites outside the nucleolus, and during metaphase form novel silver-stainable secondary constrictions, termed pseudo-NORs, that are morphologically similar to NORs.

Introduction

In most eukaryotes, the sole function of Pol I (RNA polymerase I) is the transcription of genes encoding 18 S, 5.8 S and 28 S rRNAs. Like Pol II and Pol III, it requires auxiliary factors that mediate promoter recognition, PIC (pre-initiation complex) formation and transcription initiation. PIC formation on ribosomal gene (rDNA) promoters requires the synergistic action of UBF (upstream binding factor) [1] and the promoter selectivity factor, termed SL1 in humans [2]. SL1 is composed of TBP (TATA-binding protein) and the TBP-associated factors$_I$ TAF$_I$110, TAF$_I$63 and TAF$_I$48 [3]. SL1 interacts

[1]To whom correspondence should be addressed (email brian.mcstay@cancer.org.uk).

with promoter DNA in a highly sequence-specific manner, as illustrated by the species-specific nature of rDNA transcription [4]. PICs recruit an initiation competent subfraction of Pol I, defined by the presence of TIF (transcription intermediary factor)-IA/Rrn3 [5,6]. Although clearly involved in promoter function, UBF performs additional roles. After a brief review of the chromosomal organization of human rDNA, we will focus here on the role of UBF in organizing ribosomal gene chromatin.

NOR (nucleolar organizer regions)

The human rDNA repeat is 43 kb in length, 13 kb of which is transcribed to yield the 47 S pre-rRNA. The remaining 30 kb is termed the IGS (intergenic spacer) (Figure 1). Approximately 400 copies of the rDNA repeat are distributed among the short arms of the human acrocentric chromosomes (13, 14, 15, 21 and 22) [7], each of which consists of three bands: p11, p12 and p13. Band p12 contains on average 3 Mb of tandem rDNA repeats [8] and is termed the NOR, while p11 and p13 are composed of satellite sequences packaged as heterochromatin. During metaphase, NORs that were transcriptionally active in the previous interphase form prominent chromosomal features termed secondary constrictions, which appear as achromatic gaps when stained with Giemsa or AT-specific fluorochromes such as DAPI (4′,6-diamidino-2-phenylindole) or quinacrine [9]. Electron tomography revealed that chromatin within secondary constrictions is 10-fold less condensed than the rest of the chromosome [10]. Classically, secondary constrictions are positive in silver staining, and are

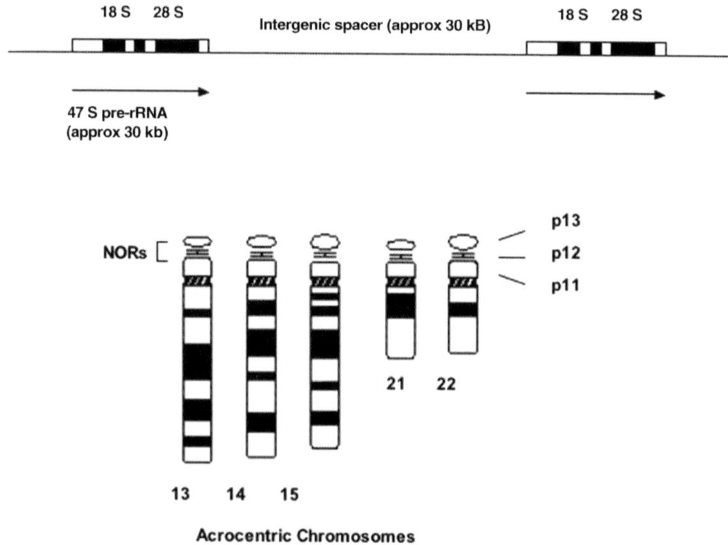

Figure 1 Structure of human ribosomal gene repeat and the chromosomal location of NORs. See text for further details.

often termed AgNORs [11]. Association of the transcription machinery with rDNA during mitosis correlates with the presence of a secondary constriction [12,13]. It is generally considered, but not proven, that some component of the transcription machinery, or indeed the act of transcription itself, is responsible for the undercondensation of NORs during metaphase.

Binding of UBF to rDNA

UBF belongs to the sequence-non-specific class of HMG (high-mobility-group) box proteins [14] (Figure 2). Unlike other HMG box proteins, UBF dimerizes in solution through sequences in its N-terminus [15]. UBF contains multiple HMG boxes and, like other HMG proteins, interacts with the minor groove of DNA and binds to structured nucleic acids such as cruciforms, or four-way junctions [16]. Like other HMG proteins, UBF induces bending upon binding to non-structured sequences. Moreover, the tandem HMG boxes enable a single dimer of UBF to organize promoter DNA into a 360° loop [16,17]. This so-called architectural ability of UBF, its abundance (approx. 5×10^5 molecules per cell) and its low sequence specifity in DNA binding were a strong indicator of other roles for UBF. Further support for this contention comes from the observation that mammalian cells contain approximately equal amounts of two UBF forms as a result of splice site variation. UBF2 differs from UBF1 in missing 37 residues

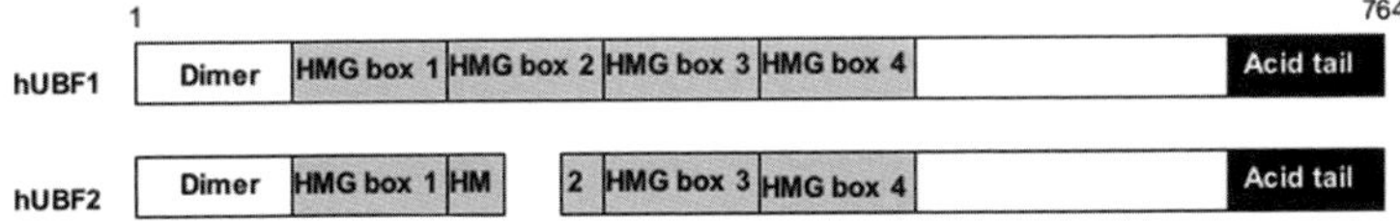

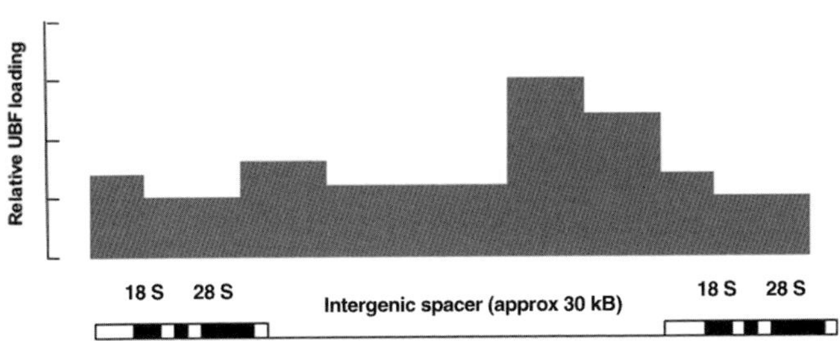

Figure 2 Structure of UBF and its distribution across the human rDNA repeat. UBF1 and UBF2 are shown in cartoon form. UBF1 comprises 764 amino acids with four HMG box DNA-binding motifs, a dimerization domain at its N-terminus and a highly acidic C-terminus. In the splice variant UBF2, 37 amino acids are deleted from HMG box 2. The relative distribution of UBF across the human rDNA repeat is shown in the lower panel. This graph was derived from previously published ChIP data [26].

from the second HMG box [18]. This deletion renders UBF2 non-functional at the promoter [19]. These characteristics suggested that UBF has a more general structural role on active NORs.

UBF can bind *in vitro* to repeated elements that function as rDNA-specific transcriptional enhancers [20,21]. Experiments with *in vitro* systems demonstrate that UBF is required for enhancer function [22]. Interestingly, UBF from heterologous species and UBF2 can support enhancer but not promoter, function. Further *in vitro* experiments have shown that UBF can bind across the entire *Xenopus* IGS [23] and to sequences downstream of the transcriptional start site [24]. However, it should be pointed out that UBF also binds to poly(dI-dC) and vector DNA. Consequently, *in vitro* binding studies do not provide reliable evidence for extensive binding of UBF to the rDNA repeat *in vivo*. In order to address the distribution of UBF *in vivo*, we developed a ChIP (chromatin immunoprecipitation) assay that utilizes an enriched nucleolar chromatin fraction [25]. We demonstrated for the first time that UBF binds *in vivo* to multiple sites distributed across the entire human rDNA repeat (transcribed sequences and IGSs) [25,26] (Figure 2). Despite this diversity of binding sites, UBF binds selectively to rDNA, as it is localized exclusively to nucleoli during interphase and at NORs during mitosis [27]. We have proposed a model in which rDNA repeats contain specialized high-affinity binding sequences that bind UBF in a co-operative manner and induce spreading of UBF on to adjacent lower-affinity binding sites across the rDNA repeat [25].

Effects of UBF binding

The next question was to determine the consequences of this extensive UBF binding. Since UBF remains bound to NORs during mitosis, we were particularly interested in determining if extensive binding of UBF is responsible for both the undercondensation and silver staining of NORs in metaphase. The strategy we employed to address this question was to generate cell lines in which large arrays of a heterologous UBF-binding sequence were integrated into ectopic sites on human chromosomes [26]. The sequences employed, XEn, were blocks of 60/81 bp repeats present in the IGS of *Xenopus* rDNA (Figure 3A). These are well characterized UBF-binding sites [28] and function in a UBF-dependent manner as transcriptional enhancers [22]. Despite a lack of sequence identity with human rDNA, it was already known that human UBF binds to these elements *in vitro* [29]. We have characterized eight independent cells lines in which arrays of XEn ranging from 100–2000 kb have been integrated at various chromosomal sites. In Figures 3(B) and 3(C) we show data from clone 3D, in which 1400 kb was integrated into the long (q) arm of human chromosome 10. Combined immuno-FISH (fluorescence *in situ* hybridization) was utilized to demonstrate that these arrays bind UBF during interphase, independent of nuclear location and chromosomal integration site (Figure 3B). Analysis of chromosomes demonstrated that UBF remains bound to XEn arrays during metaphase and is associated with the presence of a novel secondary constriction, particulary evident in reverse DAPI banding (Figure 3C). The XEn array in

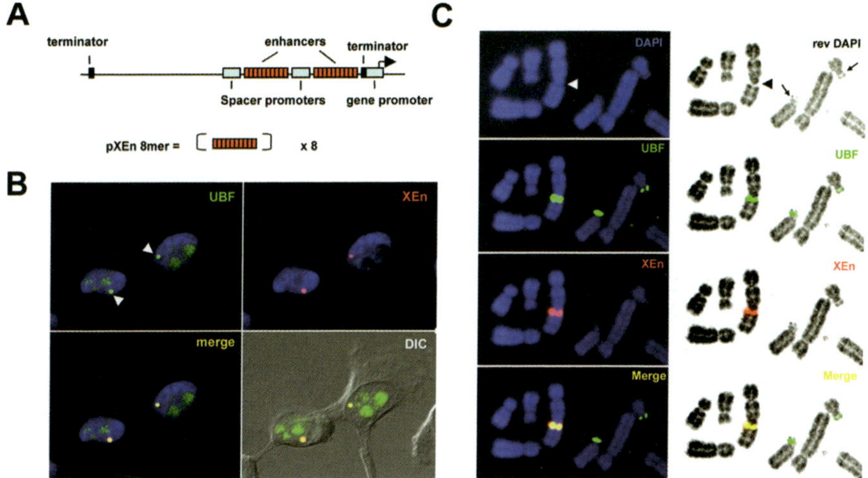

Figure 3 UBF binds to ectopic binding site arrays and results in formation of a novel secondary constriction termed a pseudo-NOR. (**A**) The plasmid p*XEn*8mer was constructed which contains eight blocks of the enhancer element or eighty 60/81 bp repeats (red boxes). The p*XEn*8mer insert was transfected into a human fibrosarcoma cell line, HT1080, together with a blasticidin resistance marker. (**B**) Combined immuno-FISH was performed on interphase cells of clone 3D. UBF was visualized using FITC-conjugated affinity-purified antibodies. *XEn* arrays were visualized with a spectrum red-labelled probe. DIC, differential interference contrast (**C**) Combined immuno-FISH was performed on metaphase spreads prepared from clone 3D. UBF and XEn were visualized as above. Left panels show standard DAPI staining; right panels show enhanced reversed DAPI staining.

this clone represents approx. 1% of the DNA present in this chromosomal arm. Notably, the novel secondary constriction associated with these sequences represents a significantly larger fraction of the contour length of the metaphase chromosome. Presumably this is a consequence of its undercondensed state.

The second primary characteristic of metaphase NORs is that they are positive in silver staining. It was generally assumed that this was due to the presence of an abundant argyophillic protein. The acidic C-terminus of UBF is a prime candidate for silver binding. The novel secondary constriction on chromosome 10q in clone 3D is clearly positive in silver staining (Figure 4A). Note also that the XEn array is clearly visualized by silver staining during interphase (Figure 4B). We conclude from these experiments that recruitment of UBF to *XEn* arrays results in formation of a chromatin structure with all the primary characteristics of NORs. We refer to these structures as pseudo-NORs, since they do not organize formation of a nucleolus. The precise role of UBF in forming secondary constrictions is not yet clear to us, but we can now propose a number of conclusions. First, as pseudo-NORs are inactive, due to the absence of promoter sequences, transcription in the previous interphase

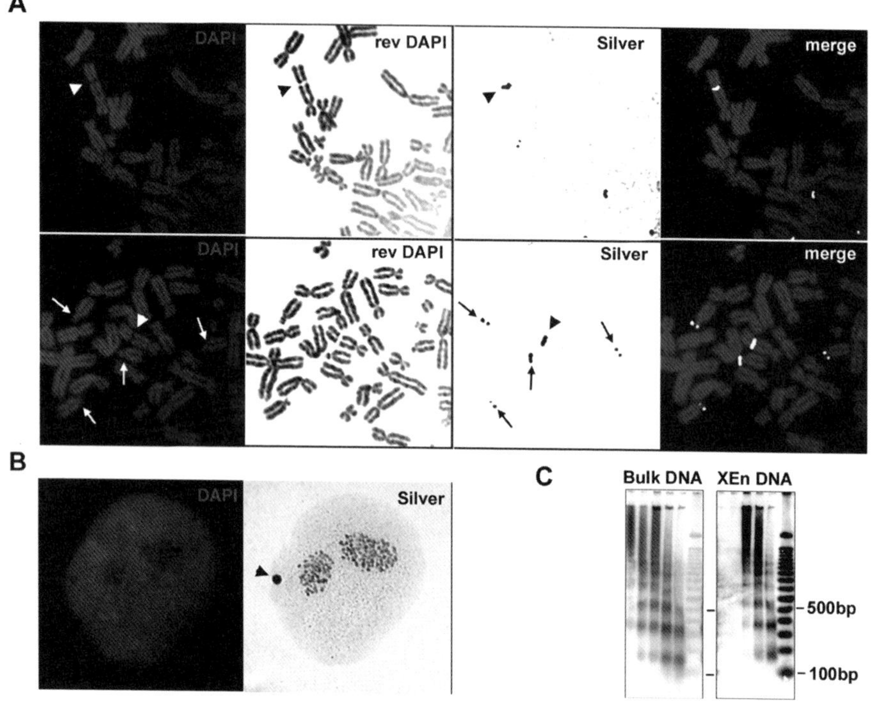

Figure 4 Pseudo-NORs are positive in silver staining and packaged as nucleosomes. (**A**) Metaphase spreads prepared from clone 3D were both DAPI-stained and-silver stained. Pseudo-NORs are indicated by arrowheads, and true NORs by arrows. (**B**) Staining of a clone 3D interphase nucleus by DAPI and silver. The pseudo-NOR is indicated by an arrowhead. (**C**) Nuclei from clone 3D were digested with an increasing concentration of micrococcal nuclease. DNA was recovered, Southern-blotted and probed with human repetitive DNA sequences (COT**1**) DNA to visualize bulk DNA (left panel) or XEn DNA (right panel).

is not a pre-requisite. Secondly, the ability to form secondary constrictions is intrinsic to rDNA. The *de novo* formation of pseudo-NORs at ectopic sites demonstrates that sequences distal and proximal to NORs are not required. Finally, binding sites for other components of the Pol I transcription machinery are not required, as UBF-binding sites are sufficient to specify this structure.

Micrococcal nuclease digests of nuclei prepared from clone 3D clearly demonstrate a classical nucleosomal ladder associated with XEn sequences (Figure 4C). The nature of these nucleosomes is under investigation. The key question, however, is how does extensive UBF binding result in undercondensation and formation of a secondary constriction during metaphase? One possibility is that UBF binding to nucleosomal DNA inhibits binding of linker histones required for chromatin condensation. In this regard, it is worth pointing out that UBF can bind *in vitro* to nucleosomes with high affinity and displace linker histone H1 [30].

Conclusions and perspectives

Although highly conserved among vertebrate organisms, UBF is not present in invertebrates, fungi or plants. A role for HMG box proteins in rDNA chromatin organization does, however, appear to have been conserved throughout evolution. Hmo1p, a nucleolar protein with a single HMG box, has some sequence similarity to UBF and strongly enhances rDNA transcription in yeast [31]. This may explain why secondary constrictions can be observed associated with the NORs of non-vertebrate organisms.

We have observed that, during interphase, pseudo-NORs sequester a significant fraction of every component of the Pol I transcription machinery, even to sites outside nucleoli and independent of both promoter sequences and transcription [26]. These results have forced us to re-evaluate how the Pol I machinery is recruited to nucleoli and rDNA. Moreover, it appears that maintenance of rDNA chromatin in an undercondensed state throughout mitosis facilitates the rapid onset of transcription in the subsequent G_1 stage of the cell cycle.

We thank Sam Raggett for technical assitance. This work was funded by a programme grant from the MRC to B.M.

References

1. Jantzen, H.M., Admon, A., Bell, S.P. and Tjian, R. (1990) Nature (London) **344**, 830–836
2. Learned, R.M., Cordes, S. and Tjian, R. (1985) Mol. Cell. Biol. **5**, 1358–1369
3. Comai, L., Tanese, N. and Tjian, R. (1992) Cell **68**, 965–976
4. Heix, J. and Grummt, I. (1995) Curr. Opin. Genet. Dev. **5**, 652–656
5. Bodem, J., Dobreva, G., Hoffmann-Rohrer, U., Iben, S., Zentgraf, H., Delius, H., Vingron, M. and Grummt, I. (2000) EMBO Rep. **1**, 171–175
6. Miller, G., Panov, K.I., Friedrich, J.K., Trinkle-Mulcahy, L., Lamond, A.I. and Zomerdijk, J.C. (2001) EMBO J. **20**, 1373–1382
7. Henderson, A.S., Warburton, D. and Atwood, K.C. (1972) Proc. Natl. Acad. Sci. U.S.A. **69**, 3394–3398
8. Sakai, K., Ohta, T., Minoshima, S., Kudoh, J., Wang, Y., de Jong, P.J. and Shimizu, N. (1995) Genomics **26**, 521–526
9. Sumner, A.T. (1982) Cancer Genet. Cytogenet. **6**, 59–87
10. Heliot, L., Kaplan, H., Lucas, L., Klein, C., Beorchia, A., Doco-Fenzy, M., Menager, M., Thiry, M., O'Donohue, M.F. and Ploton, D. (1997) Mol. Biol. Cell **8**, 2199–2216
11. Goodpasture, C. and Bloom, S.E. (1975) Chromosoma **53**, 37–50
12. Weisenberger, D. and Scheer, U. (1995) J. Cell Biol. **129**, 561–575
13. Roussel, P., Andre, C., Comai, L. and Hernandez-Verdun, D. (1996) J. Cell Biol. **133**, 235–246
14. Thomas, J.O. and Travers, A.A. (2001) Trends Biochem. Sci. **26**, 167–174
15. McStay, B., Frazier, M.W. and Reeder, R.H. (1991) Genes Dev. **5**, 1957–1968
16. Copenhaver, G.P., Putnam, C.D., Denton, M.L. and Pikaard, C.S. (1994) Nucleic Acids Res. **22**, 2651–2657
17. Bazett-Jones, D.P., Leblanc, B., Herfort, M. and Moss, T. (1994) Science **264**, 1134–1137
18. O'Mahony, D.J. and Rothblum, L.I. (1991) Proc. Natl. Acad. Sci. U.S.A. **88**, 3180–3184
19. Kuhn, A., Voit, R., Stefanovsky, V., Evers, R., Bianchi, M. and Grummt, I. (1994) EMBO J. **13**, 416–424
20. Pikaard, C.S., Pape, L.K., Henderson, S.L., Ryan, K., Paalman, M.H., Lopata, M.A., Reeder, R.H. and Sollner, W.B. (1990) Mol. Cell. Biol. **10**, 4816–4825
21. Pikaard, C.S., Smith, S.D., Reeder, R.H. and Rothblum, L. (1990) Mol. Cell. Biol. **10**, 3810–3812

22. McStay, B., Sullivan, G.J. and Cairns, C. (1997) EMBO J. **16**, 396–405
23. Mougey, E.B., Pape, L.K. and Sollner-Webb, B. (1996) J. Biol. Chem. **271**, 27138–27145
24. Leblanc, B., Read, C. and Moss, T. (1993) EMBO J. **12**, 513–525
25. O'Sullivan, A.C., Sullivan, G.J. and McStay, B. (2002) Mol. Cell. Biol. **22**, 657–668
26. Mais, C., Wright, J.E., Prieto, J.L., Raggett, S.L. and McStay, B. (2005) Genes Dev. **19**, 50–64
27. Roussel, P., Andre, C., Masson, C., Geraud, G. and Hernandez, V.D. (1993) J. Cell Sci. **104**, 327–337
28. Pikaard, C.S., McStay, B., Schultz, M.C., Bell, S.P. and Reeder, R.H. (1989) Genes Dev. **3**, 1779–1788
29. Bell, S.P., Pikaard, C.S., Reeder, R.H. and Tjian, R. (1989) Cell **59**, 489–497
30. Kermekchiev, M., Workman, J.L. and Pikaard, C.S. (1997) Mol. Cell. Biol. **17**, 5833–5842
31. Gadal, O., Labarre, S., Boschiero, C. and Thuriaux, P. (2002) EMBO J. **21**, 5498–5507

Biochem. Soc. Symp. **73**, 85–96
(Printed in Great Britain)

9

Nutrient-regulated gene expression in eukaryotes

Richard J. Reece[1], Laila Beynon, Stacey Holden, Amanda D. Hughes, Karine Rébora and Christopher A. Sellick

Faculty of Life Sciences, University of Manchester, The Michael Smith Building, Oxford Road, Manchester M13 9PT, U.K.

Abstract

The recognition of changes in environmental conditions, and the ability to adapt to these changes, is essential for the viability of cells. There are numerous well characterized systems by which the presence or absence of an individual metabolite may be recognized by a cell. However, the recognition of a metabolite is just one step in a process that often results in changes in the expression of whole sets of genes required to respond to that metabolite. In higher eukaryotes, the signalling pathway between metabolite recognition and transcriptional control can be complex. Recent evidence from the relatively simple eukaryote yeast suggests that complex signalling pathways may be circumvented through the direct interaction between individual metabolites and regulators of RNA polymerase II-mediated transcription. Biochemical and structural analyses are beginning to unravel these elegant genetic control elements.

Introduction

The genome of an organism contains within its DNA sequence all of the information required to define not only cell type, but also the ability to respond to a variety of external conditions and signals. For example, the genome of the simple eukaryotic yeast *Saccharomyces cerevisiae* contains 12 million base pairs of DNA, split into 16 separate chromosomes and comprising approximately 6000 different genes. The protein products encoded by these genes rarely act individually, and whole pathways are often regulated in a co-ordinated fashion in response to a particular signal. The complex processes of the cell depend on the differential expression of sets of genes in particular cell types, at a particular time, and/or

[1]To whom correspondence should be addressed (email Richard.Reece@manchester.ac.uk).

under certain environmental conditions. Efficient genetic switches within the cell permit the co-ordinated transcription of sets of genes in response to a signal.

It is becoming increasingly apparent that what were traditionally thought of as metabolic enzymes can play diverse roles in a wide range of cellular processes [1]. Similarly, metabolites appear not just to serve as substrates for enzymes. Indeed, the link between metabolic substrates and their ability to influence transcription directly is long established in prokaryotes. For example, in *Escherichia coli*, the interaction between lactose and the *lac* repressor [2], or galactose and GalR [3], in modulating the DNA-binding properties of each respective protein is well understood. Here we will discuss two eukaryotic genetic switches and the ways in which they are regulated by the action of small-molecule metabolites. Insights into the processes of metabolically regulated gene expression in eukaryotes have been gained from the study of the relatively simple genetic switches of yeast – although the mechanistic details have turned out to be somewhat complex.

Metabolic signalling through protein complex formation – *GAL* induction

The expression of the *GAL* genes of the yeast *Saccharomyces cerevisiae* is a paradigm for eukaryotic transcriptional regulation. The *GAL* genetic switch provides the transcriptional control of genes whose products are required for the metabolism of the sugar galactose – the enzymes of the Leloir pathway [4]. When yeast cells are grown in a medium without galactose, the *GAL* genes are essentially transcriptionally inert. However, if galactose is the only source of carbon available to the cell, then the *GAL* genes will become transcriptionally active – both rapidly and to a high level – and galactose is converted into glucose 1-phosphate [5]. Galactose is a comparatively poor sugar source for the cell, and yeast will preferentially metabolize other sources of carbon (e.g. glucose) to galactose even if a mixture of the two sugars is available [6]. Glucose triggers catabolite repression of the *GAL* genes, and the expression of the activator upon which they depend, Gal4p, is severely reduced when cells are grown on glucose [7]. In the presence of other carbon sources, e.g. raffinose or glycerol, Gal4p is produced in the cell, and can be found tethered upstream of the *GAL* genes, but does not bring about transcriptional activation. The transcriptional activity of DNA-bound Gal4p is inhibited through its interaction with another protein, Gal80p [8]. Activation of *GAL* genes by Gal4p is thought to involve the ordered recruitment of various protein complexes. The transcriptionally active form of DNA-bound Gal4p interacts directly with the Tra1p subunit of the SAGA (Spt/Ada/Gcn5 acetyltransferase) co-activator/histone modifying complex [9–11]. Other subunits of SAGA (Spt3p and Spt8p) interact directly with specific subunits of the RNA polymerase II mediator complex (Med2p and Pgd1p respectively) [12]. Gal4p may also interact with components of the mediator directly [13]. Finally, the mediator recruits RNA polymerase II holoenzyme to the promoter, and the initiation of transcription can occur.

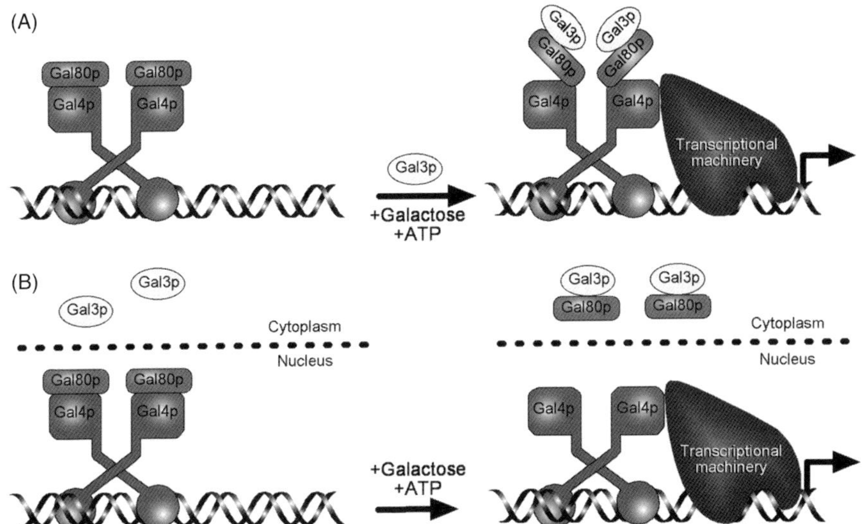

Figure 1 Models for activation of the yeast *GAL* genes. (A) A non-dissociation model, in which activation occurs through the formation of a tripartite complex of Gal4p, Gal80p and Gal3p in the presence of galactose and ATP. A conformational change in Gal80p occurs, allowing Gal4p-mediated recruitment of the transcriptional machinery. **(B)** A dissociation model, in which the interaction between Gal3p and Gal80p results in a reduction in the nuclear concentration of Gal80p, thereby allowing Gal4p to interact with the transcriptional machinery.

Activation of the transcriptional properties of Gal4p

Although the presence of galactose within the cell triggers the activation of Gal4p, neither Gal4p nor Gal80p acts as the galactose sensor. A transcriptional inducer, Gal3p, interacts with the transcriptional repressor, Gal80p, in a galactose- and ATP-dependent manner [14]. Gal3p appears to require galactose and ATP so that it can adopt a conformation to allow it to interact with Gal80p [14]. The net result of this interaction is that Gal4p becomes active, and transcription of the *GAL* genes proceeds.

The molecular mechanism by which the activation of the *GAL* genes occurs has been the focus of a much debate. Two somewhat conflicting models for its molecular mode of action have been proposed (Figure 1). It has been suggested that the induction of the *GAL* genes may occur through the association of a tripartite complex formed between Gal4p, Gal80p and Gal3p [8]. In favour of this model (Figure 1A) are the observations that (1) Gal4p purified from yeast grown in either the presence or the absence of galactose is associated with Gal80p [15], (2) artificially constructed Gal80p molecules that contain an activation domain are able to regulate transcription in the presence and absence of galactose [16], (3) the three proteins assemble *in vitro* in a gel-shift assay [8], and (4) Gal4p and Gal80p do not dissociate from each other in the presence or absence of galactose, as demonstrated recently using FRET (fluorescence resonance energy transfer) [11]. Other evidence, however, suggests that this model is incorrect, and that Gal80p dissociates from Gal4p and interacts with Gal3p in the cytoplasm of yeast cells [17]. This would result in the freeing of Gal4p from the inhibitory effects of Gal80p, allowing transcriptional activation to occur [18]. The dissociation model (Figure 1B) is supported by data indicating that Gal3p is located predominately in the cytoplasm of yeast cells [17] and that the expression of a myristoylated version of the protein (which targets it to the plasma membrane) does not unduly impair the induction of the *GAL* genes [18]. In addition, ChIP (chromatin immunoprecipitation) experiments [18] and pull-down assays [19] suggest that the Gal4p–Gal80p complex is somewhat weakened (although perhaps not completely dissociated) when cells are grown in the presence of galactose. In either of the above models, the ability of Gal3p to interact with Gal80p in a galactose-dependent fashion is essential for the transcriptional induction of the *GAL* genes. Understanding this interaction is therefore key to understanding the transcriptional activation in response to galactose.

Gal3p is highly similar to galactokinases

The analysis of Gal3p function has been greatly assisted by its similarity to Gal1p – yeast galactokinase. Gal1p and Gal3p are ~70% identical and ~90% similar at the amino acid level [20,21]. Both proteins require galactose and ATP to function: Gal3p requires these to interact with Gal80p [22], while Gal1p produces galactose 1-phosphate from galactose at the expense of ATP. Despite the high level of similarity between the two proteins, Gal3p does not possess a galactokinase activity [23]. The removal of Gal1p from yeast cells results in an inability to grow on galactose, whereas Gal3p-depleted cells can grow on galactose, but show a severe long-term adaptation phenotype. Rather than inducing the *GAL* genes within 30 min of addition of the sugar to cultures, cells lacking Gal3p only fully induce the *GAL* genes some 3–4 days later [24]. The effects of the loss of Gal3p can be overcome if Gal1p is produced at higher than wild-type levels [25]. However, yeast cells missing both *GAL1* and *GAL3* function are unable to induce the *GAL* genes at all [24]. Recent modelling data have suggested that the *GAL* genes become induced in the presence of Gal1p, but the absence of Gal3p due to stochastic changes in the cellular concentration of Gal80p [26,27]. Other galactokinases, e.g. those from *E. coli* [24], are able to substitute for the loss of Gal1p in yeast to rescue the galactokinase deficiency, but are unable to overcome the transcriptional defect in the *GAL* genes resulting from the loss of Gal3p. Combined, these data suggest that although Gal1p is primarily a galactokinase enzyme, it can also function as a weak transcriptional inducer. This conclusion has also been supported through a number of *in vitro* experiments [8,23]. Other yeast species, e.g. the milk yeast *Kluyveromyces lactis*, contain a single galactokinase-like molecule that functions both as an efficient transcriptional inducer (Gal3p-like function) and as a galactokinase enzyme (Gal1p-like function) [28].

Gal1p and Gal3p appear to have evolved from each other. Gal3p is a defective galactokinase because it contains a two-amino-acid deletion within its galactokinase signature motif [23]. The re-introduction of these residues (a serine

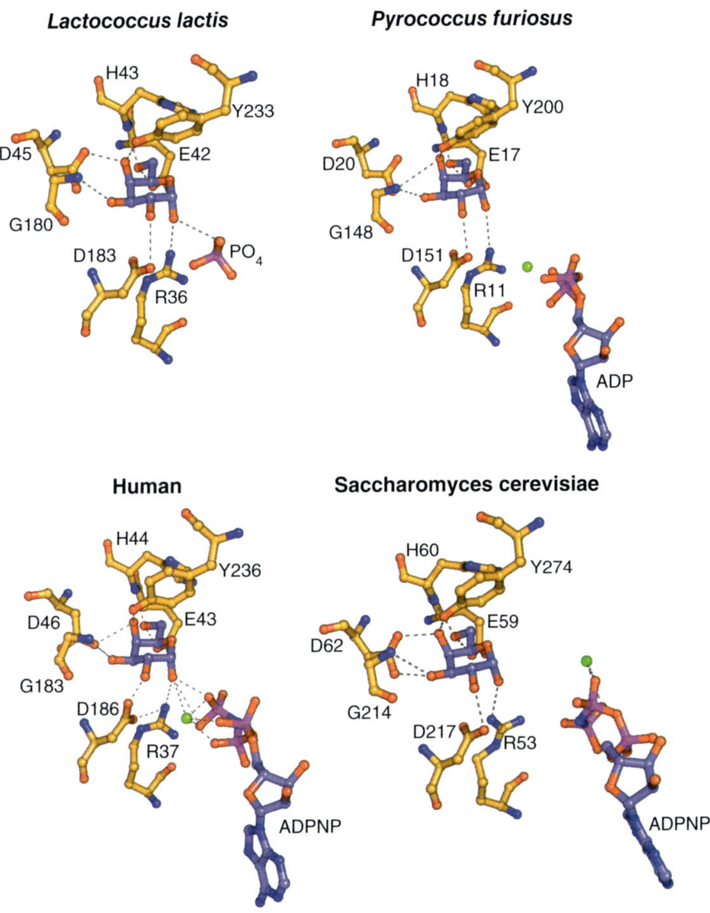

Figure 2 Conserved ligand-binding site in different galactokinases. The galactose-binding sites from the *L. lactis*, *P. furiosus*, human and *S. cerevisiae* enzymes are shown, together with the relative position of either a phosphate ion, ADP or ADPNP, present in the crystal structure as indicated. A magnesium ion (green sphere) is also present in some structures. The dashed lines indicate distances of between 2.3 and 3.3 Å between the protein and the ligands, and may indicate potential hydrogen bond interactions. Amino acids with the capacity to form hydrogen bonds with the sugar are labelled using the single letter amino acid code.

and an alanine residue starting at position 164) imparts the resulting protein with a measurable galactokinase activity *in vitro* and *in vivo* [23]. The structures of the galactokinase enzymes from *Lactococcus lactis* [29], *Pyrococcus furiosus* [30] and human [31] have recently been solved. The equivalent residue to serine-164 in both the *P. furiosus* and human galactokinase structures interacts with the magnesium ion and the β-phosphate of the nucleotide [30,31]. Despite the relative lack of overall sequence similarity between the enzymes from diverse organisms, the galactose- and nucleotide-binding sites of each crystallized protein show a high level of conservation (Figure 2). A model for the bifunctional Gal1p from *K. lactis* has been proposed based on its similarity to other sugar kinases whose structures have been solved [32]. The overall shape of the molecule within the region of the protein involved in the interaction with the ligands may indeed be similar to the galactokinases, but the likely Gal80p interaction site is not complete in these models. It therefore seems probable that Gal3p also shares similar galactose- and nucleotide-binding pockets and is likely to interact with its small molecules in a manner similar to the galactokinases. Crystals of *S. cerevisiae* Gal1p that diffract to 2.1 Å have recently been grown in the presence of galactose and the non-hydrolysable ATP analogue ADPNP (adenosine 5′-[β,γ-imido]triphosphate), and the structure has been solved [32a]. The overall structure of the protein is, as expected, highly similar to those of the previously solved galactokinases. In addition, the galactose- and ATP-binding sites are also highly conserved (Figure 2). The structure of Gal1p, and consequently the implied structure of Gal3p, is generating superb insights into the differences between the purely enzymatic proteins and those capable of transcriptional induction. The structure also allows the precise mapping of mutations in *GAL3* that give rise to a constitutive phenotype (activation of transcription in the absence of galactose). In a linear amino acid map, these mutations are spread throughout the length of the protein [33,34]; however, in the structure, they cluster at the interface between the N- and C-terminal domains of the protein. It is tempting to speculate that this represents the site of Gal80p binding.

Mutations in Gal80p that result in an impaired interaction with Gal3p are also spread throughout the length of the protein [35,36]. Gal80p has been suggested to interact with Gal4p as a dimeric protein [37], but its interaction with Gal3p occurs as a monomer only in the presence of galactose and ATP [38]. The influence of Gal3p and Gal4p upon the monomer/dimer interaction of Gal80p is not fully characterized, and the structure of Gal80p is not currently known. Modelling the primary amino sequence of Gal80p on to the previously solved structure of the sequence-related, but functionally distinct, glucose–fructose oxidoreductase from *Zymomonas mobilis* [39] suggested that the mutations in Gal80p that affect its interaction with Gal3p may cluster on one surface of the protein, while mutations that affect the interaction with Gal4p reside on another surface [36]. Caution should be exercised in interpreting a homology model based on two distantly related proteins (15% amino acid identity; 30% similarity), but the similarity between Gal80p and an oxidoreductase may suggest that Gal80p is also a dinucleotide-binding protein. This could provide yet another layer of complexity to the control of expression of the *GAL* genes.

An additional impediment to the biochemical analysis of Gal3p is that the interaction between galactose and Gal3p, or Gal1p, appears to be somewhat weak in nature. The *GAL* genetic switch is activated maximally *in vitro* using galactose at a concentration 1.5 mM [8], and the K_m for galactose of Gal1p is 1.2 mM [21]. The weakness of this interaction may be a controlling influence on the overall activation of the *GAL* genes *in vivo*. The weak interaction between Gal3p and galactose may mean that a relatively high concentration of the metabolite must be reached inside the cell before the *GAL* genes are switched on fully. The *GAL* system thus offers an intriguing, and complex, interplay between a transcriptional activator, regulator proteins and small molecules. The recent structural analysis of related proteins is providing a mechanism to dissect the molecular details of this elegant switch.

Direct modulation of transcription factor function – Put3p

S. cerevisiae can utilize a variety of compounds as potential sources of nitrogen. When preferred nitrogen sources, such as ammonia or glutamate, are available, the transcription of genes involved in the utilization of poorer nitrogen sources is repressed [40]. Proline, a comparatively poor nitrogen source, is transported into the cell by the general amino acid permease Gap1p and by the proline-specific transporter Put4p (Figure 3). Both of these are located in the plasma membrane, and the activity of each is regulated in response to the quality of the nitrogen available to the cell [41,42]. For its utilization as a nitrogen source, intracellular proline is converted into glutamate in the mitochondria by the products of the *PUT1* and *PUT2* genes, encoding proline oxidase and Δ^1-pyrroline-5-carboxylate dehydrogenase respectively [43]. The presence of proline, but the absence of preferred nitrogen sources, results in the high-level transcriptional induction of both *PUT1* and *PUT2* [44]. This induction of transcription is regulated on several levels. In response to the overall quality of nitrogen available to the cell, Nil1p and Ure2p – two GATA factor proteins – regulate the expression of both *PUT1* and *PUT2* [45,46]. However, neither Nil1p nor Ure2p is essential for the growth of yeast on proline as the sole source

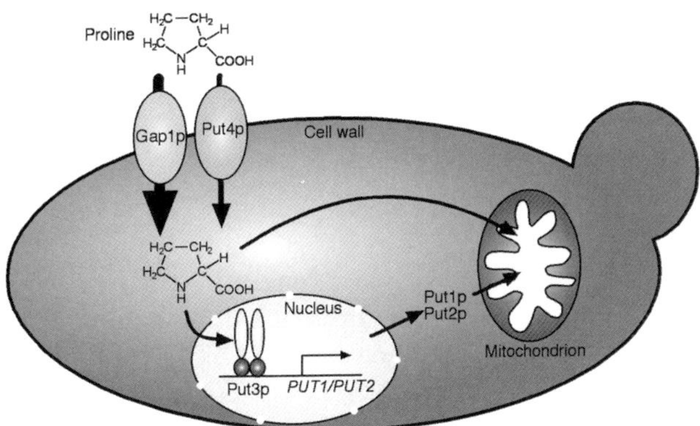

Figure 3 Proline utilization by yeast cells. Proline enters the cell via one of two transporters – Gap1p or Put4p. Once inside the cell, proline is enzymatically converted into glutamate in the mitochondria. Additionally, proline probably enters the nucleus to convert Put3p into a transcriptionally active form. The presence of proline, but only as the sole source of nitrogen, results in induction of Put3p, which activates transcription of the proline utilization genes, *PUT1* and *PUT2*. The enzymes encoded by these genes, Put1p and Put2p, are transported into the mitochondria, where they convert proline into glutamate.

of nitrogen, whereas the specific regulator of the proline utilization pathway, Put3p, is absolutely required for growth on proline [45].

Put3p architecture

Put3p is a transcriptional activator that has an overall architecture typical of the $Zn(II)_2Cys_6$ binuclear cluster family of proteins [47]. It possesses an N-terminal DNA-binding and dimerization domain [48] and an acidic activation domain at its C-terminus [49]. The protein is bound to its target DNA sites upstream of the *PUT1* and *PUT2* genes irrespective of whether those genes are transcriptionally active [50]. Put3p is a phosphoprotein, and the extent of phosphorylation has been correlated with transcriptional activity [51]. However, phosphorylation does not appear to be dependent on the presence of proline. Rather, Put3p is differentially phosphorylated in response to the quality of the nitrogen source available to cells. This was further confirmed by the observation that treatment of yeast cells with rapamycin, which mimics nitrogen-limiting conditions, resulted in the hyperphosphorylation of Put3p [46].

Conformational changes in Put3p

The activation of transcription by Put3p in response to proline appears to be mediated by a conformational change within the protein. The addition of proline to immunoprecipitated Put3p resulted in increased resistance to digestion by endoproteinase Glu-C [52]. Additionally, GST (glutathione S-transferase)–Put3p fusion protein immunoprecipitated from proline-grown yeast cells had increased resistance to thrombin-mediated cleavage of the GST tag compared with the same protein prepared from ammonia-grown cells [52]. Combined, these data suggest that Put3p undergoes a conformational change in the presence of proline. It is conceivable that this conformational change results in unmasking of the activation domain, thereby allowing the recruitment of the transcriptional machinery.

Transcriptional activation induced by proline

In vivo footprinting clearly demonstrated that Put3p binds to its DNA recognition sites in a proline-independent manner [53]. Therefore transcriptional activation by Put3p cannot be regulated through the control of its DNA-binding properties, but instead changes in the transcriptional activity of DNA-tethered Put3p must occur in response to proline. *In vitro* transcription experiments demonstrated that the activity of Put3p is induced in the presence of proline itself [54].

Direct interaction between Put3p and proline

The ability of proline to modulate the activity of Put3p *in vitro* suggests that the metabolite is directly affecting the activity of the transcriptional regulator. Proline affinity chromatography demonstrated that the recognition of proline by Put3p is indeed via a direct physical interaction. When Put3p was applied to a column bearing immobilized proline attached through its carboxy group, the protein was able to bind the matrix and could be eluted with proline [54]. Put3p was eluted from the column at a proline concentration of approx. 25 mM, which suggests that the interaction between Put3p and proline is weak [54].

However, this is a level of the metabolite comparable with that required for the activation of *PUT1* and *PUT2 in vivo* [51], and to elicit a response using *in vitro* transcription assays [54]. If a similar column was produced in which the proline was attached through its imino-group nitrogen atom, then no Put3p binding could be detected. This result, and other data, suggested that that the pyrrolidine ring of proline is vital for recognition by Put3p. Pyrrolidine itself was found to be able to activate Put3p in an *in vitro* transcription assay [54], while the addition of hydroxy groups to the pyrrolidine ring ablated its ability to modulate the activity of Put3p. Therefore Put3p appears to recognize and interact with a simple pyrrolidine ring, and the carboxy group found in proline may be removed or modified without unduly affecting the interaction with the protein.

The interaction between Put3p and proline is sufficient to bring about a transcriptional response. However, the position and structure of the proline-binding site within Put3p are not known. The activation domain of Put3p is not proline-responsive and, when fused to a heterologous DNA-binding domain, it activates transcription in a constitutive manner [49]. This suggests that regulation of the transcriptional activity of Put3p occurs elsewhere in the protein. We are currently striving to identify the proline-binding site within Put3p.

Model for transcriptional activation by Put3p

At present, the model for transcriptional activation by Put3p is that a dimeric, DNA-bound form of Put3p 'senses' the presence of proline within a cell (Figure 4). The presence of proline, but only as the sole source of nitrogen, results in the activation of transcription of *PUT1* and *PUT2*. The conversion of Put3p from a transcriptionally inert form into a transcriptionally active form is thought to occur via a conformational change in the protein. It is possible that this conformational change results in unmasking of the activation domain in the presence of proline, thus allowing the recruitment of the transcriptional machinery.

Phosphorylation may also play a role in regulating the overall transcriptional activity of Put3p. It has been proposed that phosphorylation of Put3p is required for transcriptional activation and that underphosphorylated forms of the protein are unable to activate transcription [51]. However, analysis of the phosphorylation profile of recombinant purified Put3p that activated transcription *in vitro* revealed that the protein was unphosphorylated [54]. This suggests that proline is able to act on unphosphorylated Put3p to convert it into a transcriptional activator *in vitro*, and raises the question as to the role of phosphorylation in regulating transcriptional activation by Put3p. It is possible that phosphorylated forms of Put3p are able to respond to lower levels of proline; alternatively, phosphorylation of Put3p may be involved in the global regulation of the proline utilization pathway by the GATA factors. The mechanism by which the overall quality of the nitrogen source available to a yeast cell is detected is still largely unknown. In addition, the molecular mechanism by which this information is relayed to the *PUT* genes is unclear. A complex interplay between nitrogen transporters, GATA factors and Put3p is likely to be involved.

Other metabolically regulated transcriptional switches

The systems described above represent just two well-characterized transcriptional programmes that respond directly to changes in the intracellular levels of metabolites. There are, however, others where activation is controlled by metabolites. For example, (1) the expression of genes required for pyrimidine biosynthesis is regulated by orotic acid (a biosynthetic intermediate of the pyrimidine pathway), which controls the activity of the transcriptional regulator Ppr1p [55]; (2) the formation of an active transcriptional complex between Bas1p and Bas2p to control *de novo* AMP biosynthesis is dependent on 5′-phosphoribosyl-5-aminoimidazole-4-*N*-succinocarboxamide [56]; (3) leucine biosynthesis is regulated by the action of α-isopropyl malate on Leu3p [57]; (4) lysine biosynthesis is regulated by a presumed interaction between α-aminoadipate semialdehyde and Lys14p [58]; (5) arginine biosynthesis is regulated by the interaction between arginine and Arg81p, which promotes the DNA-binding properties of the Arg81p–Mcm1p complex [59]; and (6) in response to weak acid stress, War1p activates certain stress response genes through the likely interaction between the protein and weak acids [60,61]. In each of the above cases, the small molecule is presumed to modulate the transcriptional properties of the activator protein by some mechanism yet to be determined.

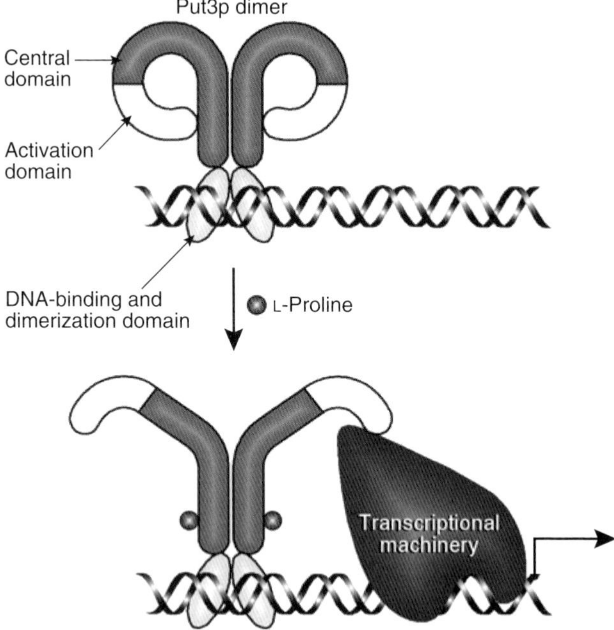

Figure 4 Proline-specific transcriptional activation by Put3p. A DNA-tethered Put3p dimer acts as a proline sensor. When the proline concentration is sufficiently high, proline interacts directly with Put3p. The proline-binding domain of Put3p is likely to be within the central domain of the protein. The direct interaction results in a conformational change in Put3p that unmasks the activation domain and allows the protein to activate transcription.

Conclusion

The direct modulation of transcription factor function by small-molecule metabolites is well established in prokaryotes. As we have shown here, this mechanism of transcriptional regulation is also common in single-celled eukaryotes. Yeasts contain many of the signalling pathways found in higher eukaryotes, e.g. MAPK (mitogen-activated protein kinase) cascades [62–64], but lack the metabolic homoeostasis systems present in higher, multicellular, eukaryotes. In higher eukaryotes, the interactions between, say, hormones and transcription factors, are understood in detail [65], but it would appear that direct modulation of transcription factor function by metabolites is largely lacking. Single-celled organisms have generally less control over their metabolic content, and this appears to have been exploited to provide a series of elegant and precise genetic controls. Understanding the diverse ways in which metabolites can directly influence and regulate transcription is vital if transcription and metabolism are to be understood in a fully integrated fashion.

We are grateful to Hazel Holden and David Timson for discussions and their comments on, and suggestions for, this chapter. Work in the authors' laboratory is funded by The Wellcome Trust and the BBSRC.

References

1. Kim, J.W. and Dang, C.V. (2005) Trends Biochem. Sci. **30**, 142–150
2. Lewis, M., Chang, G., Horton, N.C., Kercher, M.A., Pace, H.C., Schumacher, M.A., Brennan, R.G. and Lu, P. (1996) Science **271**, 1247–1254
3. Chatterjee, S., Ghosh, K., Dhar, A. and Roy, S. (2002) Proteins **49**, 554–559
4. Reece, R.J. (2000) Cell. Mol. Life Sci. **57**, 1161–1171
5. Yarger, J.G., Halvorson, H.O. and Hopper, J.E. (1984) Mol. Cell. Biochem. **61**, 173–182
6. Broach, J.R. (1979) J. Mol. Biol. **131**, 41–53
7. Griggs, D. and Johnston, M. (1991) Proc. Natl. Acad. Sci. U.S.A. **88**, 8597–8601
8. Platt, A. and Reece, R.J. (1998) EMBO J. **17**, 4086–4091
9. Larschan, E. and Winston, F. (2001) Genes Dev. **15**, 1946–1956
10. Bhaumik, S.R. and Green, M.R. (2001) Genes Dev. **15**, 1935–1945
11. Bhaumik, S.R., Raha, T., Aiello, D.P. and Green, M.R. (2004) Genes Dev **18**, 333–343
12. Roberts, S.M. and Winston, F. (1997) Genetics **147**, 451–465
13. Larschan, E. and Winston, F. (2005) Mol. Cell. Biol. **25**, 114–123
14. Zenke, F.T., Engles, R., Vollenbroich, V., Meyer, J., Hollenberg, C.P. and Breunig, K.D. (1996) Science **272**, 1662–1665
15. Parthun, M.R. and Jaehning, J.A. (1992) Mol. Cell. Biol. **12**, 4981–4987
16. Leuther, K.K. and Johnston, S.A. (1992) Science **256**, 1333–1335
17. Peng, G. and Hopper, J.E. (2000) Mol. Cell. Biol. **20**, 5140–5148
18. Peng, G. and Hopper, J.E. (2002) Proc. Natl. Acad. Sci. U.S.A. **99**, 8548–8553
19. Sil, A.K., Alam, S., Xin, P., Ma, L., Morgan, M., Lebo, C.M., Woods, M.P. and Hopper, J.E. (1999) Mol. Cell. Biol. **19**, 7828–7840
20. Schell, M.A. and Wilson, D.B. (1977) J. Biol. Chem. **252**, 1162–1166
21. Timson, D.J. and Reece, R.J. (2002) Biochimie **84**, 265–272
22. Yano, K.I. and Fukasawa, T. (1997) Proc. Natl. Acad. Sci. U.S.A. **94**, 1721–1726
23. Platt, A., Ross, H.C., Hankin, S. and Reece, R.J. (2000) Proc. Natl. Acad. Sci. U.S.A. **97**, 3154–3159
24. Bhat, P.J., Oh, D. and Hopper, J.E. (1990) Genetics **125**, 281–291
25. Bhat, P.J. and Hopper, J.E. (1992) Mol. Cell. Biol. **12**, 2701–2707
26. Verma, M., Bhat, P.J. and Venkatesh, K.V. (2004) Biotechnol. Appl. Biochem. **39**, 89–97
27. Bhat, P.J. and Venkatesh, K.V. (2005) FEBS Lett. **579**, 597–603
28. Meyer, J., Walker-Jonah, A. and Hollenberg, C.P. (1991) Mol. Cell. Biol. **11**, 5454–5461

29. Thoden, J.B. and Holden, H.M. (2003) J. Biol. Chem. **278**, 33305–33311
30. Hartley, A., Glynn, S.E., Barynin, V., Baker, P.J., Sedelnikova, S.E., Verhees, C., de Geus, D., van der Oost, J., Timson, D.J., Reece, R.J. and Rice, D.W. (2004) J. Mol. Biol. **337**, 387–398
31. Thoden, J.B., Timson, D.J., Reece, R.J. and Holden, H.M. (2005) J. Biol. Chem. **280**, 9662–9670
32. Menezes, R.A., Amuel, C., Engels, R., Gengenbacher, U., Labahn, J. and Hollenberg, C.P. (2003) J. Mol. Biol. **333**, 479–492
32a. Thoden, J.B., Sellick, C.A., Timson, D.J., Reece, R.J. and Holden, H.M. (2005) J. Biol. Chem., in the press
33. Blank, T.E., Woods, M.P., Lebo, C.M., Xin, P. and Hopper, J.E. (1997) Mol. Cell. Biol. **17**, 2566–2575
34. Suzuki-Fujimoto, T., Fukuma, M., Yano, K.-I., Sakurai, H., Vonika, A., Johnston, S.A. and Fukasawa, T. (1996) Mol. Cell. Biol. **16**, 2504–2508
35. Nogi, Y., Matsumoto, K., To-e, A. and Oshima, Y. (1977) Mol. Gen. Genet. **152**, 137–144
36. Pilauri, V., Bewley, M., Diep, C.Q. and Hopper, J.E. (2005) Genetics **169**, 1903–1914
37. Melcher, K. and Xu, H.E. (2001) EMBO J. **20**, 841–851
38. Timson, D.J., Ross, H.C. and Reece, R.J. (2002) Biochem. J. **363**, 515–520
39. Kingston, R.L., Scopes, R.K. and Baker, E.N. (1996) Structure **4**, 1413–1428
40. Brandriss, M.C. and Magasanik, B. (1979) J. Bacteriol. **140**, 498–503
41. Jauniaux, J.C. and Grenson, M. (1990) Eur. J. Biochem. **190**, 39–44
42. Jauniaux, J.C., Vandenbol, M., Vissers, S., Broman, K. and Grenson, M. (1987) Eur. J. Biochem. **164**, 601–606
43. Brandriss, M.C. (1987) Genetics **117**, 429–435
44. Marczak, J.E. and Brandriss, M.C. (1991) Mol. Cell. Biol. **11**, 2609–2619
45. Xu, S., Falvey, D.A. and Brandriss, M.C. (1995) Mol. Cell. Biol. **15**, 2321–2330
46. Saxena, D., Kannan, K.B. and Brandriss, M.C. (2003) Eukaryotic Cell **2**, 552–559
47. Todd, R.B. and Andrianopoulos, A. (1997) Fungal Genet. Biol. **21**, 388–405
48. Swaminathan, K., Flynn, P.J., Reece, R.J. and Marmorstein, R. (1997) Nat. Struct. Biol. **4**, 751–759
49. des Etages, S.G., Falvey, D.A., Reece, R.J. and Brandriss, M.C. (1996) Genetics **142**, 1069–1082
50. Selleck, S.B. and Majors, J.E. (1987) Mol. Cell. Biol. **7**, 3260–3267
51. Huang, H.L. and Brandriss, M.C. (2000) Mol. Cell. Biol. **20**, 892–899
52. des Etages, S.G., Saxena, D., Huang, H.L., Falvey, D.A., Barber, D. and Brandriss, M.C. (2001) Mol. Microbiol. **40**, 890–899
53. Axelrod, J.D., Majors, J. and Brandriss, M.C. (1991) Mol. Cell. Biol. **11**, 564–567
54. Sellick, C.A. and Reece, R.J. (2003) EMBO J. **22**, 5147–5153
55. Flynn, P.J. and Reece, R.J. (1999) Mol. Cell. Biol. **19**, 882–888
56. Rébora, K., Desmoucelles, C., Borne, F., Pinson, B. and Daignan-Fornier, B. (2001) Mol. Cell. Biol. **21**, 7901–7912
57. Kohlhaw, G.B. (2003) Microbiol. Mol. Biol. Rev. **67**, 1–15
58. El Alami, M., Feller, A., Pierard, A. and Dubois, E. (2002) Mol. Microbiol. **43**, 1629–1639
59. Amar, N., Messenguy, F., El Bakkoury, M. and Dubois, E. (2000) Mol. Cell. Biol. **20**, 2087–2097
60. Schüller, C., Mamnun, Y.M., Mollapour, M., Krapf, G., Schuster, M., Bauer, B.E., Piper, P.W. and Kuchler, K. (2004) Mol. Biol. Cell **15**, 706–720
61. Kren, A., Mamnun, Y.M., Bauer, B.E., Schuller, C., Wolfger, H., Hatzixanthis, K., Mollapour, M., Gregori, C., Piper, P. and Kuchler, K. (2003) Mol. Cell. Biol. **23**, 1775–1785
62. Gustin, M.C., Albertyn, J., Alexander, M. and Davenport, K. (1998) Microbiol. Mol. Biol. Rev. **62**, 1264–1300
63. Posas, F., Takekawa, M. and Saito, H. (1998) Curr. Opin. Microbiol. **1**, 175–182
64. Banuett, F. (1998) Mol. Biol. Rev. **62**, 249–274
65. Bassett, J.H., Harvey, C.B. and Williams, G.R. (2003) Mol. Cell. Endocrinol. **213**, 1–11

Biochem. Soc. Symp. **73**, 97–108
(Printed in Great Britain)

10

Chromatin-remodelling factors and the maintenance of transcriptional states through DNA replication

Sofia Aligianni and Patrick Varga-Weisz[1]

Marie Curie Research Institute, The Chart, Oxted RH8 0TL, Surrey, U.K.

Abstract

At the replication fork, nucleosomes, transcription factors and RNA polymerases are stripped off the DNA, the DNA double strands are unzipped and DNA methylation marks may be erased. Therefore DNA replication is both a 'curse' and 'bliss' for the epigenome, as it disrupts its stability by causing chromatin perturbations, yet it offers an opportunity to initiate changes in chromatin architecture and gene expression patterns, especially during development. Thus the DNA replication site is a critical point for regulation. It has become apparent that there is a close functional relationship between those factors that regulate transcriptional competence and the DNA replication programme. In this review we discuss novel insights into how chromatin-remodelling factors at replication sites are involved in both the maintenance and regulation of transcriptional states.

Introduction

In all eukaryotic cells, histones H2A, H2B, H3 and H4 spool the DNA in almost two turns to form the structural building block of chromatin, the nucleosome. Linker histones of the H1 family stabilize this structure by binding to the entry–exit site of the DNA on the nucleosome. Chromatin is involved in regulatory processes by limiting access to DNA. Two major classes of enzymes regulate chromatin structure. The first includes factors that covalently modify histones, such as histone acetyltransferases, which are usually associated with transcriptional activation, and histone deacetylases (HDACs), which often

[1]To whom correspondence should be sent, at present address: Babraham Institute, Babraham, Cambridge CB2 4AT, U.K. (email patrick.varga-weisz@bbsrc.ac.uk).

mediate gene silencing (reviewed by Ehrenhofer-Murray [1]). The second class employs alterations in positional and structural features of nucleosomes catalysed by ATP hydrolysis, thus organizing chromatin architecture [2]. Chromatin structure, histone modifications and, in many organisms, DNA methylation are important determinants of the activity of specific genes. These attributes constitute an epigenetic 'code' that allows the development of multicellular organisms using one single genetic blueprint. Such transcriptional states have to be maintained and transmitted through many cell generations to determine cell lineages during development.

The replication fork is incompatible with ongoing transcription and nucleosomes, and destabilizes chromatin over a distance of 650–1100 base pairs [3,4] (reviewed by Annunziato [5]). In that respect, DNA replication is a key 'hurdle' that the genome has to face in order to maintain epigenetic information. This is because its very mechanism evicts all transcriptional components from the DNA and clears epigenetic marks in a repeated manner during every cell cycle. For instance, it has been shown that, following replication, the newly formed chromatin is composed of both 'old' and 'new' nucleosomes; classical experiments have revealed that the histone H3/H4 cores are fully transmitted from the parental chromatin, whereas the remaining complement of nucleosomes is composed of *de novo*-synthesized histones (reviewed by Annunziato [5]). These new histones carry different modifications compared with the parental histones. Therefore epigenetic marks established by specific histone modifications will be diluted out over subsequent cell generations if there are no mechanisms to ensure their propagation from the parental to the daughter chromatin strands. DNA methylation is another important mediator of epigenetic information during DNA replication, because hemi-methylated DNA can direct the remethylation of newly synthesized DNA. However, epigenetic phenomena exist in organisms with no or very little DNA methylation, and it has been argued that DNA methylation regulates primarily very specific aspects of gene activities, such as transposon and endogenous retrovirus silencing, gene imprinting and X-inactivation [6]. Thus other mechanisms must exist.

Feedback mechanisms involving interactions between specifically modified histones and proteins that recognize the modifications and recruit the modifying enzymes have been proposed to be of central importance for the maintenance of epigenetic information. For example, histone H3 methylated at lysine-9 (Lys^9) provides a binding site for HP1 (heterochromatin protein 1), an important determinant of heterochromatin structure. HP1, in turn, interacts with the H3 Lys^9 methyltransferase SUV3-9H1 and its homologues, which may propagate the methylation of more H3 Lys^9 in the vicinity of the original mark (reviewed by Grewal and Elgin [7]). However, there is little evidence so far that such mechanisms operate at the replication site, and DNA replication is a process that conceivably stretches such an information preservation system to its limits. Yet both DNA replication and transcriptional inheritance operate within a chromatin context, thus making chromatin modifications an absolute requirement for stable genetic and epigenetic inheritance. In this review, we attempt to provide insights into the chromatin-remodelling mechanisms to which the replication programme is subject, and how these influence transcriptional inheritance.

DNA replication *per se* can 'license' transcriptional regulation

Given that DNA replication is coupled to a profound perturbation of chromatin, it is not surprising that the cell uses this opportunity to regulate some genes, and there are several striking examples to illustrate this. *In vitro* experiments have shown that transcriptional activation can be mechanistically linked to DNA replication [8]. In mouse embryos, the release of transcriptional arrest that occurs between the end of oocyte maturation and the two-cell stage requires DNA replication [9]. Another example is found in the regulation of the *HoxB* gene cluster, a linear array of genes that are activated in a sequential manner during early development. The *Hox* genes encode homologous transcription factors that are involved in patterning of various tissues in animals. The relationship between the activation of the genes in this cluster and DNA replication has been studied recently in two systems: during embryogenesis in the frog *Xenopus*, and during the retinoic acid-induced differentiation of pluripotent mouse P19 cells [10]. In the *Xenopus* system, inhibition of DNA replication before the mid-blastula transition (the time at which zygotic transcription starts) abolishes transcriptional induction of the *HoxB* genes. If replication is inhibited one or two cycles after the mid-blastula transition, the expression pattern is altered. In P19 cells, the activation of the *HoxB* genes requires progression through a single S phase. Furthermore, activation of most of the *Hox* genes in the *HoxB* cluster is sequential (co-linear) during S phase in synchronized cells. In these experiments, the *HoxB* locus replicated early, but co-linear activation occurred later in S phase. Therefore the replication process does not immediately activate transcription, but 'licenses' the genes for subsequent activation. Chromatin replication may also be involved in the efficient silencing of genes, as shown for thyroid-receptor-mediated repression [11]. Together, these findings suggest that DNA replication may 'license' transcriptional regulation.

Replication timing and transcriptional activity

DNA synthesis is organized in such a way that specific regions of the genome are replicated at precise times, following a set 'replication programme', which ensures that the whole genome is duplicated only once (reviewed by McNairn and Gilbert [12]). Light microscopic studies on cultured cells indicate that transcription and replication are mutually exclusive and happen in distinct, yet related, domains [13]. At the individual gene level, the timing of DNA replication and transcription has been examined in rDNA genes in HeLa cells using light microscopy [14]. Most rDNA foci were transcriptionally active in early S phase without exhibiting active replication sites. Cells in mid-to-late S phase showed a reversal of those functions: transcription was shut down where replication of individual rDNA loci commenced. These observations suggest that there might be a switching mechanism that turns off transcription for the replication of these higher-order chromatin domains. Yet another light microscopic study on cultured cells points to a close spatial relationship between transcription and DNA replication, at least in the early stages of S phase [15]. This may be in line with a longstanding notion that there is a link between replication timing and the transcriptional activity of specific loci in higher eukaryotes [12]. A striking example to illustrate this correlation is found with the mammalian X chromosomes in female cells, where one X chromosome is active and largely early-replicating, whereas the other is silenced and late-replicating. This replication profile does not seem to be determined by differences in origin usage or firing, but rather by a delay in the time of their firing caused by the heterochromatin environment in the inactive X chromosome [16].

Recent technological advances with DNA microarray approaches has allowed the genome-wide investigation of the relationship between transcriptional status and replication timing. A mapping study in *Drosophila* revealed that there is a strong general correlation between early replication and active gene expression [17]. In the same system, the co-ordination of replication timing and transcriptional activity has been demonstrated further by showing that RNA polymerase II is more densely localized in early-replicating sequences [18]. This correlation is not at the level of individual loci, but rather over extended chromosomal domains. Other links between replication initiation and transcription have been described in fission yeast, where active origins reside close to promoter regions [19]

Analysis of human chromosome 22 also indicated that there is a correlation between transcriptional activity and early replication [20, 20a]. However, a subset of late-replicating regions was found to be highly transcribed; the mechanisms that may underlie this exception have yet not been identified [20]. Two recent studies have explored the link between gene expression and the timing of DNA replication using mouse embryonic stem cells that can be induced to differentiate in culture. It was shown that as cells commit to differentiation they undergo programmed changes in chromatin organization and gene expression. These changes coincide with a switch in replication timing, from early to late or late to early, according to transcriptional activity, of a set of genes encoding transcription factors critical for early embryonic development or lineage specification [21]. A study from the Gilbert laboratory showed that this switching in replication timing upon the differentiation of mouse embryonic stem cells into neural precursors is restricted to genes within AT-rich/LINE (long interspersed nuclear element)-rich isochors [22]. In addition, early-replicating domains appear to be generally more CG-rich and LINE-poor, and have a high gene density [20, 20a]. Therefore it has been suggested that sequence features provide a regulatory mechanism by which the timing of gene replication is defined. Such a mechanism could co-regulate the availability of activators (early-S) or repressors (late-S) and their subsequent incorporation into chromatin. These observations link replication timing to the evolution of metazoan genomes and their struggle to keep mobile, parasitic DNA, such as transposons, under control [22]. Evidence for the temporal partition of regulators is provided from experiments that demonstrated that DNA in human cells is more likely to be transcribed when microinjected into early rather than late S-phase cells [23]. Taken together, the correlation between replication timing and transcriptional activity in higher eukaryotes indicates a temporal segregation of the replication of domains with different transcriptional states. This segregation, in turn, may help to maintain the transcriptional status of specific domains. This temporal correlation of transcription activity and DNA replication timing may be a unique feature of higher eukaryotes, as no link has been found between replication timing and transcriptional states in budding yeast [24].

Origins of replication: determinants and victims of chromatin structure

The multi-step process of genomic replication commences with the assembly of pre-replication complexes at discrete sites, called origins of replication. In bacteria, bacteriophages, animal viruses and budding yeast, these regions are defined DNA sequences. In *Schizosaccharomyces pombe*, as well as in higher eukaryotes, replication origins are less well defined. The ORC (origin recognition complex) that identifies and binds to origins is composed of six subunits (Orc1–6), all of which are required for cell viability (see [25] for a review on DNA replication initiation). ORC assembly is followed by binding of Cdc6p and Cdt1p and subsequent recruitment of MCM (mini-chromosome maintenance) helicase factors. Replication origins are important not only because of their ability to initiate replication fork formation and subsequent DNA replication, but also because their chromatin environment can be modulated, so that origin initiation can be regulated and thus, potentially, kept 'in tune' with transcriptional demands. Importantly, ORC proteins are determinants of chromatin structures.

In yeast, multiple connections have been identified between proteins that initiate DNA synthesis and those that mediate gene silencing or heterochromatin formation (reviewed by Ehrenhofer-Murray [1]). For example, ORC proteins and DNA polymerase α interact with heterochromatin proteins. The dynamic interplay of the replication machinery with other cellular functions is illuminated by findings on the participation of ORC subunits in tasks downstream of replication, such as chromatin assembly, heterochromatin regulation and chromosome segregation [26–30]. Collectively, these findings suggest that ORC subunits assist in the establishment of chromatin structures, which in turn will establish functional domains for sister-chromatid cohesion and gene silencing.

Not only are the origins of replication determinants of chromatin structure, but they are themselves subject to regulation through chromatin modifications. In *Drosophila* and *Xenopus*, initiation of DNA replication is regulated developmentally, whereby after the mid-blastula transition, replication switches from a random localization to preferential localization at promoters or intergenic regions [31,32]. Transcription activators can select (specify) and activate replication origins for initiation in viral systems, yeast and metazoans, and they perform this task, at least in part, by mediating chromatin remodelling ([33–35] and references therein). A non-viral vector that replicates episomally in mammalian cells requires active transcription into a chromosomal S/MAR (scaffold/matrix attachment region) sequence to stimulate autonomous replication from an SV40 (simian virus 40) replication origin [36]. This shows that active transcription can stimulate initiation of replication and indicates a role for nuclear architecture in this process. Transcription factors and their associated co-activators have also been found to be targeted to replication sites throughout S phase (e.g. the androgen receptor and its co-activator hZimp10), but the role of this targeting is not clear [37].

Further recent work underscores the links between chromatin modifications, regulation of origin function and subsequent transcriptional regulation. In the budding yeast *Saccharomyces cerevisiae*, a screen for suppressors of a *cdc6-4* mutant defective in pre-replication complex assembly led to the identification of the HDAC protein Sir2p [38]. Sir2p removes acetyl groups from lysine residues within the histone H3 and H4 N-termini and is conserved from yeast to mammals. The findings indicate that Sir2p negatively regulates pre-replication complex formation by enhancing the establishment of heterochromatic structures at pre-replication sites. Sir2p may achieve this by deacetylating origin-proximal nucleosomes, but the possibility cannot be excluded that Sir2p also targets other chromatin proteins. Deletion of another HDAC in budding yeast, *RPD3*, was shown to advance the replication timing of origins, and this was linked to the increase in histone acetylation near origins [39].

HDACs also regulate origin activity in metazoans: chromatin acetylation was shown to be critical for a developmental transition in origin specificity in *Drosophila* [40] and in human cells in culture [41]. Interestingly, at the Epstein–Barr virus origin of plasmid replication, histone H3 acetylation decreased during G_1 phase, coinciding with nucleosome remodelling and MCM3 loading, and preceding the onset of DNA replication [42]. The ATP-dependent nucleosome-remodelling factor SNF2H (sucrose non-fermenting 2 homologue; see below) was found to be targeted to this site together with HDAC2, and this complex appears to co-ordinate the G_1-specific chromatin remodelling leading to DNA replication initiation. All of these results suggest that ORC subunits, transcription factors, chromatin structure and histone modifications are important determinants of origin activity, which in turn influences transcriptional inheritance.

Histone chaperones: deliverers of histones and more

Histone chaperones bind histones in order to assist and regulate their interactions with DNA and other molecules, and in this way they are essential in chromatin assembly (reviewed in [43,44]). The analysis of several histone chaperones suggests that they are involved in the maintenance of chromatin states (reviewed in [44,45]). One example is provided by CAF-1 (chromatin assembly factor-1), a histone chaperone that delivers histones H3 and H4 in DNA synthesis-coupled chromatin assembly. CAF-1 interacts with the central replication molecule PCNA (proliferating-cell nuclear antigen), which could explain how it is targeted to the replication site [46]. CAF-1 is not essential in budding yeast, but has a role in maintaining silenced chromatin structures (reviewed in [45,47]). It is not clear how CAF-1 is involved in this process, but the finding that CAF-1 interacts with HP1 suggests a mechanism by which CAF-1 delivers heterochromatin components during chromatin assembly [48]. Recent studies in cultured mammalian cells shed further light on the role of CAF-1 in heterochromatin propagation during DNA replication. Pericentric heterochromatin is the highly condensed chromatin around the centromeres. Its inheritance in mouse cells has been shown to follow a highly dynamic process involving CAF-1 and HP1 proteins [49]. This study showed that CAF-1 forms complexes either with histones H3/H4 or with HP1α/γ. Stable interaction of HP1 with heterochromatin involves an unknown RNA moiety and histone H3 methylation at Lys^9, but CAF-1 is able to dynamically recruit HP1α/γ to the pericentromeric heterochromatin during its replication in the absence of this RNA or histone H3 Lys^9 methylation. Given that CAF-1 interacts with either histones or HP1, one could imagine a sequential interaction of CAF-1 with these proteins to ensure both chromatin assembly and the heritability of HP1-containing heterochromatin domains. One wonders if these complexes may also have a role in a disassembly reaction to facilitate heterochromatin replication.

Histone chaperones such as Asf1 (anti-silencing function 1, for H3/H4) or NAP-1 (nucleosome assembly protein-1, for H2A/H2B) assist CAF-1 at the replication fork, but have also been shown to have roles in determining chromatin structure and gene regulation in a replication-independent manner. Disruption of Asf1 enhances the silencing defects of CAF-1 deletion, which points to a synergistic role of CAF-1 and Asf1 in promoting silent chromatin [50,51]. However, Asf1 seems to counteract chromatin condensation globally and is required for the activation of specific genes by catalysing nucleosome disruption [52–54]. Asf1 is also involved in transcription-coupled chromatin assembly by catalysing the incorporation of the histone H3 variant H3.3 [55]. Thus Asf1 could epigenetically mark transcriptionally active sites. In *Drosophila*, Asf1 mutations de-repress silenced genes. On the other hand, Asf1 has also been shown to be associated with decondensed and transcriptionally active regions on polytene chromosomes and to interact with the chromatin-remodelling factor Brahma, an ATP-dependent nucleosome remodelling factor involved in transcriptional regulation [56]. These findings therefore demonstrate how histone chaperones are not only mere shuttles of histones to replication forks in order to assemble new chromatin; they function in parallel as histone-exchange media elsewhere in the genome.

Histone-modifying and nucleosome-remodelling factors at the replication site

Specific histone-modifying and nucleosome-remodelling factors have been found to be targeted to replication sites, which suggests that they may be involved in facilitating DNA replication through chromatin and/or the assembly of specific chromatin structures [1,12,45]. The specific targeting of HDAC2 and the nucleosome-remodelling ACF1 (ATP-utilizing chromatin assembly and remodelling factor 1)–ISWI (imitation switch) complex to replicating heterochromatin suggests that such factors act directly at the replication site to mediate or facilitate the formation of condensed chromatin [57,58]. ISWI (and its mammalian homologue SNF2H) are potent nucleosome-remodelling factors *in vitro*, whose major activity appears to be the mediation of nucleosome sliding along DNA [59]. ISWI is the central 'motor' of several complexes that have been shown to have a role in chromatin assembly *in vitro* and in early embryos of *Drosophila* and *Xenopus*, which makes these factors good candidates for some role in chromatin replication (reviewed in [60]). Several ISWI complexes have been shown to mediate the assembly of repressive chromatin structures, and have been implicated in transcriptional silencing [61–64]. One ISWI complex, called WICH [WSTF (Williams-syndrome transcription factor)/ISWI/chromatin-remodelling factor], was found to be targeted to replicating pericentromeric heterochromatin in mouse cells [65]. A substantial overlap of WSTF and replication site staining in earlier stages of S phase made it difficult to judge if there is a more general targeting of this complex to replication sites. Recently, an extraction protocol using high salt- and Triton-containing buffer allowed visualization of the general targeting of WSTF to replication foci [66]. This targeting seems to be mediated through direct interaction of WSTF with PCNA, a key factor in chromatin replication and DNA repair. Depletion of WSTF results in a state of global chromatin condensation, an increase in heterochromatin markers such as HP1 and a consequent impairment of transcriptional activity [66,67]. This change in chromatin structure upon WSTF depletion is dependent on S-phase progression [66]. These observations fit a model which suggests that chromatin remodelling by WICH right at the replication site allows the rapid rebinding of transcription regulators that have been evicted by the replication fork back to the chromatin in the daughter strands [67] (Figure 1). Because ISWI complexes, including WICH, mobilize nucleosomes, they facilitate the interaction of factors with DNA in chromatin [59]. These factors could be DNA-binding transcriptional regulators, such as activators or repressors, but also chromatin-binding proteins or chromatin-modifying enzymes. In the absence of WSTF, chromatin fibres would form immediately following DNA replication, which could subsequently promote heterochromatin assembly, resulting in the observed global chromatin condensation, increased heterochromatin protein content and transcriptional impairment. This hypothesis links chromatin remodelling by WICH at the replication site to the maintenance of epigenetic patterns through DNA replication.

In fission yeast there is no ISWI orthologue (A. Neves-Costa and P. Varga-Weisz, unpublished work). Therefore, other ATP-dependent nucleosome-remodelling factors might be involved in resetting chromatin structure after replication in this organism and others. One potential candidate could be INO80, a conserved member of the SWI/SNF family whose functions include chromatin remodelling and a 3–5′ helicase activity [68]. Budding yeast strains lacking INO80 display mis-regulated transcription and are hypersensitive to DNA-damaging agents, suggesting a possible role in DNA repair [68]. In *Arabidopsis thaliana*, INO80 is reported to have a role in homologous recombination and transcriptional regulation [69]. Studies in *S. cerevisiae* indicate a

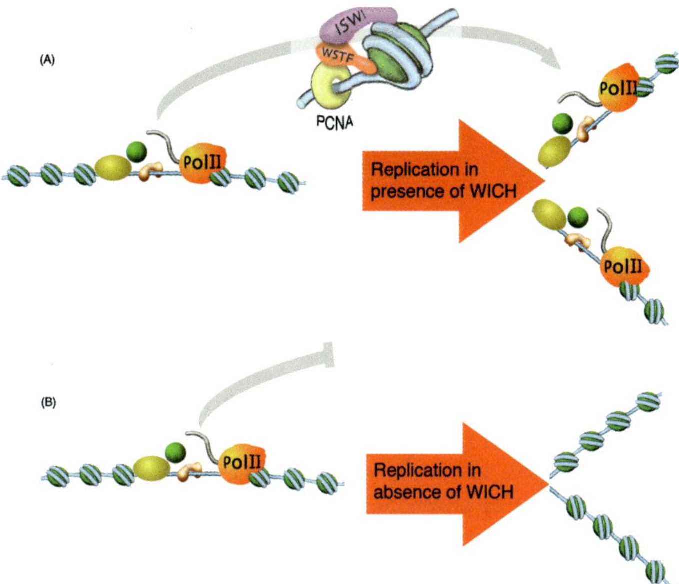

Figure 1 Chromatin-remodelling factors at the replication site as facilitators of epigenetic inheritance. A model is depicted that shows WICH bound to PCNA (**A**), facilitating DNA accessibility by moving nucleosomes on the newly replicated DNA, thereby allowing the rebinding of factors from the parental to the daughter strands. These factors may be transcription factors, as depicted here, resulting in the propagation of transcriptionally active chromatin. They may, however, also be factors that promote silencing, such as transcriptional repressors, resulting in the transmission of silent chromatin and/or heterochromatin. In the absence of WICH (**B**), no rebinding of factors can occur, allowing the rapid formation of inaccessible chromatin fibres. PolII, RNA polymerase II.

direct role for Ino80p in facilitating the repair of double-strand DNA breaks in the context of chromatin [70,71]. Given the above evidence, the role of INO80 as a chromatin modifier might be extended to replication sites, as there is a connection between DNA replication and repair, and this may account for the fact that INO80 is implicated in transcriptional regulation.

Conclusions

There is increasing evidence that the multi-step process of replication, and the chromatin organization associated with it, can be regulated on different levels, so that the elegant balance of genomic regeneration and epigenetic maintenance can be retained (Figure 2). Histone-modifying enzymes, chromatin-remodelling factors, DNA methylation and RNA components are all components of the chromatin-remodelling machinery toolbox, and they all facilitate, via a wide

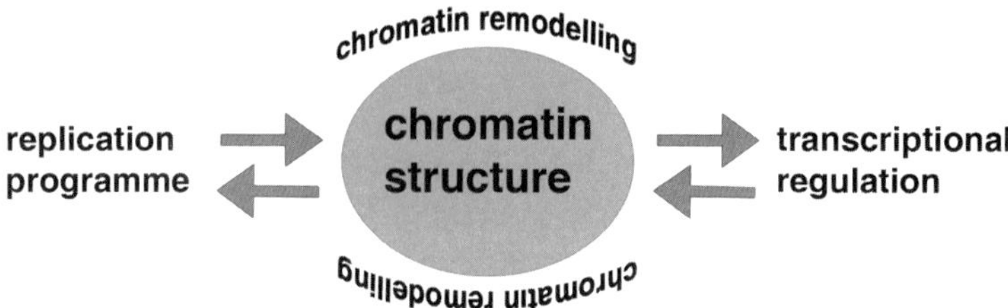

Figure 2 Chromatin structure is the global mediator that ties the replication programme together with transcriptional regulation. Replication initiation, replication timing and chromatin assembly at replication sites are determinants of epigenetic inheritance through a broad range of chromatin modulations. In return, the transcriptional states influence the replication programme by utilizing the chromatin-remodelling machinery, so that ultimately the two processes are entangled in a cycle of co-regulation.

range of mechanisms, different types of chromatin organization. Hence much scientific attention has focused on how these components are involved in the faithful transmission of epigenetic traits, but we are still largely unaware of their underlying function and regulation during DNA replication. Therefore, future research will have to address the following questions: (1) which determinants of the replication programme contribute to effective chromatin compartmentalization and thereby transcriptional maintenance? (2) how are replication sites regulated in their chromatin architecture, and how does such regulation guarantee epigenetic inheritance? (3) what is the function of chromatin-modifying and -remodelling complexes at replication forks? and, finally, (4) what are the mechanisms of propagation of specific chromatin structures?

This work was supported by Marie Curie Cancer Care. Current work in the laboratory of P.V.-W. is funded by the BBSRC, a grant from the Association of International Cancer Research, St Andrews, and support within the *Epigenome* European Network of Excellence as a Newly Established Team. We thank Bilada Bilican, Barbara Xella, Ludmila Bozhenok, Nicola Hawkes and Ana Neves-Costa for comments that improved the manuscript, and Peter Fraser for help with one figure.

References

1. Ehrenhofer-Murray, A.E. (2004) Eur. J. Biochem. **271**, 2335–2349
2. Becker, P.B. and Hörz, W. (2002) Annu. Rev. Biochem. **71**, 247–273
3. Sogo, J.M., Stahl, H., Koller, T. and Knippers, R. (1986) J. Mol. Biol. **189**, 189–204
4. Gasser, R., Koller, T. and Sogo, J.M. (1996) J. Mol. Biol. **258**, 224–239
5. Annunziato, A.T. (2005) J. Biol. Chem. **280**, 12065–12068
6. Walsh, C.P. and Bestor, T.H. (1999) Genes Dev. **13**, 26–34
7. Grewal, S.I. and Elgin, S.C. (2002) Curr. Opin. Genet. Dev. **12**, 178–187
8. Crowe, A.J., Piechan, J.L., Sang, L. and Barton, M.C. (2000) Mol. Cell. Biol. **20**, 4169–4180
9. Forlani, S., Bonnerot, C., Capgras, S. and Nicolas, J.F. (1998) Development **125**, 3153–3166
10. Fisher, D. and Mechali, M. (2003) EMBO J. **22**, 3737–3748
11. Wong, J., Shi, Y.B. and Wolffe, A.P. (1995) Genes Dev. **9**, 2696–2711

12. McNairn, A.J. and Gilbert, D.M. (2003) Bioessays **25**, 647–656
13. Wei, X., Samarabandu, J., Devdhar, R.S., Siegel, A.J., Acharya, R. and Berezney, R. (1998) Science **281**, 1502–1506
14. Pliss, A., Koberna, K., Vecerova, J., Malinsky, J., Masata, M., Fialova, M., Raska, I. and Berezney, R. (2005) J. Cell. Biochem. **94**, 554–565
15. Hassan, A.B., Errington, R.J., White, N.S., Jackson, D.A. and Cook, P.R. (1994) J. Cell Sci. **107**, 425–434
16. Gomez, M. and Brockdorff, N. (2004) Proc. Natl. Acad. Sci. U.S.A. **101**, 6923–6928
17. Schubeler, D., Scalzo, D., Kooperberg, C., van Steensel, B., Delrow, J. and Groudine, M. (2002) Nat. Genet. **32**, 438–442
18. MacAlpine, D.M., Rodriguez, H.K. and Bell, S.P. (2004) Genes Dev. **18**, 3094–3105
19. Gomez, M. and Antequera, F. (1999) EMBO J. **18**, 5683–5690
20. White, E.J., Emanuelsson, O., Scalzo, D., Royce, T., Kosak, S., Oakeley, E.J., Weissman, S., Gerstein, M., Groudine, M., Snyder, M. and Schubeler, D. (2004) Proc. Natl. Acad. Sci. U.S.A. **101**, 17771–17776
20a. Woodfine, K., Fiegler, H., Beare, D.M., Collins, J.E., McCann, O.T., Young, B.D., Debernardi, S., Mott, R., Durham, I. and Carter, N.P. (2004) Hum. Mol. Genet. **13**, 191–202
21. Perry, P., Sauer, S., Billon, N., Richardson, W. D., Spivakov, M., Warnes, G., Livesey, F.J., Merkenschlager, M., Fisher, A.G. and Azuara, V. (2004) Cell Cycle **3**, 1645–1650
22. Hiratani, I., Leskovar, A. and Gilbert, D.M. (2004) Proc. Natl. Acad. Sci. U.S.A. **101**, 16861–16866
23. Zhang, J., Xu, F., Hashimshony, T., Keshet, I. and Cedar, H. (2002) Nature (London) **420**, 198–202
24. Raghuraman, M.K., Winzeler, E.A., Collingwood, D., Hunt, S., Wodicka, L., Conway, A., Lockhart, D.J., Davis, R.W., Brewer, B.J. and Fangman, W.L. (2001) Science **294**, 115–121
25. Bell, S.P. and Dutta, A. (2002) Annu. Rev. Biochem. **71**, 333–374
26. Pak, D.T., Pflumm, M., Chesnokov, I., Huang, D.W., Kellum, R., Marr, J., Romanowski, P. and Botchan, M.R. (1997) Cell **91**, 311–323
27. Prasanth, S.G., Prasanth, K.V., Siddiqui, K., Spector, D.L. and Stillman, B. (2004) EMBO J. **23**, 2651–2663
28. Lidonnici, M.R., Rossi, R., Paixao, S., Mendoza-Maldonado, R., Paolinelli, R., Arcangeli, C., Giacca, M., Biamonti, G. and Montecucco, A. (2004) J. Cell Sci. **117**, 5221–5231
29. Huang, D.W., Fanti, L., Pak, D.T., Botchan, M.R., Pimpinelli, S. and Kellum, R. (1998) J. Cell Biol. **142**, 307–318
30. Suter, B., Tong, A., Chang, M., Yu, L., Brown, G.W., Boone, C. and Rine, J. (2004) Genetics **167**, 579–591
31. Hyrien, O., Maric, C. and Mechali, M. (1995) Science **270**, 994–997
32. Sasaki, T., Sawado, T., Yamaguchi, M. and Shinomiya, T. (1999) Mol. Cell. Biol. **19**, 547–555
33. Mysiak, M.E., Wyman, C., Holthuizen, P.E. and van der Vliet, P.C. (2004) Nucleic Acids Res. **32**, 6218–6225
34. Bodmer-Glavas, M., Edler, K. and Barberis, A. (2001) Nucleic Acids Res. **29**, 4570–4580
35. Danis, E., Brodolin, K., Menut, S., Maiorano, D., Girard-Reydet, C. and Mechali, M. (2004) Nat. Cell Biol. **6**, 721–730
36. Jenke, A.C., Stehle, I.M., Herrmann, F., Eisenberger, T., Baiker, A., Bode, J., Fackelmayer, F.O. and Lipps, H.J. (2004) Proc. Natl. Acad. Sci. U.S.A. **101**, 11322–11327
37. Sharma, M., Li, X., Wang, Y., Zarnegar, M., Huang, C.Y., Palvimo, J.J., Lim, B. and Sun, Z. (2003) EMBO J. **22**, 6101–6114
38. Pappas, Jr, D.L., Frisch, R. and Weinreich, M. (2004) Genes Dev. **18**, 769–781
39. Vogelauer, M., Rubbi, L., Lucas, I., Brewer, B.J. and Grunstein, M. (2002) Mol. Cell **10**, 1223–1233
40. Aggarwal, B.D. and Calvi, B.R. (2004) Nature (London) **430**, 372–376
41. Kemp, M.G., Ghosh, M., Liu, G. and Leffak, M. (2005) Nucleic Acids Res. **33**, 325–336
42. Zhou, J., Chau, C.M., Deng, Z., Shiekhattar, R., Spindler, M.P., Schepers, A. and Lieberman, P.M. (2005) EMBO J. **24**, 1406–1417
43. Akey, C.W. and Luger, K. (2003) Curr. Opin. Struct. Biol. **13**, 6–14
44. Loyola, A. and Almouzni, G. (2004) Biochim. Biophys. Acta **1677**, 3–11
45. Varga-Weisz, P. (2005) in Epigenetics and Chromatin, vol. 38 (Jeanteur, P., ed.), pp. 1–30,

Springer Verlag, Heidelberg
46. Shibahara, K. and Stillman, B. (1999) Cell **96**, 575–585
47. Ridgway, P. and Almouzni, G. (2000) J. Cell Sci. **113**, 2647–2658
48. Murzina, N., Verreault, A., Laue, E. and Stillman, B. (1999) Mol. Cell **4**, 529–540
49. Quivy, J.P., Roche, D., Kirschner, D., Tagami, H., Nakatani, Y. and Almouzni, G. (2004) EMBO J. **23**, 3516–3526
50. Tyler, J.K., Adams, C.R., Chen, S.R., Kobayashi, R., Kamakaka, R.T. and Kadonaga, J. T. (1999) Nature (London) **402**, 555–560
51. Krawitz, D.C., Kama, T. and Kaufman, P.D. (2002) Mol. Cell. Biol. **22**, 614–625
52. Adkins, M.W. and Tyler, J.K. (2004) J. Biol. Chem. **279**, 52069–52074
53. Adkins, M.W., Howar, S.R. and Tyler, J.K. (2004) Mol. Cell **14**, 657–666
54. Zabaronick, S.R. and Tyler, J.K. (2005) Mol. Cell. Biol. **25**, 652–660
55. Tagami, H., Ray-Gallet, D., Almouzni, G. and Nakatani, Y. (2004) Cell **116**, 51–61
56. Moshkin, Y.M., Armstrong, J.A., Maeda, R.K., Tamkun, J.W., Verrijzer, P., Kennison, J.A. and Karch, F. (2002) Genes Dev. **16**, 2621–2626
57. Rountree, M.R., Bachman, K.E. and Baylin, S.B. (2000) Nat. Genet. **25**, 269–277
58. Collins, N., Poot, R.A., Kukimoto, I., Garcia-Jimenez, C., Dellaire, G. and Varga-Weisz, P.D. (2002) Nat. Genet. **32**, 627–632
59. Becker, P.B. (2002) EMBO J. **21**, 4749–4753
60. Corona, D.F. and Tamkun, J.W. (2004) Biochim. Biophys. Acta **1677**, 113–119
61. Goldmark, J.P., Fazzio, T.G., Estep, P.W., Church, G.M. and Tsukiyama, T. (2000) Cell **103**, 423–433
62. Yasui, D., Miyano, M., Cai, S., Varga-Weisz, P. and Kohwi-Shigematsu, T. (2002) Nature (London) **419**, 641–645
63. Fyodorov, D.V., Blower, M.D., Karpen, G.H. and Kadonaga, J.T. (2004) Genes Dev. **18**, 170–183
64. Li, J., Santoro, R., Koberna, K. and Grummt, I. (2005) EMBO J. **24**, 120–127
65. Bozhenok, L., Wade, P.A. and Varga-Weisz, P. (2002) EMBO J. **21**, 2231–2241
66. Poot, R.A., Bozhenok, L., van den Berg, D.L., Steffensen, S., Ferreira, F., Grimaldi, M., Gilbert, N., Ferreira, J. and Varga-Weisz, P.D. (2004) Nat. Cell Biol. **6**, 1236–1244
67. Poot, R.A., Bozhenok, L., van den Berg, D.L., Hawkes, N. and Varga-Weisz, P.D. (2005) Cell Cycle **4**, 543–546
68. Shen, X., Mizuguchi, G., Hamiche, A. and Wu, C. (2000) Nature (London) **406**, 541–544
69. Fritsch, O., Benvenuto, G., Bowler, C., Molinier, J. and Hohn, B. (2004) Mol. Cell **16**, 479–485
70. van Attikum, H., Fritsch, O., Hohn, B. and Gasser, S.M. (2004) Cell **119**, 777–788
71. Morrison, A.J., Highland, J., Krogan, N.J., Arbel-Eden, A., Greenblatt, J.F., Haber, J.E. and Shen, X. (2004) Cell **119**, 767–775

Biochem. Soc. Symp. **73**, 109–119
(Printed in Great Britain)

11

Nucleosome dynamics

Chris Stockdale, Michael Bruno, Helder Ferreira, Elisa Garcia-Wilson, Nicola Wiechens, Maik Engeholm, Andrew Flaus and Tom Owen-Hughes[1]

Division of Gene Regulation and Expression, School of Life Sciences, University of Dundee, Dundee DD1 5EH, Scotland, U.K.

Abstract

In the 30 years since the discovery of the nucleosome, our picture of it has come into sharp focus. The recent high-resolution structures have provided a wealth of insight into the function of the nucleosome, but they are inherently static. Our current knowledge of how nucleosomes can be reconfigured dynamically is at a much earlier stage. Here, recent advances in the understanding of chromatin structure and dynamics are highlighted. The ways in which different modes of nucleosome reconfiguration are likely to influence each other are discussed, and some of the factors likely to regulate the dynamic properties of nucleosomes are considered.

Introduction

Although nucleosomes must fulfil the requirement to compact DNA, they must also possess the flexibility to allow genetic processes to occur. In the following sections, the main spontaneous rearrangements that the nucleosome is known to undergo are described.

Modes of dynamic chromatin reconfiguration

Site exposure

Although nucleosomes are resistant to complete dissociation at temperatures as high as 70°C [1], they can undergo a variety of conformational transitions. Thermal denaturation studies of nucleosome core particles have provided evidence of intermediate, structurally distinct states prior to complete melting. Biophysical studies using combinations of circular dichroism, calorimetry,

[1]To whom correspondence should be addressed (email t.a.owenhughes@dundee.ac.uk).

ultracentrifugation and exonucleases concluded that the histone–DNA contacts at the edge of the nucleosome are disrupted at temperatures of 40–45°C, leading to the dissociation of approx. 20–25 bp of DNA from each edge [2,3]. In a subsequent phase above 72°C, the central 100 bp of DNA is released and denatured, accompanied by protein denaturation [4]. Subsequently, it was observed that transcription factors bound preferentially to DNA near the edge of nucleosomes, and that one DNA-binding protein could assist the binding of another [5,6]. A major breakthrough involved the use of restriction enzymes to probe nucleosome accessibility, as this enabled quantitative models for DNA accessibility to be formulated [7–10]. More recently, fluorescence-based techniques have been used to monitor DNA unpeeling from nucleosomes [11,12]. These studies showed directly that DNA from the outer turns of the nucleosome is unwrapped for short (millisecond) periods of time, followed by re-wrapping, such that DNA is accessible for 4–20% of the time.

These studies have been used to support a site exposure model in which the progressive dissociation of DNA from the histone octamer allows access to locations successively further into the nucleosome at progressively reduced rates [13]. However, recent direct observations of accessibility using a single restriction enzyme to probe equivalent sites in nucleosomal DNA are in fact more consistent with the original biphasic model, whereby the outer turns of the nucleosome are considerably more accessible than the central region of DNA [14]. It is also possible to interpret the published data obtained by Widom and colleagues using different enzymes for different sites as being consistent with biphasic behaviour (Figure 1). Accessibility to restriction enzymes drops fairly steadily over the outermost turns of DNA associated with the histone octamer. However, over a region spanning approx. 45 bp from the nucleosome dyad, measurements are considerably more variable, but exhibit much less dependence on the distance from the dyad, measurements are considerably more variable. The variability in measured enzyme accessibility is such that, in some cases, internal sites are orders of magnitude more accessible than those located outside them. Although this may be due in part to differences in restriction enzymes and the rotational orientation of sites, the overall trend is qualitatively similar to the biphasic one observed by Wu and Travers [14]. This raises the possibility that, while unpeeling may occur on the outer extremities of the nucleosome, a different process may be required to gain access to DNA closer to the dyad.

It is likely that some DNA-binding proteins are better suited to engage DNA within the central region of the nucleosome region. For example, TBP (TATA box binding protein) and NF1 (nuclear factor 1) appear to bind nucleosomal DNA poorly, but NF-κB (nuclear factor-κB), nuclear hormone receptors such as glucocorticoid receptor, and FoxA bind nucleosomal DNA relatively well [15–18].

Nucleosome repositioning

A broad base of observations support the concept that nucleosomes can shuttle spontaneously between different locations on DNA ([19–23]; reviewed in [24,25]). Although the ability of nucleosomes to reposition appears to be a fairly

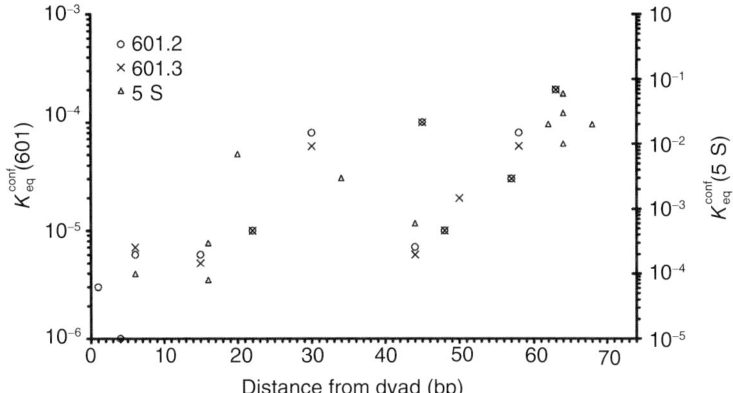

Figure 1 Accessibility of nucleosomes to restriction enzymes. Shown is a plot of site accessibility equilibrium (K^{conf}_{eq}) values for distance from the centre of the enzyme recognition site to the nucleosome pseudodyad, using data for 5 S, 601.2 and 601.3 mononucleosomes from Polach and Widom [7], Anderson and Widom [9] and Anderson et al. [10] respectively.

general phenomenon [19], the physiological relevance of thermal nucleosome movement is often questioned. One reason for this is that the temperatures required for thermal redistribution are frequently higher than 37°C. However, the temperature required to reposition a nucleosome is affected significantly by the underlying DNA sequences, and studies of nucleosome mobility have favoured a subset of sequences that direct the positioning of nucleosomes at defined locations. Because an inherent property of these sequences is that they are refractory to nucleosome redistribution, the temperatures required are likely to be higher than average. Consistent with this, it is possible for nucleosomes to be relocated at temperatures lower than 37°C from strong positioning sequences at a reduced rate, or when DNA fragments are manipulated to reduce the strength of positioning (A. Flaus and T. Owen-Hughes, unpublished work). A second reason to question the physiological relevance of thermal nucleosome redistribution is that physiological concentrations of bivalent cations inhibit nucleosome sliding [21]. However, this inhibition is not absolute, and significant thermal nucleosome redistribution can be observed in the presence of bivalent ions [26]. On balance, it seems likely that most nucleosomes will be engaged in vibrational movement. This can result in translational repositioning, the extent of which will be constrained by the structural properties of the underlying DNA [27,28], the binding of transcription factors [29,30] or nucleosome-binding proteins [31], and the action of ATP-dependent chromatin remodellers and polymerases [32].

Histone exchange

Dimers of histones H2A and H2B do not remain stably associated with the H3–H4 tetramer at physiological salt concentrations [33]. The wrapping of DNA around the octamer is therefore required to maintain nucleosome integrity. *In vivo*, there is evidence that histone dimers are prone to dissociation from the nucleosome [34–36]. Similarly, histone dimers are prone to dissociation from nucleosomes *in vitro*. In concentrated solutions of chromatin, this exchange may be difficult to detect, as reassociation with the resulting dimer-depleted chromatin may be occurring at a similar rate. However, in dilute solutions nucleosomes are not stable [37,38].

Figure 2 shows that low, but detectable, levels of histone dimer exchange occur spontaneously during thermal incubation. In that study it was possible to monitor nucleosome repositioning in parallel. As most of the nucleosomes were

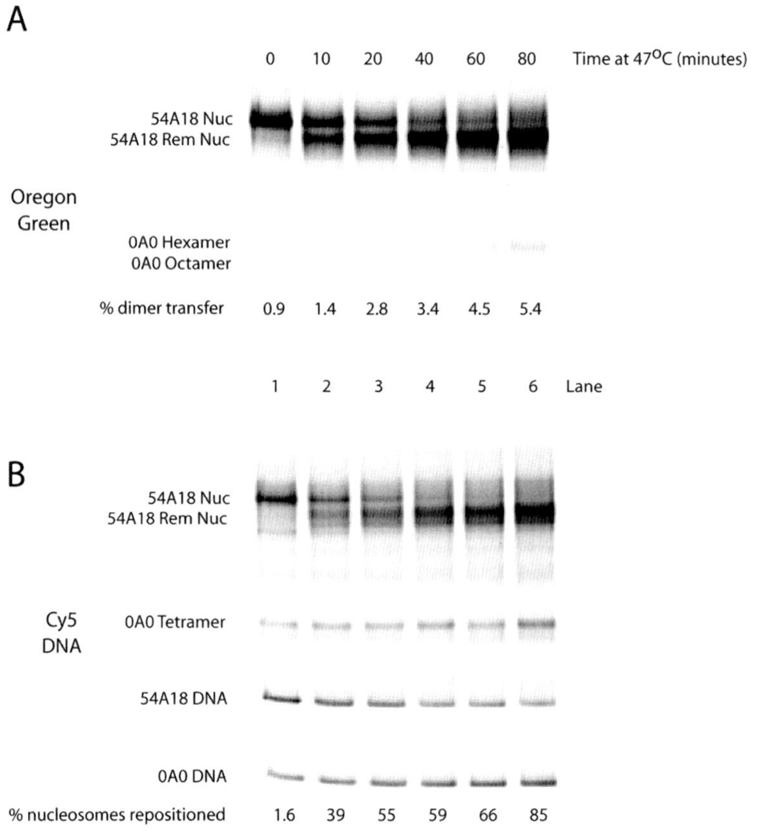

Figure 2 Detection of spontaneous histone H2A–H2B dimer exchange. Nucleosomes (Nuc; 200 nM) comprising Cy5-labelled 54A18 DNA were incubated in the presence of a 4-fold molar excess of tetramer on Cy5-labelled 0A0 DNA at 47°C for increasing times. The upper image (**A**) shows the Oregon Green-labelled histones, and the lower image (**B**) shows the Cy5 signal from DNA. The efficiency of dimer transfer from the remodelled (Rem) 54A18 nucleosomes on to 0A0/tetramer and the percentage of nucleosome repositioned were calculated. Methods were as described previously [99].

repositioned prior to the detection of significant dimer exchange, nucleosome repositioning appeared to be more favourable than loss of histone dimers under these conditions. Nonetheless, the loss or exchange of histone dimers between nucleosomes should be recognized as one of the spontaneous transitions that the nucleosome can undergo.

Other structural rearrangements of the nucleosome

The consistent organization of histones in different nucleosome structures may contribute to the perception that this arrangement is inviolable. However, there is evidence that even the arrangement of histones within the H3–H4 tetramer is somewhat dynamic. Allfrey and colleagues [39] noticed that the cysteine residues in histone H3 became more accessible in transcriptionally active chromatin. Although the structural basis for this remains unclear, there are other observations that suggest that rearrangement of the tetramer may be possible [40–42].

Chromatin fibre formation

It has been appreciated for some time that linear arrays of nucleosomes can condense spontaneously to form more compact structures. This condensation reaction is sensitive to the concentrations of univalent and particularly bivalent cations in solution [43]. Until recently there was considerable controversy as to the arrangement of nucleosomes within such fibres. However, it is now apparent that nucleosomes are arranged as a two-start helix in these 30 nm fibres, at least in an idealized situation [44]. The formation of chromatin fibres is likely to affect access to the underlying DNA. Consistent with this, a correlation has been observed between the level of condensation and the propensity for gene expression [45]. What is not clear at present is whether the organization of chromatin fibres generated *in vitro* is similar to that of those that occur inside nuclei, where additional proteins may contribute to the stabilization of chromatin fibres.

Relationship between different aspects of chromatin dynamics

Although five different modes by which chromatin can be dynamically rearranged have been described above, many of these alterations may be interrelated. The unpeeling of DNA from the edge of nucleosomes may provide a starting point for the formation of loops, which could provide a means for nucleosome movement [46]. The finding that a subset of remodelling enzymes closely related to the yeast SWI/SNF (sucrose non-fermenting) complex can reposition nucleosomes to locations at which substantial amounts of DNA are removed from the edge of nucleosomes [23,47,48] provides a pathway by which nucleosome movement may promote dimer exchange. Conversely, removal of histone dimers by equilibration with the histone chaperone NAP-1 (nucleosome assembly protein-1) has been observed to promote nucleosome mobility [49]. Similarly, it is plausible that the loss of histone dimers might contribute to site exposure.

All of the modes by which the structure of individual nucleosomes may be altered may also influence the ability of arrays of nucleosomes to form chromatin fibres. As the arrangement of nucleosomes within chromatin fibres is ordered, the orientation and length of linker DNA are likely to have optimum values that favour fibre condensation. Alterations to mononucleosome structure are likely to influence these parameters. An example for which this may be especially relevant is in the characterization of Sin mutations that suppress the requirement for the SWI/SNF complex *in vivo*. Nucleosomes bearing these mutations have dramatically increased thermal mobility [26]. Some of these mutations have also been found to inhibit the ability of nucleosomes to form chromatin fibres [50]. It is possible that the increased mobility of nucleosomes may reduce the time that nucleosomes are correctly aligned for fibre formation. This could occur either if nucleosomes slide translationally away from locations necessary for folding or if parts of the nucleosome structure participating in folding take up non-productive orientations.

Means of regulating nucleosome dynamics

Post-translational modifications

Amino acid residues within the tail domains of the nucleosome have long been recognized as targets for post-translational modification [51]. Modifications of these highly basic tails have moderate effects on chromatin at the level of the nucleosome, since they protrude from the nucleosome core [52]. Modifications may also act to create or destroy docking sites for factors functioning downstream. Protein domains common to chromatin-associated factors which are capable of binding differently modified amino acids have been identified. For example bromodomains seem to be acetyl-lysine-binding modules [54–56], several chromodomains bind methyl-lysine residues [57–61], and tudor domains have been identified with either methyl-lysine or methylarginine binding specificities [62,63].

The histone tails may also play an important role in organizing higher-order structures by contacting linker DNA or components of adjacent nucleosome cores. An example of this is seen in the first high-resolution crystal structure of a nucleosome, where the histone H4 tail contacts an acidic patch within the globular region of an H2A–H2B dimer from a neighbouring nucleosome [64]. Modifications that alter the charge of a residue (e.g. lysine acetylation) may affect these interactions. Recently, lysine acetylation has been shown to impede incorporation of nucleosomes into higher-order fibres in living cells [65]. Consistent with this, the histone H4 tail plays a critical role in mediating condensation of nucleosome arrays *in vitro* [66], raising the possibility that histone H4 modifications may specifically influence chromatin folding. Historically, most investigations into the role of histone tails in nucleosome dynamics have involved removal of all histone tails by trypsin digestion. However, if different tails have different and possibly opposing roles, then important effects may have been overlooked.

A recent study using sophisticated mass spectrometry identified many sites of post-translational modification within the globular domains of the histones [67]. By their nature, chemically reactive and thus modifiable amino acids are hydrophilic, and therefore they tend to occur on the external surface of the histone octamer. As a consequence, within the nucleosome these modifications will be in close proximity to the path of the DNA. This has led to the suggestion that post-translational modifications of the histone octamer globular core will affect the association of the DNA with this structure [68].

Post-translational modification of DNA by CpG methylation may provide another means of manipulating nucleosome dynamics. DNA methylation can affect the affinity of histone octamers for DNA, as well as altering translational positioning on DNA [69,70].

Histone variants

Almost all eukaryotes have several genes for each of the core histones. These fall into two categories. The main histone genes with replication-dependent expression are in multiply repeated arrays which, although highly similar, do exhibit a low level of gene-to-gene variation. A second set of more significant sequence variants have more specialized and replication-independent expression profiles, such that the proteins that they encode can be used in specific circumstances. Interestingly, histones H3 and H2A seem to be subject to the most specialization, whereas H2B and H4 are in general invariant [71]. Little is known about how incorporation of these variants affects the dynamic properties of chromatin. Incorporation of H2A.Z into a histone octamer increases the rate of its thermal repositioning [26] and influences chromatin fibre formation [72]. It is unclear how these altered properties relate to the role of this variant *in vivo*, which is to protect against the spread of silenced chromatin in budding yeast [73]; it may also assist in the recruitment of HP1 (heterochromatin protein 1) in higher eukaryotes [74].

The macroH2A variant is found in the silenced X chromosome of female mammals [75]. It has a large C-terminal extension which can function *in vitro* to further impede the binding of transcription factors to nucleosomal DNA [76]. In contrast, H2A.Bbd has a truncation of the C-terminus of the histone fold region and is completely excluded from inactive X chromosomes [77]. *In vitro*, nucleosomes containing this variant organize only 118 bp of DNA [78], and can be transcribed through more easily than those containing canonical histones [79]. These properties are consistent with a role in active transcription, which was first suggested from the co-localization of H2ABbd with regions of H4 acetylation [77].

ATP-dependent remodelling

Nucleosome structure can also be changed non-covalently to alter the accessibility of the associated DNA. One group of enzymes that cause these changes are ATP-dependent chromatin remodellers belonging to the SFII (superfamily II) helicase-like proteins. These are members of the Snf2 family, which is named after the first remodeller to be identified. The Snf2 family can be subdivided further into subfamilies that cause distinct alterations to nucleosome structure. For example, enzymes of the ISWI (imitation switch) subfamily cause nucleosome repositioning [30,80,81], but are unable to significantly alter

DNA accessibility within the boundaries of the nucleosome [30,48] or cause histone dimer exchange [82]. In contrast, all three of these properties have been attributed to Swi2/Snf2-type enzymes [48,82,83]. Snf2 family members have roles in regulating such diverse processes as DNA methylation [84,85], transcription intiation [86], elongation [87], termination [88] and homologous recombination [89] within the context of chromatin (reviewed in [90]).

Association of nucleosome-binding proteins

Linker histones, including histone H1 and H5, bind to the nucleosome, protecting DNA outside of the nucleosome core and promoting compaction of nucleosomal arrays (reviewed in [43]). In general, linker histones are found to associate with regions of low transcriptional activity [91], but this association is dynamic [92]. The presence of histone H1 has also been observed to restrict both inherent and ATP-driven nucleosome repositioning [31,93] and to increase the association of histone tails with linker DNA [94], and so may provide a means of regulating nucleosome dynamics. Other nucleosome-binding proteins, including HP1 [74], HMGN (high-mobility group nucleosomal binding domain) proteins [95], polycomb proteins [96] or PARP [poly(ADP-ribose) polymerase] [97], may provide additional means of modulating nucleosome dynamics. Proteins known to associate with DNA-free histones, such as NAP-1, CAF-1 (chromatin assembly factor-1) and Asf1 (antisilencing function 1), may act differently to facilitate the assembly or removal of nucleosomes from DNA [98].

Summary

It is clear that the targeted modification of histones acts to influence the function of nucleosomes as a platform for the recruitment of proteins required for gene regulation. However, this should not detract from the likelihood that a subset of regulatory pathways target the nucleosome to alter its dynamic properties, and that this could have a role in altering the accessibility of the underlying template.

This work was supported by a Wellcome Trust Senior Fellowship (T.O.-H.), Wellcome Trust Studentships (C.S., H.F., M.E.) and a BBSRC Studentship (M.B.).

References

1. Poon, N.H. and Seligy, V.L. (1980) Exp. Cell Res. **128**, 333–341
2. Riley, D. and Weintraub, H. (1978) Cell **13**, 281–293
3. Simpson, R.T. (1979) J. Biol. Chem. **254**, 10123–10127
4. Weischet, W.O., Tatchell, K., Van Holde, K.E. and Klump, H. (1978) Nucleic Acids Res. **5**, 139–160
5. Taylor, C.A., Workman, J.L., Schuetz, T.J. and Kingston, R.E. (1991) Genes Dev. **5**, 1285–1298
6. Adams, C.C. and Workman, J.L. (1995) Mol. Cell. Biol. **15**, 1405–1421
7. Polach, K.J. and Widom, J. (1995) J. Mol. Biol. **254**, 130–149
8. Polach, K.J. and Widom, J. (1996) J. Mol. Biol. **258**, 800–812
9. Anderson, J.D. and Widom, J. (2000) J. Mol. Biol. **296**, 979–987
10. Anderson, J.D., Thastrom, A. and Widom, J. (2002) Mol. Cell. Biol. **22**, 7147–7157

11. Li, G., Levitus, M., Bustamante, C. and Widom, J. (2005) Nat. Struct. Mol. Biol. **12**, 46–53
12. Li, G. and Widom, J. (2004) Nat. Struct. Mol. Biol. **11**, 763–769
13. Widom, J. (2001) Q. Rev. Biophys. **34**, 269–324
14. Wu, C. and Travers, A. (2004) Nucleic Acids Res. **32**, e122
15. Angelov, D., Lenouvel, F., Hans, F., Muller, C.W., Bouvet, P., Bednar, J., Moudrianakis, E.N., Cadet, J. and Dimitrov, S. (2004) J. Biol. Chem. **279**, 42374–42382
16. Cirillo, L.A. and Zaret, K.S. (1999) Mol. Cell **4**, 961–969
17. Archer, T.K., Lefebvre, P., Wolford, R.G. and Hager, G.L. (1992) Science **255**, 1573–1576
18. Imbalzano, A.N., Kwon, H., Green, M.R. and Kingston, R.E. (1994) Nature (London) **370**, 481–485
19. Meersseman, G., Pennings, S. and Bradbury, E.M. (1992) EMBO J. **11**, 2951–2959
20. Beard, P. (1978) Cell **15**, 955–967
21. Pennings, S., Meersseman, G. and Bradbury, E.M. (1991) J. Mol. Biol. **220**, 101–110
22. Flaus, A. and Richmond, T.J. (1998) J. Mol. Biol. **275**, 427–441
23. Flaus, A. and Owen-Hughes, T. (2003) Mol. Cell. Biol. **23**, 7767–7779
24. Workman, J.L. and Kingston, R.E. (1998) Annu. Rev. Biochem. **67**, 545–579
25. Widom, J. (1998) Annu. Rev. Biophys. Biomol. Struct. **27**, 285–327
26. Flaus, A., Rencurel, C., Ferreira, H., Wiechens, N. and Owen-Hughes, T. (2004) EMBO J. **23**, 343–353
27. Virstedt, J., Berge, T., Henderson, R.M., Waring, M.J. and Travers, A.A. (2004) J. Struct. Biol. **148**, 66–85
28. Mai, X., Chou, S. and Struhl, K. (2000) Mol. Cell. Biol. **20**, 6668–6676
29. Kang, J.G., Hamiche, A. and Wu, C. (2002) EMBO J. **21**, 1406–1413
30. Kassabov, S.R., Henry, N.M., Zofall, M., Tsukiyama, T. and Bartholomew, B. (2002) Mol. Cell. Biol. **22**, 7524–7534
31. Ura, K., Hayes, J.J. and Wolffe, A.P. (1995) EMBO J. **14**, 3752–3765
32. Studitsky, V.M., Walter, W., Kireeva, M., Kashlev, M. and Felsenfeld, G. (2004) Trends Biochem. Sci. **29**, 127–135
33. Eickbush, T.H. and Moudrianakis, E.N. (1978) Biochemistry **17**, 4955–4964
34. Jackson, V. (1990) Biochemistry **29**, 719–731
35. Kimura, H. and Cook, P.R. (2001) J. Cell Biol. **153**, 1341–1353
36. Thiriet, C. and Hayes, J.J. (2005) Genes Dev. **19**, 677–682
37. Godde, J.S. and Wolffe, A.P. (1995) J. Biol. Chem. **270**, 27399–27402
38. Claudet, C., Angelov, D., Bouvet, P., Dimitrov, S. and Bednar, J. (2005) J. Biol. Chem. **280**, 19958–19965
39. Allegra, P., Sterner, R., Clayton, D.F. and Allfrey, V.G. (1987) J. Mol. Biol. **196**, 379–388
40. Mozziconacci, J. and Victor, J.M. (2003) J. Struct. Biol. **143**, 72–76
41. Hamiche, A., Carot, V., Alilat, M., De Lucia, F., O'Donohue, M.F., Revet, B. and Prunell, A. (1996) Proc. Natl. Acad. Sci. U.S.A. **93**, 7588–7593
42. Tagami, H., Ray-Gallet, D., Almouzni, G. and Nakatani, Y. (2004) Cell **116**, 51–61
43. Hansen, J.C. (2002) Annu. Rev. Biophys. Biomol. Struct. **31**, 361–392
44. Dorigo, B., Schalch, T., Kulangara, A., Duda, S., Schroeder, R.R. and Richmond, T.J. (2004) Science **306**, 1571–1573
45. Gilbert, N., Boyle, S., Fiegler, H., Woodfine, K., Carter, N.P. and Bickmore, W.A. (2004) Cell **118**, 555–566
46. Flaus, A. and Owen-Hughes, T. (2003) Biopolymers **68**, 563–578
47. Kassabov, S.R., Zhang, B., Persinger, J. and Bartholomew, B. (2003) Mol. Cell **11**, 391–403
48. Fan, H.Y., He, X., Kingston, R.E. and Narlikar, G.J. (2003) Mol. Cell **11**, 1311–1322
49. Park, Y.J., Chodaparambil, J.V., Bao, Y., McBryant, S.J. and Luger, K. (2005) J. Biol. Chem. **280**, 1817–1825
50. Horn, P.J., Crowley, K.A., Carruthers, L.M., Hansen, J.C. and Peterson, C.L. (2002) Nat. Struct. Biol. **9**, 167–171
51. van Holde, K.E. (1988) Chromatin, Springer-Verlag, New York
52. Lee, D.Y., Hayes, J.J., Pruss, D. and Wolffe, A.P. (1993) Cell **72**, 73–84
53. Kornberg, R.D. and Lorch, Y. (1999) Curr. Opin. Genet. Dev. **9**, 148–151
54. Dhalluin, C., Carlson, J.E., Zeng, L., He, C., Aggarwal, A.K. and Zhou, M.M. (1999) Nature (London) **399**, 491–496
55. Hudson, B.P., Martinez-Yamout, M.A., Dyson, H.J. and Wright, P.E. (2000) J. Mol. Biol. **304**, 355–370

56. Jacobson, R.H., Ladurner, A.G., King, D.S. and Tjian, R. (2000) Science **288**, 1422–1425
57. Bannister, A.J., Zegerman, P., Partridge, J.F., Miska, E.A., Thomas, J.O., Allshire, R.C. and Kouzarides, T. (2001) Nature (London) **410**, 120–124
58. Lachner, M., O'Carroll, D., Rea, S., Mechtler, K. and Jenuwein, T. (2001) Nature (London) **410**, 116–120
59. Cao, R., Wang, L., Wang, H., Xia, L., Erdjument-Bromage, H., Tempst, P., Jones, R.S. and Zhang, Y. (2002) Science **298**, 1039–1043
60. Plath, K., Fang, J., Mlynarczyk-Evans, S.K., Cao, R., Worringer, K.A., Wang, H., de la Cruz, C.C., Otte, A.P., Panning, B. and Zhang, Y. (2003) Science **300**, 131–135
61. Pray-Grant, M.G., Daniel, J.A., Schieltz, D., Yates, III, J.R. and Grant, P.A. (2005) Nature (London) **433**, 434–438
62. Sprangers, R., Groves, M.R., Sinning, I. and Sattler, M. (2003) J. Mol. Biol. **327**, 507–520
63. Huyen, Y., Zgheib, O., Ditullio, Jr, R.A., Gorgoulis, V.G., Zacharatos, P., Petty, T.J., Sheston, E.A., Mellert, H.S., Stavridi, E.S. and Halazonetis, T.D. (2004) Nature (London) **432**, 406–411
64. Luger, K., Mader, A.W., Richmond, R.K., Sargent, D.F. and Richmond, T.J. (1997) Nature (London) **389**, 251–260
65. Toth, K.F., Knoch, T.A., Wachsmuth, M., Frank-Stohr, M., Stohr, M., Bacher, C.P., Muller, G. and Rippe, K. (2004) J. Cell Sci. **117**, 4277–4287
66. Dorigo, B., Schalch, T., Bystricky, K. and Richmond, T.J. (2003) J. Mol. Biol. **327**, 85–96
67. Freitas, M.A., Sklenar, A.R. and Parthun, M.R. (2004) J. Cell. Biochem. **92**, 691–700
68. Cosgrove, M.S., Boeke, J.D. and Wolberger, C. (2004) Nat. Struct. Mol. Biol. **11**, 1037–1043
69. Davey, C., Pennings, S. and Allan, J. (1997) J. Mol. Biol. **267**, 276–288
70. Davey, C.S., Pennings, S., Reilly, C., Meehan, R.R. and Allan, J. (2004) Nucleic Acids Res. **32**, 4322–4331
71. Malik, H.S. and Henikoff, S. (2003) Nat. Struct. Biol. **10**, 882–891
72. Fan, J.Y., Gordon, F., Luger, K., Hansen, J.C. and Tremethick, D.J. (2002) Nat. Struct. Biol. **9**, 172–176
73. Meneghini, M.D., Wu, M. and Madhani, H.D. (2003) Cell **112**, 725–736
74. Fan, J.Y., Rangasamy, D., Luger, K. and Tremethick, D.J. (2004) Mol. Cell **16**, 655–661
75. Costanzi, C. and Pehrson, J.R. (1998) Nature (London) **393**, 599–601
76. Angelov, D., Molla, A., Perche, P.Y., Hans, F., Cote, J., Khochbin, S., Bouvet, P. and Dimitrov, S. (2003) Mol. Cell **11**, 1033–1041
77. Chadwick, B.P. and Willard, H.F. (2001) J. Cell Biol. **152**, 375–384
78. Bao, Y., Konesky, K., Park, Y.J., Rosu, S., Dyer, P.N., Rangasamy, D., Tremethick, D.J., Laybourn, P.J. and Luger, K. (2004) EMBO J. **23**, 3314–3324
79. Angelov, D., Verdel, A., An, W., Bondarenko, V., Hans, F., Doyen, C.M., Studitsky, V.M., Hamiche, A., Roeder, R.G., Bouvet, P. and Dimitrov, S. (2004) EMBO J. **23**, 3815–3824
80. Hamiche, A., Sandaltzopoulos, R., Gdula, D.A. and Wu, C. (1999) Cell **97**, 833–842
81. Langst, G., Bonte, E.J., Corona, D.F. and Becker, P.B. (1999) Cell **97**, 843–852
82. Bruno, M., Flaus, A., Stockdale, C., Rencurel, C., Ferreira, H. and Owen-Hughes, T. (2003) Mol. Cell **12**, 1599–1606
83. Whitehouse, I., Flaus, A., Cairns, B.R., White, M.F., Workman, J.L. and Owen-Hughes, T. (1999) Nature (London) **400**, 784–787
84. Dennis, K., Fan, T., Geiman, T., Yan, Q. and Muegge, K. (2001) Genes Dev. **15**, 2940–2944
85. Jeddeloh, J.A., Stokes, T.L. and Richards, E.J. (1999) Nat. Genet. **22**, 94–97
86. Lomvardas, S. and Thanos, D. (2001) Cell **106**, 685–696
87. Corey, L.L., Weirich, C.S., Benjamin, I.J. and Kingston, R.E. (2003) Genes Dev. **17**, 1392–1401
88. Alen, C., Kent, N.A., Jones, H.S., O'Sullivan, J., Aranda, A. and Proudfoot, N.J. (2002) Mol. Cell **10**, 1441–1452
89. Alexiadis, V. and Kadonaga, J.T. (2002) Genes Dev. **16**, 2767–2771
90. Becker, P.B. and Horz, W. (2002) Annu. Rev. Biochem. **71**, 247–273
91. Kamakaka, R.T. and Thomas, J.O. (1990) EMBO J. **9**, 3997–4006
92. Misteli, T., Gunjan, A., Hock, R., Bustin, M. and Brown, D.T. (2000) Nature (London) **408**, 877–881
93. Ramachandran, A., Omar, M., Cheslock, P. and Schnitzler, G.R. (2003) J. Biol. Chem. **278**, 48590–48601
94. Zheng, C. and Hayes, J.J. (2003) J. Biol. Chem. **278**, 24217–24224

95. Catez, F., Brown, D.T., Misteli, T. and Bustin, M. (2002) EMBO Rep. **3**, 760–766
96. King, I.F., Francis, N.J. and Kingston, R.E. (2002) Mol. Cell. Biol. **22**, 7919–7928
97. Kim, M.Y., Mauro, S., Gevry, N., Lis, J.T. and Kraus, W.L. (2004) Cell **119**, 803–814
98. Loyola, A. and Almouzni, G. (2004) Biochim. Biophys. Acta **1677**, 3–11
99. Bruno, M., Flaus, A. and Owen-Hughes, T. (2004) Methods Enzymol. **375**, 211–228

Biochem. Soc. Symp. **73**, 121–129
(Printed in Great Britain)

12

Convergence of the SUMO and MAPK pathways on the ETS-domain transcription factor Elk-1

Shen-Hsi Yang and Andrew D. Sharrocks[1]

Faculty of Life Sciences, University of Manchester, Michael Smith Building, Oxford Road, Manchester M13 9PT, U.K.

Abstract

The ETS-domain transcription factor Elk-1 is regulated by phosphorylation in response to activation of the MAPK (mitogen-activated protein kinase) pathways. This phosphorylation triggers a series of molecular events that convert Elk-1 from a transcriptionally silent state into a highly active state and then back to a basal level. At the same time, activation of the ERK (extracellular-signal-regulated kinase) MAPK pathway leads to loss of modification of Elk-1 by SUMO (small ubiquitin-related modifier). As SUMO imparts repressive properties on Elk-1, ERK-mediated SUMO loss leads to de-repression at the same time as the ERK pathway promotes activation of Elk-1. Thus a two-step mechanism is employed to convert Elk-1 into its fully activated state. Here, the molecular events underlying these changes in Elk-1 status, and the role of PIASxα [protein inhibitor of activated STAT (signal transducer and activator of transcription) xα] as a co-activator that facilitates this process, are discussed.

Introduction

Extracellular signals are sensed and converted into cellular responses by relaying the information within the cell via signalling pathways. One of the major outputs of these signalling pathways is an alteration in the gene expression profiles in the cell that ultimately dictates a change in cellular function or phenotype. The MAPK (mitogen-activated protein kinase) pathways are some of the best studied

[1]To whom correspondence should be addressed (email a.d.sharrocks@manchester.ac.uk).

signalling routes to the nucleus, and different MAPK pathways can respond to distinct signals (reviewed in [1,2]). For example, the ERK (extracellular-signal-regulated kinase) pathway responds to mitogens, whereas the JNK (c-Jun N-terminal kinase) and p38 pathways respond to stress stimuli. Upon activation, the terminal kinases in these cascades then phosphorylate transcriptional regulators, either directly or indirectly through downstream kinases (reviewed in [3]). These phosphorylation events trigger a number of distinct molecular events, leading to either activation or inactivation of their target proteins, and hence to either enhanced or reduced expression respectively of their target genes.

In addition to regulation via phosphorylation, transcriptional regulators can also be controlled by other post-translational modifications, such as modification by SUMO (small ubiquitin-related modifier) (reviewed in [4–6]). Sumoylation can affect many different molecular activities, including DNA binding, protein stability, nuclear–cytoplasmic shuttling, subnuclear localization, and protein–protein interactions. Ultimately, these molecular events lead to changes in the transactivation capacity of the transcriptional regulator, and one common event is that sumoylation often imparts repressive properties on its substrate.

The ETS-domain transcription factor Elk-1 is regulated both by MAPK pathways and by SUMO modification, and therefore provides a useful model for understanding how these two pathways converge to regulate the activity of a single transcription factor. This review summarizes our current knowledge of the molecular mechanisms underlying how these pathways influence the activation/inactivation cycle that controls the transcriptional output from Elk-1.

The TCF (ternary complex factor) ETS-domain transcription factors

Elk-1 is a member of the TCF subfamily of ETS-domain transcription factors, which also contains SAP-1 [SRF (serum response factor) accessory protein 1; also known as Elk-4] and SAP-2/Net (also known as Elk-3) (reviewed in [7–9]). These proteins all share several conserved domains, including the ETS DNA-binding domain, the B-box SRF interaction domain, the C-terminal TAD (transcriptional activation domain), and a MAPK docking module composed of the D-domain and an FxF motif that flanks the TAD (Figure 1). However, each subfamily member also contains distinct functional motifs. In the case of Elk-1, this is reflected by the presence of the R-motif, which contains several SUMO modification sites [10–12]. The TAD is the target of the MAPK pathways, and phosphorylation causes its activation and an overall enhancement of the transcriptional activation activity of Elk-1. In contrast, both the ETS domain and the R-motif exhibit distinct repressive properties.

Elk-1 and the other TCFs play an important role in controlling the transient expression of immediate-early genes such as c-*fos* following cellular stimulation and MAPK pathway activation (reviewed in [7–9]). This gene activation involves three distinct stages: the target genes are initially inactive, followed by a rapid activation phase, and subsequently down-regulation back to basal levels. Elk-1 plays a role in controlling all three of these phases. Biologically, the roles of the

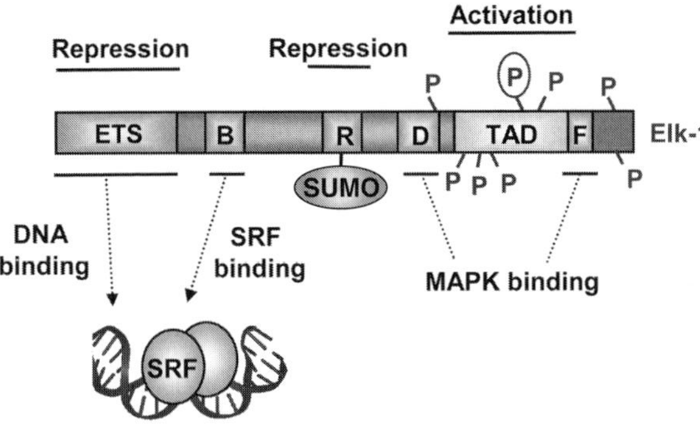

Figure 1 Domain structure of Elk-1. The ETS domain and B-box are involved in promoter recruitment by binding to DNA and the transcription factor SRF respectively. The D and F domains constitute a MAPK binding module. The locations of MAPK phosphorylation sites (P) are shown, and the critical residue Ser^{383} within the TAD is circled. Elk-1 is sumoylated within the R-motif repression domain. The areas shown to be involved in either the transcriptional repressive or activating activities associated with Elk-1 are indicated.

TCFs are currently unclear due to potential functional redundancy, typified by the minimal phenotype observed in the *elk-1* knockout mouse [13]. However, it is likely that these and other phenotypes, such as the T cell defects seen in *sap-1* knockout animals, will be compounded in double knockouts where *elk-1* and *sap-1* are deleted together [14].

Regulation of Elk-1 transactivation by MAPK-mediated phosphorylation

Elk-1 can be targeted and phosphorylated in response to activation of the three major MAPK pathways, i.e. ERK, JNK and p38 (reviewed in [7–9]). Elk-1 is phosphorylated at multiple sites in its C-terminal region; of these, Ser^{383} has been shown to play a critical role in enhancing transactivation. Phosphorylation of Elk-1 triggers a number of sequential molecular events that lead first to its activation and then to its subsequent inactivation (Figure 2). TCFs such as Elk-1 are apparently constitutively bound in a complex with SRF to promoters found in genes such as c-*fos in vivo* [15]. Thus it is assumed that the transactivation potential of Elk-1 is controlled while engaged with the promoter, although an exchange mechanism or rapid dynamic dissociation/reassociation with the promoter cannot be excluded.

Upon phosphorylation, Elk-1 undergoes a conformational change, which in turn leads to changes in contacts with pre-bound p300/CBP [CREB (cAMP response element-binding protein)-binding protein] and the activation of the

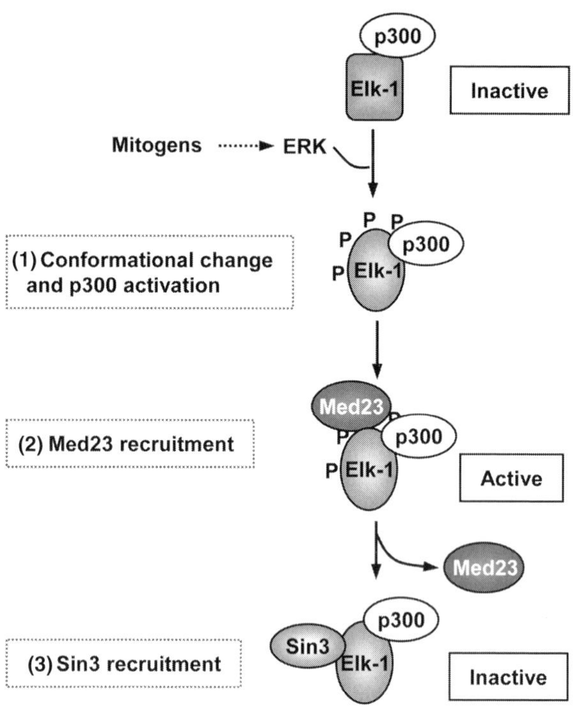

Figure 2 ERK-mediated Elk-1 activation pathway. Elk-1 is constitutively bound to p300/CBP. ERK-mediated phosphorylation causes a conformational change in Elk-1 and a change in p300 interactions. Med23 is then recruited to achieve full activation of the transactivation capacity of Elk-1. Subsequently, Med23 is lost, Sin3 is recruited and Elk-1 is dephosphorylated to return it to an inactive state.

HAT (histone acetyltransferase) activity of this co-activator (Figure 2) [16,17]. Elk-1 phosphorylation also leads to the recruitment of the mediator complex by interaction with the Med23/Sur2 subunit (Figure 2) [18]. Mediator recruitment enhances RNA polymerase II promoter engagement but, importantly, also increases the rate of transcriptional initiation by RNA polymerase II [19]. It is currently not clear whether p300/CBP activation and mediator recruitment occur in a specific temporal manner, but the resulting increase in HAT activity and transcription initiation are thought to work together to promote Elk-1 activation and the rapid induction of its target genes. The subsequent down-regulation of Elk-1 activity is accompanied by recruitment of the co-repressor Sin3 (Figure 2) [20], Med23 loss and dephosphorylation of its TAD. It is currently unclear how loss of Med23 is controlled, but the dephosphorylation of Elk-1 appears to be a likely trigger. Current evidence suggests that p300/CBP appears to be engaged with the promoter throughout the activation/inactivation process, and thus seems to be maintained on the promoter in an inactive state once Elk-1 returns to its basal state [17].

Down-regulation of Elk-1 activity by the SUMO pathway

The recruitment of the Sin3 co-repressor appears to be a specific event which follows phosphorylation-dependent activation of Elk-1 and is responsible for the shutdown phase of the Elk-1 activation process. However, Elk-1 must be kept inactive to prevent inappropriate triggering of target gene expression in the absence of mitogenic signalling. A clue as to how this might happen was provided by the observation that sumoylation of Elk-1 within the R-motif reduces its transactivation capacity [11]. Subsequently, sumoylation was also shown to reduce the rate of Elk-1 nucleocytoplasmic shuttling [12]. This transcriptional repression is mediated, at least in part, by the SUMO-dependent recruitment of HDAC-2 (histone deacetylase-2) to Elk-1-regulated promoters (Figure 3A) [21]. Importantly, Elk-1 sumoylation is not a constitutive modification, but is regulated in response to ERK MAPK signalling. Upon activation of the ERK pathway, Elk-1 sumoylation is lost and hence de-repression occurs (Figures 3A and 3B) [11]. This de-repression is accompanied by phosphorylation-dependent increases in transcriptional activation, leading to a two-step activation mechanism, which allows Elk-1 to go from a repressed state, through a de-repressed intermediate, to a fully active state (Figure 3B). Part of this active state is likely to result from an alteration of the balance between HAT and HDAC activities associated with Elk-1, as decreased HDAC-2 activity and increased p300/CBP HAT activity are triggered by ERK MAPK pathway activation.

PIASxα facilitates the de-repression of Elk-1 activity

The sumoylation pathway consists of a series of steps involving proteins that play analogous roles in the ubiquitin pathway (reviewed in [6]). The E1 SUMO-activating enzyme, SAE1/2, and an E2 SUMO-conjugating enzyme, Ubc9, constitute the basic pathway that is sufficient for substrate modification by SUMO (Figure 4A). However, E3 ligases are thought to play important roles in enhancing substrate sumoylation *in vivo*. One of the PIAS [protein inhibitor of activated STAT (signal transducer and activator of transcription)] family, PIASxα, is an example of such a protein [22]. However, recent studies identified PIASxα as a co-activator protein for Elk-1 rather than as a co-repressor, as might be expected if it caused enhanced Elk-1 sumoylation [23]. PIASxα appears to activate Elk-1 activity in an E3 ligase-independent manner, but only does so if the SUMO acceptor sites in Elk-1 are intact. The molecular explanation for this is that the co-activator function of PIASxα is due to it facilitating SUMO and HDAC-2 loss from Elk-1 in response to activation of the ERK pathway (Figure 4B). Furthermore, p300/CBP is also SUMO-modified and, as observed with Elk-1, this modification also imparts HDAC-dependent repressive properties

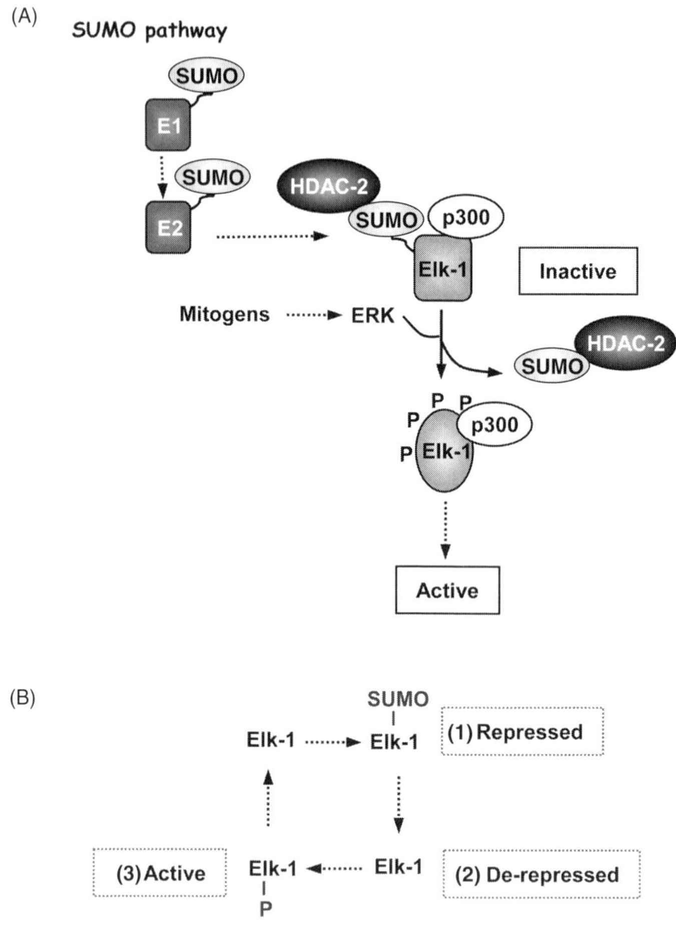

Figure 3 ERK pathway regulation of Elk-1 sumoylation. (**A**) Elk-1 is maintained in an inactive form through modification by the SUMO pathway. This in turn permits recruitment of the repressive HDAC-2. Activation of the ERK pathway leads to the loss of SUMO and associated HDAC-2, and triggers the subsequent phosphorylation-dependent Elk-1 activation steps. (**B**) Interconversion of Elk-1 between different states of activation. Sumoylated Elk-1 is inactive (1), whereas desumoylated Elk-1 is de-repressed (2). Phosphorylation of Elk-1 then coverts Elk-1 into a transcriptionally active state (3). It is currently unclear how Elk-1 re-achieves its sumoylated repressed state.

[24]. It appears that PIASxα also plays a role in promoting HDAC loss from p300, thereby potentiating its effect on Elk-1 activation [23]. Thus PIASxα plays a pivotal role in the ERK-mediated de-repression process and is an integral part of the activation pathway that converts Elk-1 from a transcriptional repressor into a transcriptional activator protein.

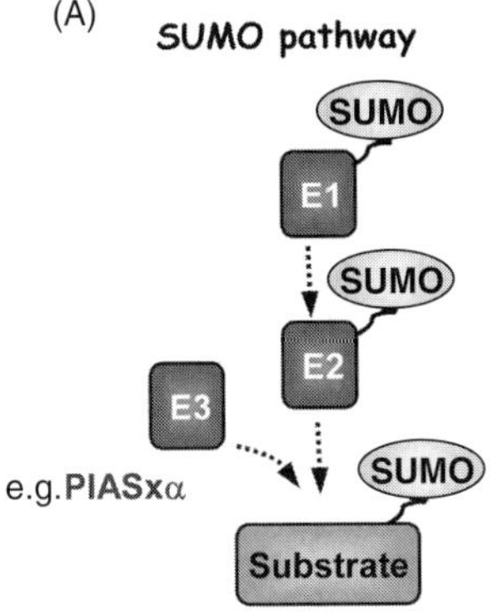

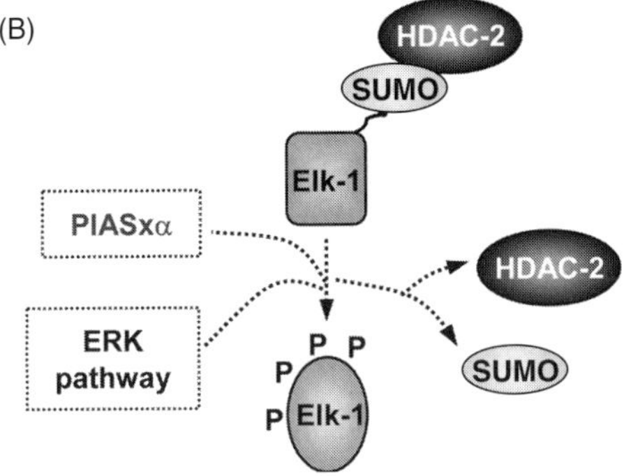

Figure 4 PIASxα acts in an E3-dependent manner to enhance substrate sumoylation and as an E3-independent Elk-1 co-activator. (**A**) The core SUMO pathway consists of the E1 (SAE1/2) and E2 (Ubc9) enzymes. E3 ligases such as PIASxα enhance Ubc9-mediated SUMO conjugation on to substrate proteins. (**B**) PIASxα as an Elk-1 co-activator. PIASxα is required for the loss of SUMO and HDAC-2 from Elk-1 upon stimulation of the ERK MAPK pathway and thus de-repression of Elk-1 activity.

Conclusions and perspectives

We now have a reasonable understanding of the main molecular events that occur during the conversion of Elk-1 from an inert transcription factor into a highly active transcriptional activator. In particular, we now know how the MAPK and SUMO pathways interact to control the transcriptional activation potential of Elk-1. This in turn allows us to understand how signals can be transmitted through the MAPK pathways to permit a specific nuclear response via a specific transcription factor complex. However, despite these advances, there are still many key mechanistic questions to be answered, and undoubtedly additional important regulatory proteins to be identified.

First, this review has focused on Elk-1, but two additional highly related TCF transcription factors, SAP-1 and SAP-2/Net, exist. It is already clear that although superficially very similar, they are controlled in distinct manners. For example, all three are targets of the ERK pathway, yet they exhibit a hierarchy in their transcriptional activation potentials, with Elk-1 being the most and SAP-2/Net the least potent (reviewed in [9]). Indeed, it is unclear whether SAP-1 and SAP-2/Net function through phosphorylation-dependent mediator recruitment in an analogous manner to Elk-1. In addition, while both Elk-1 and SAP-2/Net have been shown to be targets of the SUMO pathway, SAP-2/Net is modified through a different repression domain not found in Elk-1 [the NID (net inhibitory domain)] [25]. It is currently unclear how this modification is regulated, and whether it also functions through promoting HDAC-mediated transcriptional repression. Future studies will help to reveal how different TCFs are controlled at the molecular level.

Secondly, we have focused mainly on the output of the ERK MAPK pathway. Although Ser^{383} plays a critical role in the activation of Elk-1 in response to all MAPK pathways, the overall pattern of phosphorylation differs, and it is not clear whether this has any additional consequences (reviewed in [7–9]). Furthermore, the interplay between the stress-activated MAPK pathways and Elk-1 sumoylation is currently unknown. Similarly, the subtleties of activation of the other TCFs by different MAPK pathways have not been studied in detail. Thus, although some functional connections can probably be extrapolated from one transcription factor or MAPK pathway to another, it is dangerous to assume that all combinations will produce the same result. Indeed, it is entirely possible that the response to mitogenic or stress-activated MAPK pathways might be modified by additional proteins that bind to Elk-1 and are themselves differentially influenced by these pathways, thereby permitting specific readouts to be achieved.

Finally, we are only just beginning to unravel the role and mechanism of SUMO pathway modification of Elk-1 function (and indeed other TCFs). For example, we do not know if other co-repressors are involved in addition to HDAC-2. We also do not know how the ERK pathway causes desumoylation of Elk-1, or how PIASxα facilitates this process. Similarly, it is unclear how the SUMO-modified state of Elk-1 is established, and why this is apparently present at substoichiometric levels and yet has profound functional consequences. The latter problem is a conundrum that is faced by the majority of researchers in the SUMO field, and remains one of the major challenges that will be the subject of future studies.

The work in the authors' laboratory on Elk-1 is currently supported by the Wellcome Trust and the AICR.

References

1. Raman, M. and Cobb, M.H. (2003) Curr. Biol. **13**, R886–R888
2. Roux, P.P. and Blenis, J. (2004) Microbiol. Mol. Biol. Rev. **68**, 320–344
3. Yang, S.H., Sharrocks, A.D. and Whitmarsh, A.J. (2003) Gene **320**, 3–21
4. Verger, A., Perdomo, J. and Crossley, M. (2003) EMBO Rep. **4**, 137–142
5. Gill, G. (2004) Genes Dev. **18**, 2046–2059

6. Hay, R.T. (2005) Mol. Cell **18**, 1–12
7. Sharrocks, A.D. (2002) Biochem. Soc. Trans. **30**, 1–9
8. Shaw, P.E. and Saxton, J. (2003) Int. J. Biochem. Cell Biol. **35**, 1210–1226
9. Buchwalter, G., Gross, C. and Wasylyk, B. (2004) Gene **324**, 1–14
10. Yang, S.H., Bumpass, D.C., Perkins, N.D. and Sharrocks, A.D. (2002) Mol. Cell. Biol. **22**, 5036–5046
11. Yang, S.H., Jaffray, E., Hay, R.T. and Sharrocks, A.D. (2003) Mol. Cell **12**, 63–74
12. Salinas, S., Briancon-Marjollet, A., Bossis, G., Lopez, M.A., Piechaczyk, M., Jariel-Encontre, I., Debant, A. and Hipskind, R.A. (2004) J. Cell Biol. **165**, 767–773
13. Cesari, F., Brecht, S., Vintersten, K., Vuong, L.G., Hofmann, M., Klingel, K., Schnorr, J.J., Arsenian, S., Schild, H., Herdegen, T. et al. (2004) Mol. Cell. Biol. **24**, 294–305
14. Costello, P.S., Nicolas, R.H., Watanabe, Y., Rosewell, I. and Treisman, R. (2004) Nat. Immunol. **5**, 289–298
15. Herrera, R.E., Shaw, P.E. and Nordheim, A. (1989) Nature (London) **340**, 68–70
16. Yang, S.H., Shore, P., Willingham, N., Lakey, J.H. and Sharrocks, A.D. (1999) EMBO J. **18**, 5666–5674
17. Li, Q.-J., Yang, S.-H., Maeda, Y., Sladek, F.M., Sharrocks, A.D. and Martins-Green, M. (2003) EMBO J. **22**, 281–291
18. Stevens, J.L., Cantin, G.T., Wang, G., Shevchenko, A., Shevchenko, A. and Berk, A.J. (2002) Science **296**, 755–758
19. Wang, G., Balamotis, M.A., Stevens, J.L., Yamaguchi, Y., Handa, H. and Berk, A.J. (2005) Mol. Cell **17**, 683–694
20. Yang, S.H., Vickers, E., Brehm, A., Kouzarides, T. and Sharrocks, A.D. (2001) Mol. Cell. Biol. **21**, 2802–2814
21. Yang, S.H. and Sharrocks, A.D. (2004) Mol. Cell **13**, 611–617
22. Kotaja, N., Karvonen, U., Janne, O.A. and Palvimo, J.J. (2002) Mol. Cell. Biol. **22**, 5222–5234
23. Yang, S.-H. and Sharrocks, A.D. (2005) EMBO J. **24**, 2161–2171
24. Girdwood, D., Bumpass, D., Vaughan, O.A., Thain, A., Anderson, L.A., Snowden, A.W., Garcia-Wilson, E., Perkins, N.D. and Hay, R.T. (2003) Mol. Cell **11**, 1043–1054
25. Wasylyk, C., Criqui-Filipe, P. and Wasylyk, B. (2005) Oncogene **24**, 820–828

Biochem. Soc. Symp. **73**, 131–140
(Printed in Great Britain)

13

Non-coding RNA in transcription initiation

William O'Gorman, Kon Yew Kwek, Benjamin Thomas and Alexandre Akoulitchev[1]

Sir William Dunn School of Pathology, University of Oxford, South Parks Road, Oxford OX1 3RE, U.K.

Abstract

Diverse classes of non-coding RNAs, including snRNAs (small nuclear RNAs), play fundamental regulatory roles in gene expression. For example, 7SK RNA and the components of the splicing apparatus U1–U6 snRNAs are implicated in the regulation of transcriptional elongation. The first evidence for the involvement of RNA in the regulation of transcriptional initiation is now emerging. TFIIH (transcription factor IIH), a general transcription initiation factor, appears to associate specifically with U1 snRNA, a core splicing component. Reconstituted transcription *in vitro* demonstrates an increase in the rate of formation of the first phosphodiester bond by RNA polymerase II in presence of U1 snRNA. Reconstituted re-initiation is also stimulated by U1 snRNA. These results suggest that U1 snRNA functions in the regulation of transcription by RNA polymerase II in addition to its role in RNA processing. The implications of these data extend to the development of new technologies that will allow the identification and analysis of diverse RNA species present as regulatory components in transcription-related ribonucleoprotein complexes.

Non-coding RNAs in transcription

A number of genetic phenomena in higher eukaryotes reflect the signalling and regulatory role of RNA. These include RNAi (RNA interference), transgene silencing, position effect variegation, imprinting, DNA methylation, X-chromosome dosage compensation and co-suppression [1–8]. Interestingly, non-coding RNAs, particularly snRNAs (small nuclear RNAs), appear to play a fundamental role in regulating transcription by RNA polymerase II and processing of the transcripts (Figure 1) [9]. For example, P-TEFb (positive

[1]To whom correspondence should be addressed (email alexandre.akoulitchev@path.ox.ac.uk).

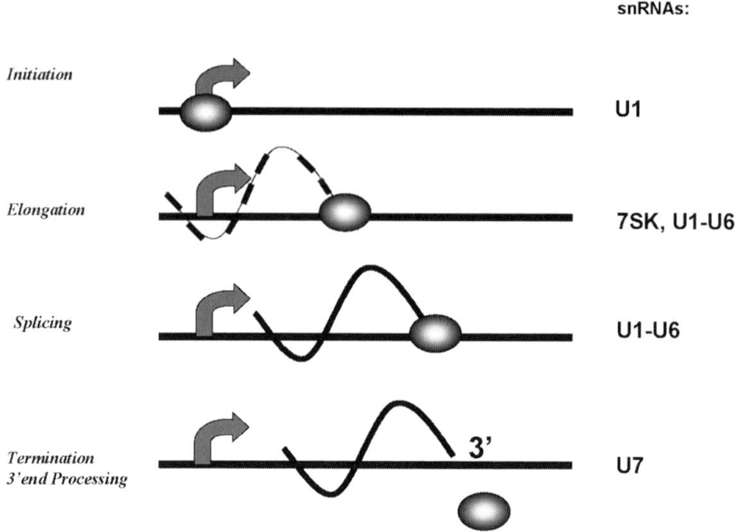

Figure 1 snRNAs in transcription by RNA polymerase II. Examples of snRNAs implicated in different stages of transcription and transcription coupled processing are shown.

transcription elongation factor b) consists of CDK9 (cyclin-dependent kinase 9) and cyclin T. CDK9 hyperphosphorylates the CTD (C-terminal domain) and controls transcription elongation [10]. Recently it has been shown that P-TEFb kinase activity can be modulated by an interaction with the non-coding RNA 7SK and the HEXIM (MAQ1) polypeptide [11,12]. As another example, U1–U6 snRNAs are implicated in splicing and also in elongation of transcription, while U7 snRNA directs the 3′ processing of histone mRNA [11–14].

Transcription initiation

Initiation of transcription by RNA polymerase II requires a minimal set of general transcription factors, and can be reconstituted *in vitro* with a minimal set of highly purified or recombinant components – the general transcription factors TFIID (transcription factor IID), TFIIB, TFIIE, TFIIF and TFIIH [15]. In such a minimal system, initiation is accurate, but remains largely insensitive to activation. In addition, this system does not support the re-initiation of transcription. Of the general transcription factors mentioned, TFIIH is of particular interest for this review. It is a general transcription factor that is implicated not only in the regulation of RNA polymerase II-mediated transcription [15–17], but also in nucleotide-excision repair and transcription-coupled repair (Figure 2) [18–20]. Functionally, TFIIH is involved in transcription at the stages of initiation [21,22], promoter escape and early elongation [23–26], formation of the re-initiation scaffolding [27] and, potentially, the formation of the gene loops [28]. Structurally, TFIIH is composed of six core subunits and a tripartite

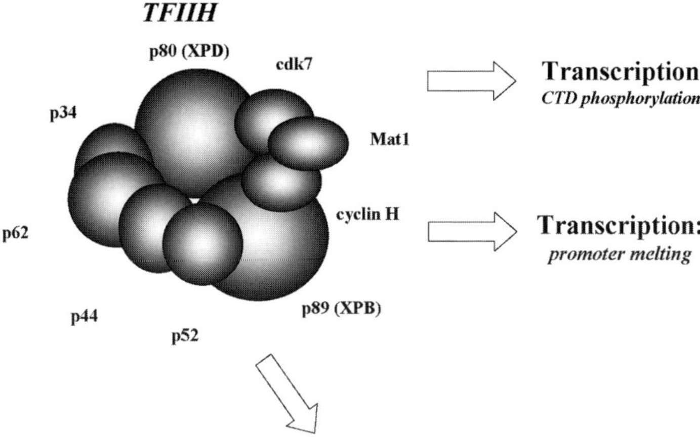

Figure 2 The general transcription factor TFIIH. The composition and multiple functions of TFIIH are shown.

kinase subcomplex, CAK (CDK-activating kinase) [29–32]. The CAK (CDK7/cyclin H/MAT1)-associated CDK7 phosphorylates the RNA polymerase II large-subunit CTD [33,34]. The phosphorylation of the CTD is a regulatory event implicated in transcription initiation, elongation [35,36] and the coupling of transcription to pre-mRNA processing [14,37,38].

An RNA associates with TFIIH

In order to determine whether any non-coding RNAs play a role in the initiation of transcription, general transcription factors can be purified and tested for potential association with RNA. This can be done by labelling with T4 RNA ligase in the presence of [5′-^{32}P]pCp[2]. Such an approach identified a distinct product of approx. 160 nucleotides present in both conventionally and affinity-purified general transcription factor TFIIH [39–41]. Treatment with RNase A was used to demonstrate the nature of the labelled nucleic acid as RNA, not DNA [41]. Following the cloning and sequencing of the TFIIH-associated RNA, it was confirmed to be identical with the gene database entry no. J00318, i.e. mature, 3′-processed U1 snRNA. Several independent assays have been used to confirm further the association of U1 with TFIIH. First, an RNase protection assay with a probe complementary to nucleotides 12–75 of U1 snRNA confirmed that our preparations of TFIIH contained U1 snRNA [41]. Secondly, TFIIH was selected successfully via its interaction with U1. A biotinylated DNA oligonucleotide, complementary to the region of U1 comprising nucleotides 12–75, was incubated with purified TFIIH and was able to specifically precipitate the p62 and CDK7 subunits of TFIIH on streptavidin-coated magnetic beads. These subunits were precipitated only when using the anti-U1 oligonucleotide, but not a control oligonucleotide [41].

Effects of U1 snRNA on abortive initiation

The functional relevance of the interaction between the U1 snRNA and TFIIH was tested in TFIIH-dependent reconstituted abortive [42] and productive transcription assays. In abortive initiation assays on the AdML (adenovirus major late) promoter, RNA polymerase II advances to the second position from the start site in the presence of ATP and [α-^{32}P]CTP, but cannot proceed further in the absence of GTP, which is required for the third position. As a result, polymerase forms the first phosphodiester bond, and an ApC dinucleotide is released as a product of abortive cycles of transcription initiation. The levels of released dinucleotides reflect the initiation potential of the pre-initiation complex. In contrast, in the productive transcription assay, all four nucleoside triphosphates are added to the reaction and polymerase is able to proceed past the second nucleotide of the template. A small fraction of RNA polymerase II molecules eventually escape the promoter and advance along the template as a ternary complex in its elongation mode. TFIIH dissociates from elongating polymerase during the synthesis of the first 30–50 nucleotides of the transcript [15].

Titration of the TFIIH samples with and without U1 snRNA in the TFIIH-dependent productive transcription assay demonstrated partial loss of productive activity, while titration of TFIIH in the abortive initiation assay indicated that initiation is reduced by more than 10-fold in the absence of U1 [41]. The role of U1 in abortive initiation was confirmed further by digesting it directly by T1 RNase or micrococcal nuclease, and by complementation with synthetic U1 snRNA [41]. Neither *Escherichia coli* tRNA nor synthetic U2 snRNA had any effect in these experiments. These results strongly suggest that U1 snRNA specifically stimulates the rate of TFIIH-dependent abortive initiation.

The role of TFIIH in transcription is not limited to abortive initiation. TFIIH has also been implicated in controlling promoter escape by RNA polymerase II [23,24], as well as productive re-initiated transcription [27]. Hahn and co-workers [27] demonstrated that, upon initiation of the first round of transcription, a TFIIH-dependent re-initiation scaffold is formed on immobilized templates, with an important role in the subsequent round of transcription. Interestingly, *in vivo* evidence links transcription efficiency and the positioning of the first intron, and in particular its splice donor site, which is recognized by U1 [43,44].

When tested in the reconstituted system, the efficiency of re-initiated transcription was dependent on the presence of U1 snRNA and the promoter proximal splice donor site (Figure 3) [41]. The β-globin intron sequences [45] containing splice donor and acceptor sites were introduced into the AdML promoter template, and the efficiency of the first and second rounds of transcription was monitored separately on the immobilized templates containing the wild-type and mutant donor/acceptor sites. While the first round of transcription displayed similar levels of transcription, the second round of transcription from the template containing the wild-type intron sequence supported up to 3-fold higher levels of transcription when compared with

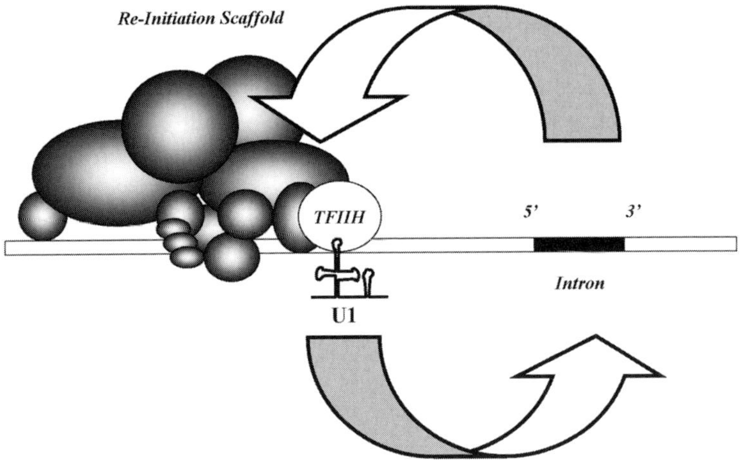

Figure 3 Promoter–proximal intron phenomenon. Shown are the re-initiation scaffold and feedback signalling via TFIIH and U1 snRNA to the promoter–proximal splice donor site and the composition and activity of the re-initiation scaffold [43,44,60].

the construct without the intron. Mutation of the donor site significantly (up to 3-fold) reduced this effect. Importantly, these high levels of transcription depended on U1 snRNA, as was shown by the targeted RNase H treatment of the re-initiation scaffold [46]. The observed effects of 3-fold regulation should be considered as significant, since the effect potentially accumulates with each round of transcription. The role of TFIIH as a re-initiation scaffold component and the effects of U1 snRNA and the splice donor site on the efficiency of re-initiation suggests that it could also be co-ordinated by the recruited components of the splicing machinery.

Analysis of the TFIIH–U1 interaction

In a further test for the specificity of the TFIIH–U1 complex, experiments with UV cross-linking established cyclin H as a subunit of TFIIH implicated in the interaction [41]. A standard approach to fine mapping of the interaction site on a polypeptide usually involves assaying a set of deletion mutants for their binding capacity. However, cyclin H constitutes an exceptional case because of its unique structural features. Previous work by Andersen et al. [46] has shown that the α-helices at the N- and C-termini of cyclin H come into a proximal position and interact as part of the functional structure. Deletion of or interference with these two helices has a drastic effect on the polypeptide structure and the activity of the kinase [46]. Therefore one requires a different method of mapping the protein–RNA interaction, without introducing any terminal deletions.

Taking into consideration the unique structural properties of cyclin H, a clear advantage is provided by the new approach to mapping protein–RNA interactions. It is based on electrospray-ionization tandem mass spectrometry, and is currently being tested in our laboratory. Following the cross-linking of the protein to RNA, the complex is subjected to tryptic digestion of the protein and affinity purification of RNA with the cross-linked peptide(s). In the final stage of the mapping, the peptide is identified by direct sequence analysis with electrospray-ionization tandem MS. The choice of cross-linking varies from UV to a chemical reversible cross-link. UV cross-linking can be carried out with significantly higher yields in the presence of nucleotide derivatives, such as 5-bromo-UTP and 5-iodo-UTP. However, one should take into account the fact that these derivatives potentially may affect the specificity and stability of the ribonucleoprotein complex. Another choice of cross-linking is chemical cross-linking, i.e. by formaldehyde. Aldehyde-based cross-linking has the advantage of reversibility and simplifies the identification of unmodified peptides. Affinity purification of RNA-containing products is an important step in this approach. Preliminary experiments demonstrate successful isolation of RNA-linked peptides with PHOS-Select Iron Affinity Gel (Sigma). We are currently conducting mapping experiments using this methodology in order to identify the site(s) of interaction of U1 with cyclin H.

In fact, the described interaction of cyclin H with U1 has an important precedent. Another transcriptional kinase of the same class, CDK9/cyclin T (P-TEFb), is involved in transcription elongation by phosphorylating the CTD of RNA polymerase II [49,50]. P-TEFb is also regulated via an interaction with non-coding RNAs. While the RNA stem–loop structure of the HIV transcript TAR, together with the Tat protein, is implicated in the activation of the P-TEFb kinase activity [51,52], the 7SK RNA–HEXIM1 complex is implicated in inactivation of this activity [11,12,53–55]. Even more importantly, in both the cases it appears that an important part in the interaction on the kinase side is played by the cyclin subunit, TAR interacts with cyclin T1 [56–59] and 7SK stabilizes the interaction of HEXIM with cyclin T [53].

The very nature of U1 snRNA as part of the splicing machinery opens up interesting possibilities for the molecular mechanism that underlies previously observed coupling between promoter strength *in vivo* [43,44,60], high levels of abortive initiation and re-initiation *in vitro* [41], and the promoter proximal position of the first intron. U1 snRNA comprises three stem–loop structures in a clover leaf conformation, followed by a 3′ single-stranded region and a fourth 3′ stem–loop [47]. Previous studies on the multiple protein interactions of U1 clearly indicated the importance of the various stem–loop structures for the specificity of all of the interactions [48]. Standard assembly of the U1 snRNP first involves recognition of U1 snRNA by the SMN (survival of motor neurons) complex [61,62]. This is followed by deposition of the Sm heptameric protein ring, hypermethylation of the snRNA 5′ cap, 3′ trimming and processing for nuclear import [48]. Once in the nucleus, the U1 snRNP contains three other specific proteins: the U1-70k protein interacts with SL1 [63], the U1-A protein interacts with the loop of SL2 [64,65], and the U1-C protein is probably attached via protein–protein interactions [66]. The three-dimensional arrangement of the U1 snRNP complex has been determined by single-particle electron cryomicroscopy [67].

Despite all of the functional biochemical data concerning the TFIIH–U1 interaction, our current understanding of the steps of assembly and interaction *in vivo* is very limited. Undoubtedly, this constitutes the most important question at this point. Interestingly, the recognition of U1 by SMN requires the same stem–loop 1 [61] that may be involved in the interaction with TFIIH (A. Akoulitchev, unpublished work). Also, analysis of the reconstituted re-initiation scaffold after a U1-dependent second round of transcription [41] reveals recruitment of the 70 kDa subunit of the U1 snRNP (A. Akoulitchev, unpublished work), which has also been identified in yeast two-hybrid analysis as a partner of Ioc4, responsible for release of RNA polymerase II into elongation [68]. Preliminary analysis of TFIIH 3′–5′ helicase activity, essential in promoter opening during initiation [22,24,25,69], also indicates that recruitment of U1 snRNP components via U1 can stimulate the helicase activity (A. Akoulitchev, unpublished work). It would therefore be very interesting to analyse the step-by-step assembly of the protein complex, which in the initial study [41] was only reflected in the conventionally purified TFIIH retaining its interaction with U1 snRNA. Of particular interest is the interaction with the SMN complex, which has also been shown to interact functionally with RNA polymerase II [70]. An important step would be to analyse the components of the U1 snRNP and potentially other polypeptides, recruited and retained in the transcriptional pre-initiation complex and re-initiation complex via interaction of TFIIH with U1 snRNA, by electrospray ionization MS.

Conclusions

TFIIH is a multi-subunit factor directly implicated in RNA polymerase II-mediated initiation [15,16,29]. Moreover, studies suggest that TFIIH plays an important role in the regulation of transcriptional re-initiation and is included in a re-initiation scaffold that forms at the promoter after the first round of transcription [27]. The nature of the TFIIH-associated RNA – U1 snRNA – provides a potential connection between the initiation of transcription and mRNA splicing [14]. The association of TFIIH with U1 stimulates the rate of initiation by RNA polymerase II. Moreover, in a re-initiation assay following the first round of transcription, the presence of the splice donor site increases the efficiency of re-initiation from the scaffold due to the presence of U1 and TFIIH. This result is particularly important in view of the earlier *in vivo* data linking the strength of eukaryotic promoters with the position of the promoter-proximal intron [43,44]. The well documented features of U1 provide an interesting insight into possible functional interconnections between transcription initiation and splicing. Interaction of the U1 snRNA with the 5′ donor splice site initiates assembly of U2, U4, U5 and U6 snRNAs and over 60 associated proteins into a spliceosome. More recent studies, however, indicate a more flexible order of assembly and a more diverse role for U1 [71]. For instance, U1 snRNA associates with a distinct, as yet uncharacterized, complex apart from the penta-snRNPs [72]. Notably, U1 snRNP can be dispensable for *in vitro* splicing when splicing regulatory proteins are present [73]. Splicing regulatory proteins have been reported to interact with the CTD of the largest subunit of RNA polymerase II, and therefore may also participate in transcriptional regulation [73,74]. All of these data suggested an additional function for U1 snRNA. Interestingly, in relation to transcriptional regulation, a selection of RNA ligands interfering with early stages of RNA polymerase II-mediated transcription gave rise to an 11-nucleotide consensus sequence complementary to U1 snRNA [75].

TFIIH plays a critical role during the initiation of transcription by RNA polymerase II, by controlling the early stages of promoter melting and formation of the open complex in an ATP-dependent manner [22,76,77]. In reconstituted systems, the levels of productive transcription are significantly reduced, or uncoupled from the levels of abortive initiation [23]. The introduction of auxiliary factors such as certain activators improves the co-ordination between these two processes, with a higher fraction of initiated RNA polymerases successfully clearing the promoter and advancing into elongation [23,78]. On the basis of the results from the abortive initiation and re-initiation assays, we would like to suggest that exposure to a promoter-proximal splice donor site via a TFIIH–U1 interaction alters the re-initiation scaffold, making initiated RNA polymerase II advance into productive elongation with higher efficiency. It remains to be elucidated whether, following the first round of transcription, the presence of the promoter-proximal intron alters the composition of the re-initiation scaffold. The detailed mechanism of the U1–TFIIH interaction and its effect on the enzymatic activities of TFIIH during initiation, promoter escape and re-initiation also remains to be determined. It will be important to identify, step by step, the interactions between the U1 snRNA and other components of the transcription and splicing machinery in the dynamic context of initiation and mRNA processing of the promoter-proximal intron.

Altogether, the results described in this chapter illustrate the dual role of U1 snRNA in the initiation of transcription and splicing. Earlier *in vivo* data demonstrated marked enhancement of transcriptional efficiency by the promoter-proximal introns [43,44]. Removal of promoter-proximal splice signals from a mammalian gene or the excision of introns from yeast genes results in a marked reduction of nascent transcription. Importantly, *in vivo* recognition of the proximal splice donor site by U1 snRNA appears to play an important role in the activation of transcriptional efficiency [60]. Altogether, these results emphasize further the importance of the interaction of U1 snRNA with TFIIH as a molecular mechanism underlying the co-ordinated control of efficient transcriptional initiation and early mRNA processing.

We thank Dr S. Murphy, Dr P. Cook, Dr O. Bensaude, Dr J. Manley, Dr D. Reinberg, Dr A. Furger and Dr N. Proudfoot for comments and discussions.

References

1. Judd, B.H. (1995) Genetics **141**, 245–253
2. Fire, A. (1999) Trends Genet. **15**, 358–363

3. Jones, D.O., Cowell, I.G. and Singh, P.B. (2000) Bioessays **22**, 124–137
4. Kelly, R.L. and Kuroda, M.I. (2000) Cell **103**, 9–12
5. Wassenegger, M. (2000) Plant Mol. Biol. **43**, 203–220
6. Morel, J.B., Mourrain, P., Beclin, C. and Vaucheret, H. (2000) Curr. Biol. **10**, 1591–1594
7. Sleutels, F., Barlow, D.P. and Lyle, R. (2000) Curr. Opin. Genet. Dev. **10**, 229–233
8. Sharp, P.A. (2001) Science **268**, 282–284
9. Eddy, S.R. (2001) Nat. Rev. Genet. **2**, 919–929
10. Price, D.H. (2000) Mol. Cell. Biol. **20**, 2629–2634
11. Nguyen, V.T., Kiss, T., Michels, A.A. and Bensaude, O. (2001) Nature (London) **414**, 322–325
12. Yang, Z., Zhu, Q., Luo, K. and Zhou, Q. (2001) Nature (London) **414**, 317–322
13. Fong, Y.W. and Zhou, Q. (2001) Nature (London) **414**, 929–933
14. Proudfoot, N.J., Furger, A. and Dye, M.J. (2002) Cell **108**, 501–512
15. Orphanides, G., Lagrange, T. and Reinberg, D. (1996) Genes Dev. **10**, 2657–2683
16. Coin, F. and Egly, J.M. (1998) Cold Spring Harbor Symp. Quant. Biol. **63**, 105–110
17. Reese, J.C. (2003) Curr. Opin. Genet. Dev. **13**, 114–118
18. Svestrup, J.Q. (2002) Nat. Rev. Mol. Cell. Biol. **3**, 21–29
19. Costa, R.M., Chigancas, V., Galhardo Rda, S., Carvalho, H. and Menck, C.F. (2003) Biochimie **85**, 1083–1099
20. van Hoffen, A., Balajee, A.S., van Zeeland, A.A. and Mullenders, L.H. (2003) Toxicology **193**, 79–90
21. Douziech, M., Coin, F., Chipoulet, J.M., Arai, Y., Ohkuma, Y., Egly, J.M. and Coulombe, B. (2000) Mol. Cell. Biol. **20**, 8168–8177
22. Kim, T.K., Ebright, R.H. and Reinberg, D. (2000) Science **288**, 1418–1422
23. Kumar, K.P., Akoulitchev, S. and Reinberg, D. (1998) Proc. Natl. Acad. Sci. U.S.A. **95**, 9767–9772
24. Moreland, R.J., Tirode, F., Yan, Q., Conaway, J.W., Egly, J.M. and Conaway, R.C. (1999) J. Biol. Chem. **274**, 22127–22130
25. Bradsher, J., Coin, F. and Egly, J.M. (2000) J. Biol. Chem. **275**, 2532–2538
26. Dvir, A., Conaway, J.W. and Conaway, R.C. (2001) Curr. Opin. Genet. Dev. **11**, 209–214
27. Yudkovsky, N., Ranish, J.A. and Hahn, S. (2000) Nature (London) **408**, 225–229
28. O'Sullivan, J.M., Tan-Wong, S.M., Morillon, A., Lee, B., Coles, J., Mellor, J. and Proudfoot, N.J. (2004) Nat. Genet. **36**, 1014–1018
29. Feaver, W.J., Svejstrup, J.Q., Henry, N.L. and Kornberg, R.D. (1994) Cell **79**, 1103–1109
30. Roy, R., Adamczewski, J.P., Seroz, T., Vermeulen, W., Tassan, J.P., Schaeffer, L., Nigg, E.A., Hoeijmakers, J.H. and Egly, J.M. (1994) Cell **79**, 1093–1101
31. Serizawa, H., Makela, T.P., Conaway, J.W., Conaway, R.C., Weinberg, R.A. and Young, R.A. (1995) Nature (London) **374**, 280–282
32. Shiekhattar, R., Mermelstein, F., Fisher, R.P., Drapkin, R., Dynlacht, B., Wessling, H.C., Morgan, D.O. and Reinberg, D. (1995) Nature (London) **374**, 283–287
33. Rossignol, M., Kolb-Cheynel, I. and Egly, J.M. (1997) EMBO J. **16**, 1628–1637
34. Yankulov, K.Y. and Bentley, D.L. (1997) EMBO J. **16**, 1638–1646
35. Palancade, B. and Bensaude, O. (2003) Eur. J. Biochem. **270**, 3859–3870
36. Hahn, S. (2004) Nat. Struct. Mol. Biol. **11**, 394–403
37. Howe, K.J. (2002) Biochim. Biophys. Acta **1577**, 308–324
38. Maniatis, T. and Reed, R. (2002) Nature (London) **416**, 499–506
39. Maldonado, E., Drapkin, R. and Reinberg, D. (1996) Methods Enzymol. **274**, 72–100
40. Le Roy, G., Drapkin, R., Weis, L. and Reinberg, D. (1998) J. Biol. Chem. **273**, 7134–7140
41. Kwek, K.Y., Murphy, S., Furger, A., Thomas, B., O'Gorman, W., Kimura, H., Proudfoot, N.J. and Akoulitchev, A. (2002) Nat. Struct. Biol. **9**, 800–805
42. Akoulitchev, S., Makela, T.P., Weinberg, R.A. and Reinberg, D. (1995) Nature (London) **377**, 557–560
43. Brinster, R.L., Allen, J.M., Behringer, R.R., Gelinas, R.E. and Palmiter, R.D. (1988) Proc. Natl. Acad. Sci. U.S.A. **85**, 836–840
44. Palmiter, R.D., Sandgren, E.P., Avarbock, M.R., Allen, D.D. and Brinster, R.L. (1991) Proc. Natl. Acad. Sci. U.S.A. **88**, 478–482
45. Efstratiadis, A., Posakony, J.W., Maniatis, T., Lawn, R.M., O'Connell, C., Spritz, R.A., DeRiel, J.K., Forget, B.G., Weissman, S.M., Slightom, J.L. et al. (1980) Cell **21**, 653–668
46. Andersen, G., Busso, D., Poterszman, A., Hwang, J.R., Wurtz, J.M., Ripp, R., Thierry,

J.C., Egly, J.M. and Moras, D. (1997) EMBO J. **16**, 958–967
47. Krol, A., Westhof, E., Bach, M., Luhrmann, R., Ebel, J.P. and Carbon, P. (1990) Nucleic Acids Res. **13**, 3803–3811
48. Will, C.L. and Luhrmann, R. (2001) Curr. Opin. Cell Biol. **13**, 290–301
49. Marshall, N.F. and Price, D.H. (1995) J. Biol. Chem. **270**, 12335–12338
50. Marshall, N.F., Peng, J., Xie, Z. and Price, D.H. (1996) J. Biol. Chem. **271**, 27176–27183
51. Herrmann, C.H. and Rice, A.P. (1995) J. Virol. **69**, 1612–1620
52. Zhu, Y., Pe'ery, T., Peng, J., Ramanathan, Y., Marshall, N., Marshall, T., Amendt, B., Mathews, M.B. and Price, D.H. (1997) Genes Dev. **11**, 2622–2632
53. Michels, A.A., Nguyen, V.T., Fraldi, A., Labas, V., Edwards, M., Bonnet, F., Lania, L. and Bensaude, O. (2003) Mol. Cell. Biol. **23**, 4859–4869
54. Yik, J.H., Chen, R., Nishimura, R., Jennings, J.L., Link, A.J. and Zhou, Q. (2003) Mol. Cell **12**, 971–982
55. Chen, R., Yang, Z. and Zhou, Q. (2004) J. Biol. Chem. **279**, 4153–4160
56. Garber, M.E., Wei, P., KewalRamani, V.N., Mayall, T.P., Herrmann, C.H., Rice, A.P., Littman, D.R. and Jones, K.A. (1998) Genes Dev. **12**, 3512–3527
57. Wei, P., Garber, M.E., Fang, S.M., Fischer, W.H. and Jones, K.A. (1998) Cell **92**, 451–462
58. Bieniasz, P.D., Grdina, T.A., Bogerd, H.P. and Cullen, B.R. (1999) Proc. Natl. Acad. Sci. U.S.A. **96**, 7791–7796
59. Zhang, J., Tamilarasu, N., Hwang, S., Garber, M.E., Huq, I., Jones, K.A. and Rana, T.M. (2000) J. Biol. Chem. **275**, 34314–34319
60. Furger, A., O'Sullivan, J.M., Binnie, A., Lee, B.A. and Proudfoot, N.J. (2002) Genes Dev. **16**, 2792–2799
61. Yong, J., Pellizzoni, L. and Dreyfuss, G. (2002) EMBO J. **21**, 1188–1196
62. Gubitz, A.K., Feng, W. and Dreyfuss, G. (2004) Exp. Cell Res. **296**, 51–56
63. Query, C.C., Bentley, R.C. and Keene, J.D. (1989) Mol. Cell. Biol. **9**, 4872–4881
64. Scherly, D., Boelens, W., van Venrooij, W.J., Dathan, N.A., Hamm, J. and Mattaj, I.W. (1989) EMBO J. **8**, 4163–4170
65. Lutz-Freyermuth, C., Query, C.C. and Keene, J.D. (1990) Proc. Natl. Acad. Sci. U.S.A. **87**, 6393–6397
66. Nelissen, R.L., Will, C.L., van Venrooij, W.J. and Luhrmann, R. (1994) EMBO J. **13**, 4113–4125
67. Stark, H., Dube, P., Luhrmann, R. and Kastner, B. (2001) Nature (London) **409**, 539–542
68. Fromont-Racine, M., Rain, J.C. and Legrain, P. (1997) Nat. Genet. **16**, 277–282
69. Tirode, F., Busso, D., Coin, F. and Egly, J.M. (1999) Mol. Cell **3**, 87–95
70. Pellizzoni, L., Charroux, B., Rappsilber, J., Mann, M. and Dreyfuss, G. (2001) J. Cell Biol. **152**, 75–85
71. Johnson, T.L. and Abelson, J. (2001) Genes Dev. **15**, 1957–1970
72. Stevens, S.W., Ryan, D.E., Ge, H.Y., Moore, R.E., Young, M.K., Lee, T.D. and Abelson, J. (2002) Mol. Cell **9**, 31–44
73. Crispino, J.D., Blencowe, B.J. and Sharp, P.A. (1994) Science **265**, 1866–1869
74. Hirose, Y. and Manley, J.L. (2000) Genes Dev. **14**, 1415–1429
75. Tian, H. (2001) FEBS Lett. **509**, 282–286
76. Holstege, F.C., Fiedler, U. and Timmers, H.T. (1997) EMBO J. **16**, 7468–7480
77. Tirode, F., Busso, D., Coin, F. and Egly, J.M. (1999) Mol. Cell **3**, 87–95
78. Liu, J., Akoulitchev, S., Weber, A., Ge, H., Chuikov, S., Libutti, D., Wang, X.W., Conaway, J.W., Harris, C.C., Conaway, R.C. et al. (2001) Cell **104**, 353–363

Biochem. Soc. Symp. **73**, 141–154
(Printed in Great Britain)

14

Activation by c-Myc of transcription by RNA polymerases I, II and III

Natividad Gomez-Roman*, Zoë A. Felton-Edkins*, Niall S. Kenneth*, Sarah J. Goodfellow*, Dimitris Athineos*, Jingxin Zhang*, Ben A. Ramsbottom*, Fiona Innes*, Theodoros Kantidakis*, Elaine R. Kerr*, Jacqueline Brodie*, Carla Grandori† and Robert J. White*[1]

*Institute of Biomedical and Life Sciences, Division of Biochemistry and Molecular Biology, University of Glasgow, Glasgow G12 8QQ, U.K. and †Division of Human Biology, Fred Hutchinson Cancer Research Center, Seattle, WA 98109-1024, U.S.A.

Abstract

The proto-oncogene product c-Myc can induce cell growth and proliferation. It regulates a large number of RNA polymerase II-transcribed genes, many of which encode ribosomal proteins, translation factors and other components of the biosynthetic apparatus. We have found that c-Myc can also activate transcription by RNA polymerases I and III, thereby stimulating production of rRNA and tRNA. As such, c-Myc may possess the unprecedented capacity to induce expression of all ribosomal components. This may explain its potent ability to drive cell growth, which depends on the accumulation of ribosomes. The activation of RNA polymerase II transcription by c-Myc is often inefficient, but its induction of rRNA and tRNA genes can be very strong in comparison. We will describe what is known about the mechanisms used by c-Myc to activate transcription by RNA polymerases I and II.

Introduction

The proto-oncogene c-*myc* is essential for embryonic development, as shown by its targeted disruption in mice [1,2]. Numerous biological activities have been attributed to its protein product, including the promotion of growth, cell cycle progression and apoptosis, as well as the suppression of differentiation

[1]To whom correspondence should be addressed (email rwhite@udcf.gla.ac.uk)

[3–11]. De-regulation of c-*myc* is associated with a broad range of cancer types [12]. For example, Burkitt's lymphomas are characterized by a chromosomal translocation that activates c-*myc* [12]. Overall, it has been estimated that c-Myc de-regulation contributes to one-seventh of all cancer deaths in the U.S.A. [5].

Despite being one of the first oncogenes discovered and the subject of intense study, c-*myc* has been described as "an enduring enigma" [9]. There is no doubt that it encodes a DNA-binding factor that regulates gene transcription. However, there is an "apparent gap between Myc's biological role and what is surmised to be its molecular function" [9]. Long lists of Myc target genes have been identified, but in most cases the effect of Myc on expression of these genes is weak and variable [4,9,10]. Indeed, Eisenman [9] complained that alongside a "mountain of biological effects, the molecular characterization of Myc – as a relatively weak transcriptional regulator of uncertain target genes – looks like a molehill". It is therefore generally agreed that identifying *bona fide* targets remains a priority for Myc research.

Transcription by pol III (RNA polymerase III) is potently activated by c-Myc

Studies of Myc had concentrated until recently exclusively on its ability to regulate transcription of protein-encoding genes by pol II. However, we found that c-Myc functions as a direct and potent activator of pol III transcription [13]. When introduced by retroviral transduction, a hydroxytamoxifen-inducible Myc–ER (oestrogen receptor) fusion gene [14] will rapidly and specifically activate pol III-dependent expression of tRNA and 5 S rRNA genes in primary human diploid fibroblasts. Activation of tRNA genes by Myc typically reaches ~12-fold or more and far exceeds that of the cyclin D2 gene, an established Myc target [14], which is induced in parallel by 3.6-fold. These effects are specific, since control genes are not induced [13]. Further evidence of specificity came from the deletion of residues 106–143 within the Myc transactivation domain, which prevents induction of both tRNA and cyclin D2 [13]. This deletion removes the highly conserved MBII (Myc box II) sequence (residues 126–144), which is essential for cell transformation [15].

The sets of genes that are induced by Myc overexpression can differ from those that respond to physiological levels of Myc [16]. However, the use of RNAi (RNA interference) and genetic knockouts showed that pol III transcription is extremely sensitive to endogenous c-Myc at physiological concentrations, in both human and rodent cells. Thus expression of pol III transcripts can be reduced by ~7-fold in c-*myc* knockout fibroblasts, and restored by introduction of exogenous c-Myc [13]. Similarly, expression of tRNA and 5 S rRNA falls substantially if RNAi is used to deplete HeLa cells of endogenous c-Myc [17,18].

The timing of tRNA induction following activation of Myc–ER parallels that of the cyclin D2 gene, which is known to be a direct target of c-Myc [14]. Thus tRNA expression increases within 3 h, whereas S phase entry takes ~17 h [14]. The alacrity of pol III activation suggests a direct effect of Myc.

This was confirmed using 2 μg/ml α-amanitin, which specifically blocks pol II transcription; any effect seen under these conditions cannot be a secondary response to induction of pol II-transcribed genes. Whereas α-amanitin prevented Myc activation of the cyclin D2 gene, expression of tRNA and 5 S rRNA was strongly induced [13]. This shows that c-Myc regulation of pol III templates does not require stimulation of protein-encoding genes. Unequivocal evidence of a direct effect came from ChIP (chromatin immunoprecipitation) studies, which showed that endogenous c-Myc associates with tRNA and 5 S rRNA genes in living fibroblasts and epithelial cells, but not with the pol II-transcribed control gene $p21^{Waf1}$ [13,17].

Although ChIP assays show clearly that endogenous c-Myc binds to tRNA and 5 S rRNA genes *in vivo*, these genes do not contain good matches to the E-box DNA sequence that is recognized by Myc. Instead, c-Myc appears to be recruited to these genes by protein–protein interactions with the pol III-specific factor TFIIIB (transcription factor IIIB) [13]. In glutathione S-transferase pull-down assays, TFIIIB binds specifically to the c-Myc transactivation domain, which is required for tRNA induction *in vivo* [13]. Furthermore, co-immunoprecipitation and co-fractionation experiments show that endogenous c-Myc associates stably and specifically with endogenous TFIIIB [13]. We believe that it is TFIIIB, rather than a DNA sequence, that attracts c-Myc to pol III-transcribed genes.

Pol I transcription is also a direct target for activation by c-Myc

It had been reported previously that levels of large rRNA respond to Myc. It was found that accumulation of large rRNA is 2.3-fold lower in c-Myc knockout fibroblasts compared with wild-type controls [19]. Furthermore, the knockout cells are unable to accumulate rRNA following serum stimulation [20]. A 45% increase in rRNA was seen following overexpression of N-Myc in neuroblastoma cells [21]. However, these studies did not determine if such changes are due to direct effects of Myc on pol I transcription; they might instead reflect changes in rRNA processing or stability, as genes involved in rRNA maturation also respond to Myc [14,16,21–25]. Indeed, c-Myc has been shown to stimulate processing of the rRNA precursor transcript into mature 5.8, 18 and 28 S rRNA [25]. That study overlooked indications that the pre-rRNA primary transcript is itself induced by Myc [25]. However, several recent papers have confirmed that this is the case [18,26–28].

Northern blotting, pulse labelling and RT-PCR (reverse transcription–PCR) analyses revealed that expression of the rRNA precursor transcript increases in response to Myc–ER induction in fibroblasts of human, rat or mouse origin, as well as in mouse granulocytes [18,26,27]. Additional evidence of transcriptional induction was obtained using nuclear run-on assays with human B cells carrying a tetracycline-inducible c-Myc construct [18]. Activation of rRNA gene transcription by c-Myc is impervious to the presence of 2 μg/ml α-amanitin [18,27], which confirms that the response is not an indirect one that is mediated through regulation of pol II-transcribed genes encoding components of the pol I machinery. This is consistent with the rapidity of pre-rRNA induction, which is observed within 3 h of c-Myc activation [18]. Additional proof that the effect of c-Myc is direct in this case comes from ChIP assays that demonstrate its presence at rRNA gene loci within living HeLa and B cells [18,27]. This is supported by immunofluorescence analyses of fibroblasts and COS-7 cells, which reveal the presence of c-Myc within nucleoli, the sites of pol I transcription *in vivo* [18,27]. Furthermore, nucleolar c-Myc co-localizes with pre-rRNA, as shown by fluorescence *in situ* hybridization (FISH) [18]. These combined data confirm that c-Myc acts directly to stimulate pol I transcription of rRNA genes in a range of human and rodent cell types.

Many of the above experiments involved overexpression of c-Myc. However, a role for endogenous c-Myc in regulating rRNA synthesis at physiological concentrations was demonstrated by RNAi [18,27]. Thus pol I transcription decreased significantly when siRNAs (small interfering RNAs) were used to reduce the concentration of endogenous c-Myc in the human cell lines U2OS and HeLa, as well as in primary human peripheral blood lymphocytes [18,27]. This effect was not seen using control siRNAs against lamin, GAPDH (glyceraldehyde-3-phosphate dehydrogenase) or luciferase [18,27]. Furthermore, the transcriptional response could be rescued by co-transfecting mRNA encoding chicken c-Myc, which is not recognized by the siRNAs [27]. These data establish that the rate of pre-rRNA synthesis by pol I is sensitive to the availability of endogenous c-Myc.

In contrast with most pol III-transcribed genes, the rRNA gene repeats contain multiple copies of the E-box DNA sequence that is known to be recognized by Myc [18,27]. Several are found within the vicinity of the promoter, a region that is occupied by c-Myc in both HeLa and B cells, as revealed by ChIP assays [18,27]. In addition, c-Myc also binds in the vicinity of a dense cluster of E-boxes downstream of the 28 S rRNA coding region [18]. However, it remains to be determined whether the E-boxes are entirely responsible for c-Myc recruitment to these sites. An indication that protein–protein interactions may also contribute comes from the observation that a c-Myc mutant that is defective in DNA binding is nevertheless retained in nucleoli, albeit less efficiently than the wild-type protein [27].

Genes for rRNA and tRNA are unusually good targets for activation by c-Myc

As mentioned in the Introduction, activation of pol II transcription by c-Myc is generally weak – typically less than 4-fold for most targets [4]. For example, Guo et al. [16] carried out a genomic screen with cDNA microarrays to identify genes that respond to c-Myc in Rat1 fibroblasts. They identified 96 protein-encoding genes that were consistently induced by >2-fold when c-Myc was restored to knockout cells and when wild-type cells were compared with the nulls; for these genes, the average induction was only 3.4-fold, and the strongest induction was 5.9-fold [16]. Coller et al. [14] used oligonucleotide microarray analysis to look for genes that were consistently induced when hydroxytamoxifen was added to primary human fibroblasts transduced with the Myc–ER construct. This approach identified 27 protein-encoding genes that were reproducibly induced by more than 2-fold; after averaging results from three identical experiments, the mean induction was 4.1-fold and the maximum was 8.9-fold [14]. Both of these studies used microarray approaches, but similar results have been obtained using other experimental techniques. For example, the gene for ODC (ornithine decarboxylase 1) is considered to be one of the best direct targets for c-Myc. In the microarray analysis by Coller et al. [14], its average induction by Myc–ER in human fibroblasts was 5.9-fold. Greasley et al. [29] tested the same gene in Rat1 fibroblasts, using nuclear run-on and Northern blot approaches; in six experiments, their Northern analyses showed an average induction of 2.3-fold in response to Myc–ER, while their run-on assays revealed a maximum ODC induction of 2.8-fold. The same group obtained comparable results using RT-PCR [30]. This Laodicean response does not reflect some intrinsic property of the ODC gene, because it was induced by over 40-fold when serum was added to the same Rat1 cells [30].

Using a combination of ChIP assays and high-density oligonucleotide arrays, it was estimated that the human genome may contain ~25 000 sites that are bound by c-Myc *in vivo* [31]. This compares with ~12 000 sites for Sp1 and 1600 sites for p53, as estimated using the same approach [31]. Of the ~25 000 c-Myc-binding sites, ~24% lie within CpG islands, which frequently encompass the promoter and first exon and intron of protein-encoding genes [31]. Indeed, 18% of the c-Myc sites were found within 1 kb of a gene's first exon [31]. Conversely, ~11% of all pol II promoters are associated with a high-affinity Myc-binding site [32]. Greasely et al. [29] conducted a screen to look for Myc-responsiveness among genes that carry consensus E-boxes within a CpG island. Of 123 such genes, only 13 showed increased expression by Northern and/or nuclear run-on analyses when Myc–ER was induced in Rat1 cells [29]. For these 13 genes, the average induction was only 2.5-fold [29].

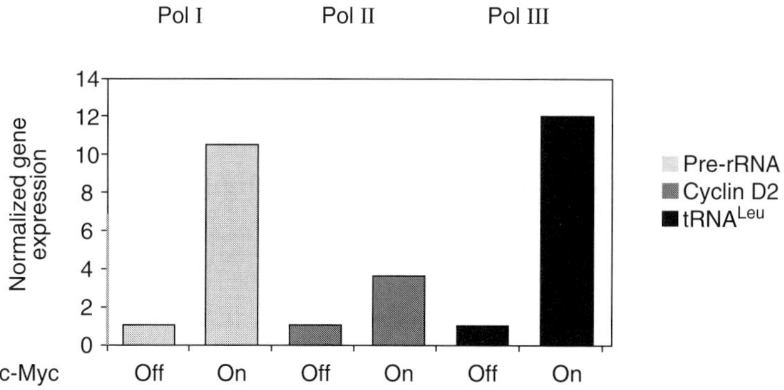

Figure 1 Changes in the relative levels of pre-rRNA, cyclin D2 mRNA and tRNA^{Leu} in response to induction of Myc–ER with hydroxytamoxifen in serum-starved human fibroblasts. Data were obtained by RT-PCR and normalized to the level of the mRNA encoding ARPP P0.

Occasional exceptions have been described. For example, Fernandez et al. [32] found a 43-fold induction in B cells of the TERT reverse transcriptase component of telomerase. Nevertheless, the consensus from many exhaustive studies makes it clear that the capacity of c-Myc to stimulate pol II transcription is "relatively wimpy" in most cases [9]. It was therefore striking that transcription by pols I and III shows very potent activation in similar assays. For example, induction of Myc–ER in primary human fibroblasts raised expression of pre-rRNA by 11-fold and expression of tRNA by 12-fold, after normalization against the mRNA for ARPP (acidic ribosomal phosphoprotein) PO [13,18]. For comparison, the same sets of samples were tested for activation of the pol II-transcribed cyclin D2 gene, one of the best characterized direct targets for c-Myc [14,29,33,34]; in this case, induction peaked at 3.6-fold (Figure 1).

As well as being extremely responsive to c-Myc overexpression, transcription by pols I and III is also very sensitive to levels of the endogenous protein. This is apparent in RNAi experiments, where partial depletion of endogenous c-Myc can cause a dramatic reduction in the expression of pre-rRNA and 5 S rRNA. When tested in parallel, cyclin D2 mRNA levels showed minimal change (Figure 2). Of the pol III-transcribed genes we have tested, almost all show hypersensitivity to c-Myc, including the $tRNA^{Leu}$, $tRNA^{Tyr}$, $tRNA^{Arg}$, 5 S rRNA, 7SL RNA and U6 snRNA (small nuclear RNA) genes. The only exception we have found is the

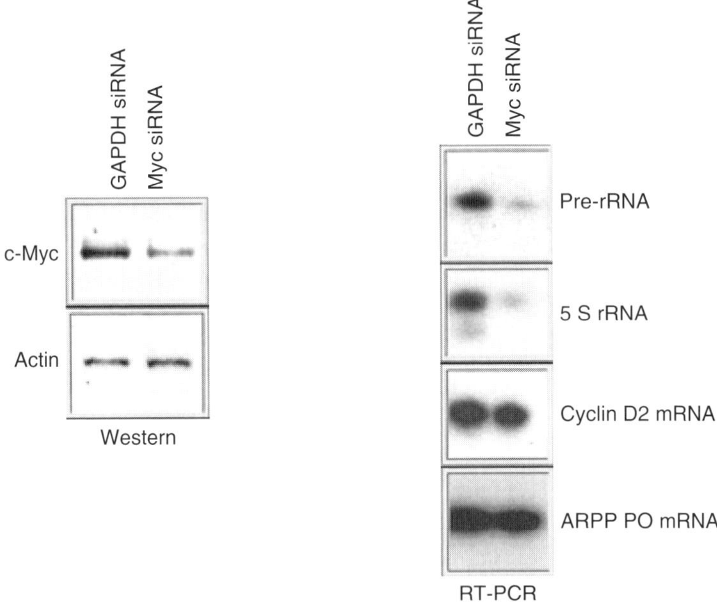

Figure 2 Changes in the relative levels of pre-rRNA, cyclin D2 mRNA and 5 S rRNA in response to partial knockdown of endogenous c-Myc. HeLa cells were transfected with siRNA against GAPDH (control) or c-Myc. Specific partial knockdown of c-Myc was confirmed by Western blot (left panel). Expression of the indicated transcripts was monitored by RT-PCR (right panel).

7SK RNA, which shows very little response. It is not yet clear why 7SK behaves differently in this respect, although it is known to be atypical in its promoter arrangement and transcription factor requirements [35–38].

Mechanisms of transcriptional activation by c-Myc

Most recent work on the mechanism(s) of transcriptional activation by c-Myc has focused on its ability to influence chromatin structure [4,9,39]. The N-terminal activation domain of c-Myc binds to the cofactor TRRAP (transformation/transactivation domain-associated protein) [40], a 400 kDa member of the ATM (ataxia telangiectasia mutated) family that forms complexes with the HATs (histone acetyltransferases) GCN5, P/CAF (p300/CBP-associated factor) and Tip60 [41,42]. These interactions allow c-Myc to recruit HAT activity to E-box sites, leading to localized acetylation of histones H3 and H4 [30,32,34,43]. Although there is a general correlation between histone acetylation and gene activation, there are cases in which transcriptional induction does not occur, despite a substantial increase in acetylated histones [30,32]. For example, Myc–ER in Rat1 cells induces an ~10-fold increase in histone H4 acetylation at the nucleolin gene without having any appreciable effect on its transcription [30]. This suggests that activation by c-Myc requires additional effects besides acetylation of histones. This might involve the remodelling of nucleosomes, since c-Myc has been shown to bind to the INI1/hSNF5 subunit of the SWI/SNF complex, as well as to the ATPase/helicases TIP48 and TIP49 [44,45]. In addition, a direct interaction has been demonstrated between c-Myc and TBP (TATA-binding protein) [46–48]. Since TBP recruitment can be sufficient to activate transcription *in vivo* [49,50], this interaction provides a plausible mechanism by which c-Myc could regulate expression of some of its targets (Figure 3).

It will be interesting to determine whether the induction of pol I and pol III transcription by c-Myc involves novel mechanisms or conforms with the paradigms that have been established for pol II. Pol III may prove to be

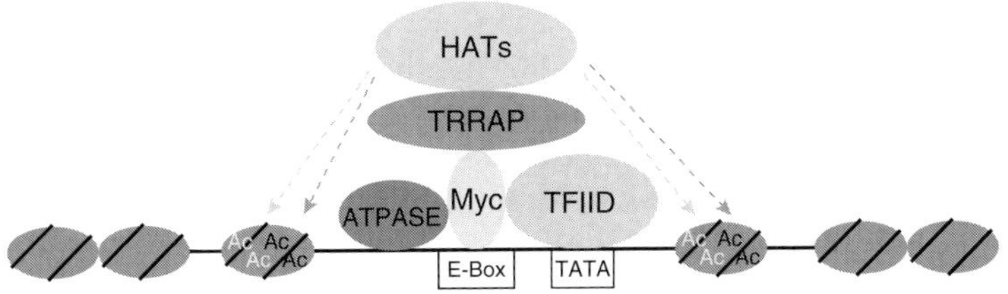

Figure 3 Model representing molecular mechanisms that may contribute to the activation of pol II transcription by c-Myc. These may include the recruitment of TFIID, via binding of c-Myc to TBP, ATP-dependent chromatin remodelling complexes, and TRRAP-containing HAT complexes leading to acetylation (Ac) of vicinal histones.

significantly different, since activation in this case appears to be independent of E-box motifs. The pol I situation seems much more familiar, in a number of respects. Here E-box sequences provide binding sites for c-Myc in the vicinity of the promoter, as well as elsewhere in the rRNA gene repeats [18,27]. Furthermore, c-Myc induction provokes marked acetylation of histones H3 and H4 at these loci in U2OS and B cells [18,27]. The increase in histone H4 acetylation is apparent around the rRNA gene promoter and termination sites, and is maximal ~1 kb upstream of the transcription start site and negligible in parts of the intergenic spacer [18,27]. However, H3 acetylation is much more localized, being strongly induced ~1 kb upstream of the start, but not at the promoter itself [18]. The significance of these changes is unclear. Psoralen cross-linking experiments have shown that the fraction of rRNA genes in an accessible chromatin configuration changes little when pol I transcription is activated by serum stimulation of Friend cells, a situation in which c-Myc is likely to contribute substantially [51]. Furthermore, we found that efficient pol I induction by c-Myc can be reproduced *in vitro* using a naked DNA template [18]. Thus nuclear extracts prepared at various times after hydroxytamoxifen addition to Myc–ER-containing fibroblasts showed increased pol I transcriptional activity that began within 90 min of Myc induction and reached ~14-fold after 7 h [18]. This robust activation in the absence of chromatin may be explained by the ability of c-Myc to bind to SL1, an essential component of the basal pol I machinery. SL1 binds to the rRNA gene promoter and is required for polymerase recruitment and transcription initiation [52–54]. Co-immunoprecipitation assays showed that it associates stably with endogenous c-Myc in human fibroblasts [18]. Pull-down assays further demonstrated that specific fragments of recombinant c-Myc can bind to an *in vitro*-translated SL1 subunit (TAF_I48, where TAF is TBP-associated factor), as well as native SL1 from HeLa cells [18]. Two distinct binding sites were detected: one in the C-terminal domain of c-Myc and the other within the MBII region of its transactivation domain [18]. The functional significance of the interaction between SL1 and c-Myc is suggested strongly by ChIP experiments, which demonstrate a substantial increase in SL1 occupancy of the rRNA gene promoter following induction of c-Myc in B cells [18]. On the basis of these observations, we suggest a model in which pol I transcriptional activation may be achieved by c-Myc binding to E-box sequences and recruiting SL1 to the promoter, thereby facilitating pre-initiation complex assembly. Histone acetylation accompanies this process, but its functional significance is unclear; it may assist rRNA gene activation *in vivo*, but is probably not essential, since robust induction can be mimicked *in vitro* in the absence of chromatin.

Components of the pol I and pol III transcription machinery can be induced by Myc

In addition to its direct effects on SL1 and TFIIIB, c-Myc can also augment a cell's capacity for rRNA and tRNA synthesis by activating class II genes encoding certain components of the pol I and pol III transcription machineries.

This was first demonstrated for the *BN51* gene, which encodes a subunit of pol III [29]. The *BN51* gene is occupied by endogenous c-Myc in a range of cell types, including human fibroblasts [32]. It contains a single consensus E-box within its first intron, a location favoured in many c-Myc targets [29]. A reporter consisting of the BN51 promoter, first exon and first intron fused to the luciferase coding region can be induced by 3-fold if c-Myc is transfected into Rat1 or 293T cells [29]. Conversely, repression was observed following transfection of the Myc antagonist Mad1 [29]. Substitution of the E-box (CACGTG→TACGTG) was sufficient to abolish these responses [29]. When the Myc–ER construct was induced by hydroxytamoxifen in Rat1 cells, transcription of the endogenous *BN51* gene increased by 2.2-fold, as assessed by nuclear run-on, whereas Northern analysis showed a 2.5-fold increase in BN51 mRNA [29]. Although these data provide convincing evidence that the *BN51* gene is regulated by c-Myc in some situations, it is not induced appreciably by Myc–ER in serum-starved primary human fibroblasts [13]; perhaps in this case an alternative factor is limiting. Genetic data point to the significance of BN51 for cell proliferation, since a temperature-sensitive BN51 mutation that compromises pol III activity results in G_1 arrest at the non-permissive temperature in a hamster cell line [55–57]. Furthermore, G_1 arrest is also caused in *Saccharomyces cerevisiae* by a temperature-sensitive mutation in the yeast homologue of BN51 [58]. These observations led Greasley et al. [29] to speculate that reduced levels of pol III may contribute to the slow growth and proliferation and prolonged G_1 and G_2 phases of c-Myc knockout Rat1 cells.

As well as this key pol III subunit, c-Myc has also been shown to induce expression of the gene encoding the pol I transcription factor UBF (upstream binding factor) [26]. E-box sequences are present in the first intron and promoter of the UBF gene, as well as further upstream, and ChIP assays show the association of endogenous c-Myc with these regions *in vivo* [26]. Furthermore, induction of Myc–ER in NIH 3T3 cells causes a rapid 2.7-fold increase in UBF mRNA [26]. Mad1 binds to the same regions as c-Myc and can repress UBF expression [26]. Deletion of the distal E-boxes abrogated regulation of a transfected UBF reporter in response to c-Myc or Mad1 [26]. These data provide a convincing case for the UBF gene as a direct target for the Myc/Mad family. Regulation of UBF can have a substantial impact on the rate of pol I transcription [52,53,59–61]. UBF induction may therefore re-inforce the direct effect of c-Myc on SL1 to ensure high rates of rRNA synthesis. That this is the case in NIH 3T3 cells is suggested by the finding that the activation of pol I transcription by c-Myc can be attenuated by using a specific siRNA to prevent induction of UBF [26].

In *Drosophila*, Myc induces genes encoding subunits of pols I and III, as well as the pol I-specific transcription factor TIF-IA (transcription intermediary factor-IA) [28]. Indeed, of the ten genes tested that encode components of the pol I or pol III machinery, eight were activated by Myc [28]. TIF-IA is believed to be a key control point in regulating pol I output that is conserved throughout evolution [52,53,61]. Overexpression of TIF-IA alone is sufficient to increase pre-rRNA expression in *Drosophila* imaginal wing discs [28]. Accordingly, Myc can stimulate pre-rRNA synthesis in wing discs and throughout third instar larvae, an effect that correlates with nucleolar expansion [28]. In contrast with

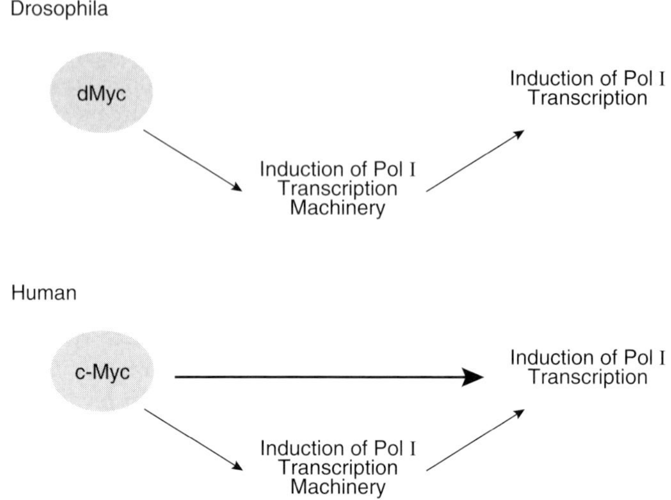

Figure 4 Stimulation of pre-rRNA synthesis by Myc. *Drosophila* Myc (dMyc) stimulates pre-rRNA synthesis indirectly by inducing pol II transcription of genes encoding components of the pol I machinery. The same is true in mammals, but here c-Myc also has a rapid, potent and direct effect upon the process of pol I transcription.

mammalian systems, the induction in fruitflies seems to be entirely indirect. Thus *Drosophila* rRNA genes do not contain E-boxes and are not bound by Myc *in vivo* [28]. Instead, pol II transcription of genes encoding pol I and TIF-IA is induced prior to the increase in pre-rRNA expression [28]. This may suggest that indirect control of pol I output by Myc evolved prior to the direct effects seen in mammals; the appearance of E-boxes and binding to SL1 may have arisen subsequently, to accelerate and amplify the effect (Figure 4). There are precedents for this type of increase in complexity. For example, the insulin receptor/phosphoinositide 3-kinase pathway acts directly on the pol I machinery in mammals [62–64], but not in fruitflies [28].

Role of pols I and III in the biological function of c-Myc

The fruitfly model provides a clear indication of the importance of rRNA synthesis for the biological effects of Myc [28]. Induction of Myc in developing wing discs stimulates growth and leads to an increase in cell size [28,65]. However, this effect can be substantially negated by using a pol I mutant to block any increase in rRNA expression [28]. Grewal et al. [28] concluded that “the growth effects of dMyc in larval wing imaginal discs require *de novo* rRNA synthesis”. They went on to suggest that “during animal development, the control of rRNA synthesis and ribosome biogenesis is an essential Myc function”.

Production of rRNA and tRNA is tightly coupled to the growth rate of cells [66,67]. Under starvation conditions, rates of rRNA and tRNA synthesis decrease, whereas total mRNA production is sustained [68–70]. There is a strict inverse relationship between cell doubling time and nucleolar size, a cytohistological indicator of pol I activity [71,72]. This is readily explained by the fact that growth (increase in mass) is directly proportional to the rate of protein accumulation [73], which is inevitable, as 80–90% of cell mass is protein [74]. Activation of translation can increase cell size [75]. Furthermore, biosynthesis and the attainment of adequate mass are essential prerequisites for cell cycle progression [75–81]. Thus a 50% reduction in translation causes fibroblasts to exit the cell cycle and quiesce [82–84]. In most cell types, the mRNA concentration exceeds that of ribosomes, and so protein accumulation depends critically on ribosome content [85,86]. As rRNA synthesis is a limiting step in ribosome production [87], high output by pols I and III is essential for cells to maintain rapid growth. Thus hypertrophic growth of cardiomyocytes can be curtailed specifically by using an antisense vector against the pol I-specific factor UBF to restrict rRNA synthesis [88]. Conversely, specific activation of pol I transcription can accelerate the growth and proliferation of HEK293 cells [89]. Levels of tRNA can also have a major impact on growth rate; for example, a 2-fold reduction in levels of initiator tRNA causes a 3-fold increase in the doubling time of *Saccharomyces cerevisiae* [90]. In *S. cerevisiae*, mutations in pol I subunits were found to cause a substantial decrease in cell size [91]. The most potent effect on size was ascribed to Sfp1, a transcription factor that directly

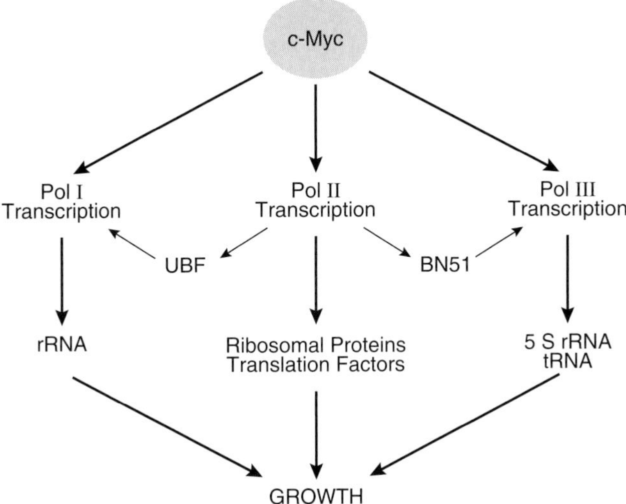

Figure 5 Growth induction by c-Myc involves direct effects on transcription by pols I, II and III, to stimulate expression of multiple components of the translation apparatus. Activation of pol I transcription raises levels of rRNA, whereas tRNA and 5 S rRNA are among the pol III products that are induced by c-Myc. A wide range of pol II-transcribed genes are targets for c-Myc, including the genes encoding UBF, BN51, translation factors and many ribosomal proteins. The resultant increase in protein synthetic capacity provides the potential for increased cell growth.

activates the genes encoding pols I and III, along with translation factors and various nucleolar components [91]. These observations link ribosome biogenesis directly to the commitment to cell division [81,91].

Activation of Myc drives a rapid increase in translation and growth, which precedes DNA replication and cell division [7,92]. Indeed, Myc can stimulate growth and protein synthesis independently of cell cycle progression [16,22,24,65,93–95]. Conversely, loss of Myc diminishes growth and protein synthesis [7,19,65]. Many of the pol II-transcribed genes that respond to c-Myc are required for cell growth, including ribosomal proteins and translation initiation factors [11,14,16,21–23,32,92,96]. Our discovery that c-Myc can also stimulate synthesis of tRNA and rRNA by pols I and III suggests an unprecedented capacity to co-ordinately induce multiple components of the protein synthetic apparatus through parallel effects on independent transcription systems (Figure 5). Since rRNA production can be rate-limiting for ribosome manufacture, protein accumulation and cell growth [61,88,89,97], the potent effect of c-Myc on rRNA synthesis may contribute substantially to its growth-promoting activity.

This work was supported by grants from Cancer Research UK, the Wellcome Trust, the Biotechnology and Biological Sciences Research Council and the National Institutes of Health.

References

1. Davis, A.C., Wims, M., Spotts, G.D., Hann, S.R. and Bradley, A. (1993) Genes Dev. **7**, 671–682
2. Trumpp, A., Refaeli, Y., Oskarsson, T., Gasser, S., Murphy, M., Martin, G.R. and Bishop, J.M. (2001) Nature (London) **414**, 768–772
3. Hueber, A.-O. and Evan, G.I. (1998) Trends Genet. **14**, 364–367
4. Cole, M.D. and McMahon, S.B. (1999) Oncogene **18**, 2916–2924
5. Dang, C.V. (1999) Mol. Cell. Biol. **19**, 1–11
6. Elend, M. and Eilers, M. (1999) Curr. Biol. **9**, R936–R938
7. Schmidt, E.V. (1999) Oncogene **18**, 2988–2996
8. Grandori, C., Cowley, S.M., James, L.P. and Eisenman, R.N. (2000) Annu. Rev. Cell Dev. Biol. **16**, 653–699
9. Eisenman, R.N. (2001) Genes Dev. **15**, 2023–2030
10. Levens, D.L. (2003) Genes Dev. **17**, 1071–1077
11. Oskarsson, T. and Trumpp, A. (2005) Nat. Cell Biol. **7**, 215–217
12. Nesbit, C.E., Tersak, J.M. and Prochownik, E.V. (1999) Oncogene **18**, 3004–3016
13. Gomez-Roman, N., Grandori, C., Eisenman, R.N. and White, R.J. (2003) Nature (London) **421**, 290–294
14. Coller, H.A., Grandori, C., Tamayo, P., Colbert, T., Lander, E.S., Eisenman, R.N. and Golub, T.R. (2000) Proc. Natl. Acad. Sci. U.S.A. **97**, 3260–3265
15. Stone, J., de Lange, T., Ramsay, G., Jakobovits, E., Bishop, J.M., Varmus, H. and Lee, W. (1987) Mol. Cell. Biol. **7**, 1697–1709
16. Guo, Q.M., Malek, R.L., Kim, S., Chiao, C., He, M., Ruffy, M., Sanka, K., Lee, N.H., Dang, C.V. and Liu, E.T. (2000) Cancer Res. **60**, 5922–5928
17. Felton-Edkins, Z.A., Kenneth, N.S., Brown, T.R.P., Daly, N.L., Gomez-Roman, N., Grandori, C., Eisenman, R.N. and White, R.J. (2003) Cell Cycle **3**, 181–184
18. Grandori, C., Gomez-Roman, N., Felton-Edkins, Z.A., Ngouenet, C., Galloway, D.A., Eisenman, R.N. and White, R.J. (2005) Nat. Cell Biol. **7**, 311–318
19. Mateyak, M.K., Obaya, A.J., Adachi, S. and Sedivy, J.M. (1997) Cell Growth Differ. **8**, 1039–1048
20. Bush, A., Mateyak, M., Dugan, K., Obaya, A., Adachi, S., Sedivy, J. and Cole, M. (1998)

Genes Dev. **12**, 3797–3802

21. Boon, K., Caron, H.N., van Asperen, R., Valentijn, L., Hermus, M.-C., van Sluis, P., Roobeek, I., Weis, I., Voute, P.A., Schwab, M. and Versteeg, R. (2001) EMBO J.. **20**, 1383–1393
22. Kim, S., Li, Q., Dang, C.V. and Lee, L.A. (2000) Proc. Natl. Acad. Sci. U.S.A. **97**, 11198–11202
23. Neiman, P.E., Ruddell, A., Jasoni, C., Loring, G., Thomas, S.J., Brandvold, K.A., Lee, R., Burnside, J. and Delrow, J. (2001) Proc. Natl. Acad. Sci. U.S.A. **98**, 6378–6383
24. Schuhmacher, M., Kohlhuber, F., Holzel, M., Kaiser, C., Burtscher, H., Jarsch, M., Bornkamm, G.W., Laux, G., Polack, A., Weidle, U.H. and Eick, D. (2001) Nucleic Acids Res. **29**, 397–406
25. Schlosser, I., Holzel, M., Murnseer, M., Burtscher, H., Weidle, U.H. and Eick, D. (2003) Nucleic Acids Res. **31**, 6148–6156
26. Poortinga, G., Hannan, K.M., Snelling, H., Walkley, C.R., Jenkins, A., Sharkey, K., Wall, M., Brandenburger, Y., Palatsides, M., Pearson, R.B. et al. (2004) EMBO J. **23**, 3325–3335
27. Arabi, A., Wu, S., Ridderstrale, K., Bierhoff, H., Shiue, C., Fatyol, K., Fahlen, S., Hydbring, P., Soderberg, O., Grummt, I. et al. (2005) Nat. Cell Biol. **7**, 303–310
28. Grewal, S.S., Li, L., Orian, A., Eisenman, R.N. and Edgar, B.A. (2005) Nat. Cell Biol. **7**, 295–302
29. Greasley, P.J., Bonnard, C. and Amati, B. (2000) Nucleic Acids Res. **28**, 446–453
30. Frank, S.R., Schroeder, M., Fernandez, P., Taubert, S. and Amati, B. (2001) Genes Dev. **15**, 2069–2082
31. Cawley, S., Bekiranov, S., Ng, H.H., Kapranov, P., Sekinger, E.A., Kampa, D., Piccolboni, A., Sementchenko, V., Cheng, J., Williams, A.J. et al. (2004) Cell **116**, 499–509
32. Fernandez, P.C., Frank, S.R., Wang, L., Schroeder, M., Liu, S., Greene, J., Cocito, A. and Amati, B. (2003) Genes Dev. **17**, 1115–1129
33. Bouchard, C., Thieke, K., Maier, A., Saffrich, R., Hanley-Hyde, J., Ansorge, W., Reed, S., Sicinski, P., Bartek, J. and Eilers, M. (1999) EMBO J. **18**, 5321–5333
34. Bouchard, C., Dittrich, O., Kiermaier, A., Dohmann, K., Menkel, A., Eilers, M. and Luscher, B. (2001) Genes Dev. **15**, 2042–2047
35. Murphy, S., Tripodi, M. and Melli, M. (1986) Nucleic Acids Res. **14**, 9243–9260
36. Wang, Z. and Roeder, R.G. (1995) Proc. Natl. Acad. Sci. U.S.A. **92**, 7026–7030
37. Yoon, J.-B., Murphy, S., Bai, L., Wang, Z. and Roeder, R.G. (1995) Mol. Cell. Biol. **15**, 2019–2027
38. Schramm, L. and Hernandez, N. (2002) Genes Dev. **16**, 2593–2620
39. Amati, B., Frank, S.R., Donjerkovic, D. and Taubert, S. (2001) Biochim. Biophys. Acta **1471**, M135–M145
40. McMahon, S.B., Van Buskirk, H.A., Dugan, K.A., Copeland, T.D. and Cole, M.D. (1998) Cell **94**, 363–374
41. Vassilev, A., Yamauchi, J., Kotani, T., Prives, C., Avantaggiati, M.L., Qin, J. and Nakatani, Y. (1998) Mol. Cell **2**, 869–875
42. Ikura, T., Ogryzko, V.V., Grigoriev, M., Groisman, R., Wang, J., Horikoshi, M., Scully, R., Qin, J. and Nakatani, Y. (2000) Cell **102**, 463–473
43. Nikiforov, M.A., Chandriani, S., Park, J., Kotenko, I., Matheos, D., Johnsson, A., McMahon, S.B. and Cole, M.D. (2002) Mol. Cell. Biol. **22**, 5054–5063
44. Cheng, S.-W.G., Davies, K.P., Yung, E., Beltran, R.J., Yu, J. and Kalpana, G.V. (1999) Nat. Genet. **22**, 102–105
45. Wood, M.A., McMahon, S.B. and Cole, M.D. (2000) Mol. Cell **5**, 321–330
46. Hateboer, G., Timmers, H.T., Rustgi, A.K., Billaud, M., van't Veer, L.J. and Bernards, R. (1993) Proc. Natl. Acad. Sci. U.S.A. **90**, 8489–8493
47. Hoang, A.T., Lutterbach, B., Lewis, B.C., Yano, T., Chou, T.-Y., Barrett, J.F., Raffeld, M., Hann, S.R. and Dang, C.V. (1995) Mol. Cell. Biol. **15**, 4031–4042
48. McEwan, I.J., Dahlman-Wright, K., Ford, J. and Wright, A.P.H. (1996) Biochemistry **35**, 9584–9593
49. Chatterjee, S. and Struhl, K. (1995) Nature (London) **374**, 820–822
50. Klages, N. and Strubin, M. (1995) Nature (London) **374**, 822–823
51. Conconi, A., Widmer, R.M., Koller, T. and Sogo, J.M. (1989) Cell **57**, 753–761
52. Moss, T. and Stefanovsky, V.Y. (2002) Cell **109**, 545–548
53. Grummt, I. (2003) Genes Dev. **17**, 1691–1702

54. Russell, J. and Zomerdijk, J.C.B.M. (2005) Trends Biochem. Sci. **30**, 87–96
55. Ittmann, M., Greco, A. and Basilico, C. (1987) Mol. Cell. Biol. **7**, 3386–3393
56. Ittmann, M., Ali, J., Greco, A. and Basilico, C. (1993) Cell Growth Differ. **4**, 503–511
57. Jackson, A.J., Ittmann, M. and Pugh, B.F. (1995) Mol. Cell. Biol. **15**, 94–101
58. Mann, C., Micouin, J.-Y., Chiannilkulchai, N., Treich, I., Buhler, J.-M. and Sentenac, A. (1992) Mol. Cell. Biol. **12**, 4314–4326
59. Hannan, K.M., Hannan, R.D. and Rothblum, L. (1998) Front. Biosci. **3**, d376–d398
60. Paule, M.R. (1998) Transcription of Ribosomal RNA Genes by Eukaryotic RNA Polymerase I, Springer-Verlag, Berlin
61. White, R.J. (2005) Nat. Rev. Mol. Cell Biol. **6**, 69–78
62. Hannan, K.M., Brandenburger, Y., Jenkins, A., Sharkey, K., Cavanaugh, A., Rothblum, L., Moss, T., Poortinga, G., McArthur, G.A., Pearson, R.B. and Hannan, R.D. (2003) Mol. Cell. Biol. **23**, 8862–8877
63. James, M.J. and Zomerdijk, J.C.B.M. (2004) J. Biol. Chem. **279**, 8911–8918
64. Mayer, C., Zhao, J., Yuan, X. and Grummt, I. (2004) Genes Dev. **18**, 423–434
65. Johnston, L.A., Prober, D.A., Edgar, B.A., Eisenman, R.N. and Gallant, P. (1999) Cell **98**, 779–790
66. Grummt, I. (1999) Prog. Nucleic Acid Res. Mol. Biol. **62**, 109–154
67. Reeder, R.H. (1999) Prog. Nucleic Acid Res. Mol. Biol. **62**, 293–327
68. Grummt, I., Smith, V.A. and Grummt, F. (1976) Cell **7**, 439–445
69. Johnson, L.F., Abelson, H.T., Green, H. and Penman, S. (1974) Cell **1**, 95–100
70. Mauck, J.C. and Green, H. (1974) Cell **3**, 171–177
71. Derenzini, M., Trere, D., Pession, A., Montanaro, L., Sirri, V. and Ochs, R.L. (1998) Am. J. Pathol. **152**, 1291–1297
72. Trere, D., Pession, A. and Derenzini, M. (1989) Exp. Cell Res. **184**, 131–137
73. Baxter, G.C. and Stanners, C.P. (1978) J. Cell. Physiol. **96**, 139–146
74. Zetterberg, A. and Killander, D. (1965) Exp. Cell Res. **39**, 22–32
75. Fingar, D.C., Salama, S., Tsou, C., Harlow, E. and Blenis, J. (2002) Genes Dev. **16**, 1472–1487
76. Killander, D. and Zetterberg, A. (1965) Exp. Cell Res. **40**, 12–20
77. Johnston, G.C., Pringle, J.R. and Hartwell, L.H. (1977) Exp. Cell Res. **105**, 79–98
78. Neufeld, T.P. and Edgar, B.A. (1998) Curr. Opin. Cell Biol. **10**, 784–790
79. Montagne, J. (2000) Mol. Cell Biol. Res. Commun. **4**, 195–202
80. Stocker, H. and Hafen, E. (2000) Curr. Opin. Genet. Dev. **10**, 529–535
81. Sudbery, P. (2002) Science **297**, 351–352
82. Brooks, R.F. (1977) Cell **12**, 311–317
83. Ronning, O.W., Lindmo, T., Pettersen, E.O. and Seglen, P.O. (1981) J. Cell. Physiol. **109**, 411–418
84. Kung, A.L., Sherwood, S.W. and Schimke, R.T. (1993) J. Biol. Chem. **268**, 23072–23080
85. Kief, D.R. and Warner, J.R. (1981) Mol. Cell. Biol. **1**, 1007–1015
86. Thomas, G. (2000) Nat. Cell Biol. **2**, E71–E72
87. Hannan, R.D. and Rothblum, L.I. (1995) Cardiovasc. Res. **30**, 501–510
88. Brandenburger, Y., Jenkins, A., Autelitano, D.J. and Hannan, R.D. (2001) FASEB J. **15**, 2051–2053
89. Zhao, J., Yuan, X., Frodin, M. and Grummt, I. (2003) Mol. Cell **11**, 405–413
90. Francis, M.A. and Rajbhandary, U.L. (1990) Mol. Cell. Biol. **10**, 4486–4494
91. Jorgensen, P., Nishikawa, J.L., Breitkreutz, B.-J. and Tyers, M. (2002) Science **297**, 395–400
92. Rosenwald, I.B. (1996) Cancer Lett. **102**, 113–123
93. Iritani, B.M. and Eisenman, R.N. (1999) Proc. Natl. Acad. Sci. U.S.A. **96**, 13180–13185
94. Schuhmacher, M., Staege, M.S., Pajic, A., Polack, A., Weidle, U.H., Bornkamm, G.W., Eick, D. and Kohlhuber, F. (1999) Curr. Biol. **9**, 1255–1258
95. Beier, R., Burgin, A., Kiermaier, A., Fero, M., Karsunky, H., Saffrich, R., Moroy, T., Ansorge, W., Roberts, J. and Eilers, M. (2000) EMBO J. **19**, 5813–5823
96. Shiio, Y., Donohoe, S., Yi, E.C., Goodlett, D.R., Aebersold, R. and Eisenman, R.N. (2002) EMBO J. **21**, 5088–5096
97. Huang, R., Wu, T., Xu, L., Liu, A., Ji, Y. and Hu, G. (2002) FASEB J. **16**, 293–301

Biochem. Soc. Symp. **73**, 155–163
(Printed in Great Britain)

15

Both normal and polyglutamine-expanded ataxin-7 are components of TFTC-type GCN5 histone acetyltransferase-containing complexes

Dominique Helmlinger*, Sara Hardy†, Adrien Eberlin†, Didier Devys*† and Làszlò Tora†[1]

*Department of Molecular Pathology, Institut de Génétique et de Biologie Moléculaire et Cellulaire, UMR 7104, CNRS/INSERM/ULP, BP 10142, 67404 Illkirch Cedex, France, and †Department of Transcription, Institut de Génétique et de Biologie Moléculaire et Cellulaire, UMR 7104, CNRS/INSERM/ULP, BP 10142, 67404 Illkirch Cedex, France

Abstract

SCA7 (spinocerebellar ataxia type 7) is a neurodegenerative disorder caused by a CAG repeat expansion in the *SCA7* gene that leads to elongation of a polyglutamine tract in ataxin-7, a protein of unknown function. Sgf73, a putative yeast orthologue of ataxin-7, has been identified as a new component of the yeast SAGA (Spt/Ada/Gcn5 acetyltransferase) multisubunit complex, a co-activator required for the transcription of a subset of RNA polymerase II-dependent genes. We show here that ataxin-7 is an integral component of mammalian SAGA-like complexes, i.e. the TFTC [TBP (TATA-binding protein)-free TAF (TBP-associated factor) complex] and the STAGA (SPT3/TAF9/GCN5 acetyltransferase) complex. In agreement with this, immunoprecipitation of ataxin-7 retained a histone acetyltransferase activity characteristic of TFTC-like complexes. Moreover, polyglutamine expansion in ataxin-7 did not affect its incorporation into TFTCs/STAGA complexes purified from cells from a SCA7 patient. We demonstrate here that ataxin-7 is the human orthologue of a the yeast SAGA Sgf73 subunit, and is a *bona fide* subunit of human TFTC-like transcriptional complexes.

[1]To whom correspondence should be addressed (email laszlo@igbmc.u-strasbg.fr)

Introduction

SCA7 (spinocerebellar ataxia type 7) is a dominantly inherited neurodegenerative disorder, characterized by late-onset neuronal loss in the cerebellum, brainstem and retina. SCA7 is caused by a CAG repeat expansion (38–460) in the SCA7 gene encoding ataxin-7, an 892-amino-acid protein of unknown function [1,2]. Recent bioinformatic studies identified a significant block of identity between human ataxin-7 and a yeast open reading frame, *YGL066w* [3,4], that was recently identified as a novel subunit of the yeast SAGA (Spt/Ada/Gcn5 acetyltransferase) complex and named Sgf73 [5,6].

The yeast SAGA complex is composed of Ada co-activators, Spt proteins, a subset of yeast TAFs [TBP (TATA-binding protein)-associated factors] and Tra1 [7,8]. Yeast SAGA acts as a transcriptional co-activator/adapter that is required at specific promoters and possesses HAT (histone acetyltransferase) activity that is mediated by Gcn5. Human homologues of yeast Gcn5 include P/CAF [p300/CBP (CREB-binding protein)-associated factor] and GCN5 [9,10], which were found in several multiprotein complexes, such as the TFTC (TBP-free TAF complex) [11,12], the STAGA (SPT3/TAF9/GCN5 acetyltransferase) complex [13] and P/CAF/GCN5 complexes [14]. These multisubunit complexes show related, but not identical, compositions, and presumably overlapping but also distinct functions. All contain TRRAP (transformation/transcription domain-associated protein), the human homologue of yeast Tra1, homologues of yeast SAGA subunits and a subset of TAFs, but clearly lack TBP [15]. However, TFTC specifically contains TAF5 and TAF6, whose yeast homologues are also found in SAGA, whereas these subunits were not found in STAGA or P/CAF/GCN5 complexes [16]. These complexes contain a GCN5 HAT activity, and were shown to preferentially acetylate histone H3 in both free and nucleosomal contexts, and to activate transcription on chromatin templates [12,17,18]. In good agreement with the functional similarity between human TFTC and yeast SAGA, their structure is evolutionarily conserved [19,20].

Here we demonstrate that ataxin-7 is an integral subunit of highly purified TFTC, and that an anti-ataxin-7 antibody co-immunoprecipitated TFTC-like complexes and a HAT activity corresponding to that of GCN5. Furthermore, we show that a polyglutamine expansion in ataxin-7 does not affect TFTC-like complex formation and HAT activity.

Materials and methods

Anti-TRRAP polyclonal and monoclonal antibodies were generated by immunization of rabbits and mice with peptides corresponding to amino acids 198–219 and 2005–2024 respectively of human TRRAP, as described in [21]. Western blots were performed as described using anti-ataxin-7 [22], anti-GCN5 [21], anti-TAF10 and anti-TBP [11] monoclonal antibodies. Immunoprecipitations were carried out using anti-ataxin-7 [22] and anti-TRRAP [21] polyclonal antibodies. Protein complexes bound to Protein A–Sepharose beads were either boiled and processed for Western blot analysis or used for HAT assays, as described in [12]. TFTC was purified as described previously [11].

Results

Ataxin-7 is a novel subunit of GCN5 HAT-containing complexes

BLAST searches using ataxin-7 as a query sequence identified a putative ataxin-7 yeast orthologue (Sgf73, encoded by *YGL066w*), which was described recently as a novel subunit of the SAGA complex [6]. Sequence comparison of ataxin-7 with Sgf73 revealed two conserved blocks: block I (residues 126–176 in human ataxin-7) and block II (residues 341–400). A third region (block III; residues 508–565) was conserved in orthologous genes from different vertebrate species (Figure 1A, and results not shown). The expandable polyglutamine motif is located N-terminal to block I in human ataxin-7 and in its vertebrate orthologues.

We thus asked whether human ataxin-7 is also a subunit of TFTC/STAGA-type complexes, which are human homologues of the yeast SAGA complex. In order to test this hypothesis, we performed co-immunoprecipitation experiments using nuclear extracts of HeLa cells. When an anti-ataxin-7 antibody was used to immunoprecipitate ataxin-7-associated proteins, we detected, by Western blot analysis, ataxin-7 and the specific TFTC subunits TRRAP and GCN5 (Figure 1B, lane 2), whereas these proteins were not detected in a control immunoprecipitation (Figure 1B, lane 4). In addition, the anti-ataxin-7 antibody also immunoprecipitated TAF10, but not TBP. Conversely, an anti-TRRAP antibody immunopurified ataxin-7 together with other human TFTC subunits, as shown by Western blot (Figure 1B, lane 3). These results suggest that ataxin-7 is associated with TRRAP in a large multiprotein complex. Importantly, an anti-TBP antibody failed to detect TBP in both ataxin-7- and TRRAP-immunopurified fractions (Figure 1B, lanes 2 and 3), suggesting that human ataxin-7 is a component of TFTC, but not of TFIID (transcription factor IID), which specifically contains TBP.

These results were confirmed further by using a highly purified TFTC preparation [11]. A monoclonal antibody against human ataxin-7 strongly and specifically detected a band of the expected size (110 kDa) in the highly purified TFTC (Figure 1C, lane 1). Taken together, these results confirmed unequivocally that ataxin-7 is a *bona fide* TFTC/STAGA subunit.

TFTCs/STAGA complexes contain a HAT activity that is mediated by GCN5 [12]. In order to confirm that endogenous ataxin-7 is an integral component of these complexes, we tested the ability of the ataxin-7-containing immunopurified complex to acetylate free histones using an *in vitro* assay. This assay was performed directly on Sepharose beads after immunoprecipitation. As shown in Figure 1(D), the ataxin-7 immunopurified fraction acetylated histone H3 very efficiently (lane 3). As a positive control, purified TFTC showed a similar pattern of histone acetylation (lane 1). In contrast, no HAT activity was detected in control immunoprecipitations (lane 4). Altogether, these results demonstrated that human ataxin-7 is a novel subunit of GCN5 HAT-containing TFTC/STAGA-type complexes *in vivo*, and thus probably represents the human orthologue of the yeast SAGA subunit Sgf73.

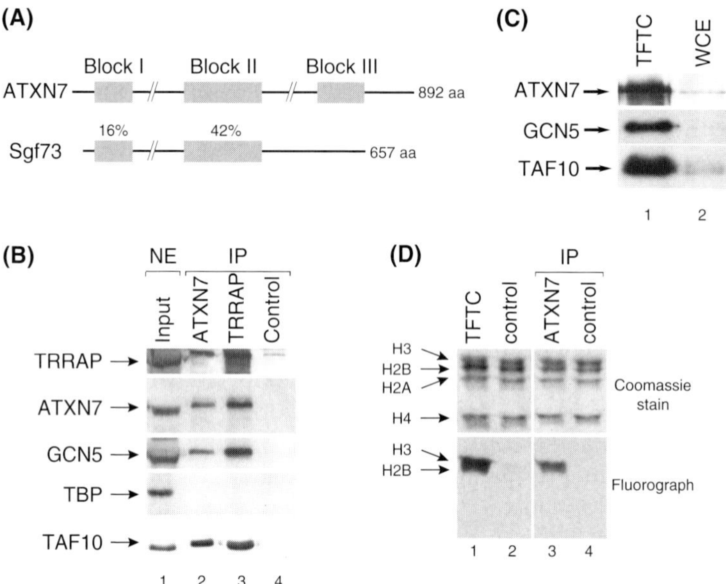

Figure 1 Ataxin-7 is an integral component of TFTC-like complexes. (**A**) Schematic representation of human ataxin-7 (ATXN7) and yeast Sgf73. For each conserved block (grey boxes), sequence identities of human ataxin-7 and yeast Sgf73 are indicated. (**B**) Crude HeLa nuclear extracts (NE) were immunoprecipitated (IP) with anti-ataxin-7 or anti-TRRAP polyclonal antibodies, or with non-immune rabbit serum as a negative control. The immunoprecipitated complexes retained on Protein A–Sepharose beads were either tested for HAT activity (see **D**) or analysed by immunoblotting with monoclonal antibodies specifically detecting TFTC/STAGA subunits, as indicated. A TFIID-specific subunit, TBP, was not found in complexes immunoprecipitated with anti-ataxin-7 or anti-TRRAP antibodies. Equal amounts of nuclear extracts (input) were also loaded. (**C**) Ataxin-7, GCN5 and TAF10 were present and significantly enriched in a highly purified TFTC fraction [11], by comparison with HeLa crude whole-cell extract (WCE). (**D**) Fluorograph of HAT assay performed using the ataxin-7 immunopurified fraction (lane 3) or the control immunoprecipitation (lane 4) shown in (**B**). The assay (lower panel) was performed by adding tritiated acetyl-CoA and purified histones (shown by Coomassie Blue staining in the upper panel) directly to immunopurified complexes retained on Sepharose beads. Highly purified TFTC was used as a positive control (lane 1) and omission of Sepharose beads as a negative control (lane 2).

Polyglutamine expansion in ataxin-7 does not affect TFTC-like complex formation and HAT activity

The mutation in SCA7 results in expansion of a polyglutamine tract in the N-terminal part of ataxin-7, which confers a gain of function to the mutant protein in SCA7 patients. SCA7 belongs to a group of neurodegenerative disorders, including Huntington's disease, that are caused by a polyglutamine expansion. A hallmark of these disorders is the aggregation of polyglutamine proteins into insoluble nuclear inclusions.

In order to determine whether the polyglutamine expansion affects the association of ataxin-7 with TFTC/STAGA complexes, we performed co-immunoprecipitation experiments using lymphoblastoid cell lines, derived from a SCA7 patient, expressing both normal (10 glutamine residues) and mutant (51 glutamine residues) ataxin-7 proteins (Figure 2A, lane 1). Both normal and mutant ataxin-7 were immunoprecipitated specifically from nuclear extracts of the SCA7 patient cells with an anti-TRRAP antibody (Figure 2A, lane 2) and with an anti-ataxin-7 antibody (lane 3). The bands corresponding to normal and expanded ataxin-7 were of comparable intensities in both the TRRAP and ataxin-7 immunoprecipitations, indicating that the two forms of

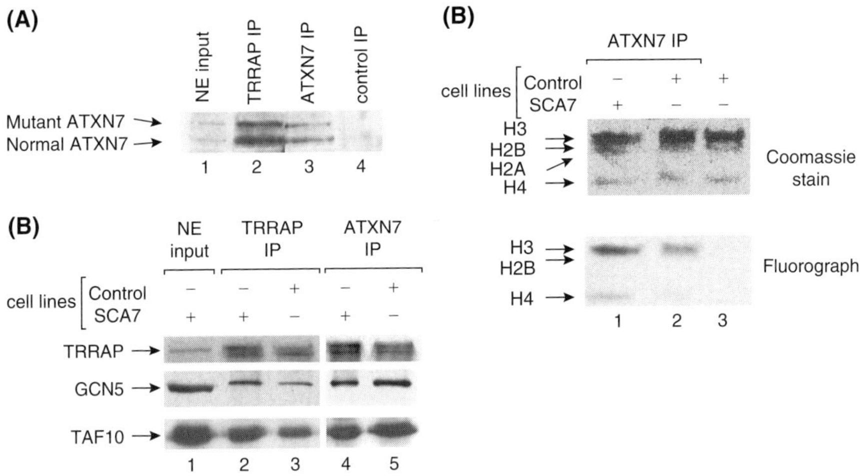

Figure 2 Polyglutamine expansion in ataxin-7 does not affect TFTC/STAGA complex formation and localization. (**A**) Nuclear extracts of SCA7 lymphoblastoid cells were immunoprecipitated (IP) with polyclonal antibodies against ataxin-7 (ATXN7) or TRRAP, or with non-immune rabbit serum as a negative control. Input SCA7 lymphoblastoid nuclear extracts (NE) was 2% (lane 1). (**B**) Nuclear extracts of control and SCA7 lymphoblastoid cells were immunoprecipitated with anti-TRRAP and anti-ataxin-7 polyclonal antibodies. Immunopurified fractions were analysed by immunoblotting with antibodies specifically detecting TFTC/STAGA subunits, as indicated. (**C**) Fluorograph of a histone acetyltransferase performed on ATXN7 immunopurified complexes obtained from cell lines derived from either control individual (lane 2) or an SCA7 patient (lane 3). The assay was corried out as described in the legend to Figure 1(D).

Table 1 Subunit composition of TBP-free HAT-containing complexes. Yeast and human TBP-free TAF HAT complexes are shown. Subunits that are consistently found in these different complexes are presented. Yeast and human proteins within the same horizontal row show homology with each other. Abbreviations: Sgf, SAGA-associated factor; Ubp, ubiquitin protease; SAP, spliceosome-associated protein; Chd, chromo-ATPase/helicase-DNA-binding domain; ND, non-defined.

	TBP-free HAT complexes	
	SAGA (yeast)	TFTC/STAGA/P/CAF (human)
HAT subunit	yGcn5	hGCN5/hPCAF
TAFs	yTAF5	hTAF5/hTAF5L
	yTAF6	hTAF6/hTAF6L
	yTAF9	hTAF9
	yTAF10	hTAF10
	yTAF12	hTAF12
SPTs, ADAs	ySpt3	hSPT3
	ySpt7	hSPT7L
	Spt8	ND
	ySpt20	ND
	yAda1	hADA1
	yAda2	hADA2
	yAda3	hADA3
Other components	Tra1	TRRAP
	Sgf11	ND
	Sgf29	ND
	Sgf73	hATXN7
	Ubp8	ND
	ND	SAP130
	Chd1	ND
	Sus1	he(y)2?

ataxin-7 were incorporated equally within TFTC/STAGA-type complexes in SCA7 cells. Furthermore, we performed anti-TRRAP and anti-ataxin-7 immunoprecipitation using nuclear extracts from the SCA7 cells used in Figure 2(A) or from control lymphoblastoid cells derived from an unaffected individual, expressing only normal ataxin-7. Western blot analysis of TRRAP- and ataxin-7-associated proteins showed that GCN5 and TAF10 were immunopurified similarly from both cell lines (Figure 2B, lanes 2 and 3). Nuclear extracts representing 10% of the amount used for immunoprecipitation were loaded on the gel (Figure 2B, lane 1), and it was revealed that endogenous TRRAP was also enriched efficiently after immunoprecipitation in both the control cells and the SCA7 cell line. These results together indicate that TFTC/STAGA subunit composition is not altered by polyglutamine expansion in ataxin-7.

We next tested the ability of the ataxin-7- or TRRAP-containing immunopurified complexes prepared from normal and SCA7 lymphoblastoid cells to acetylate free histones using an *in vitro* HAT assay. We did not detect any differences in the acetylation pattern of the histones when comparing complexes from normal and SCA7 cells (Figure 2C and results not shown). These data suggest that the HAT activity of the ataxin-7-containing (normal and mutataed) TFTC/STAGA-type complexes is not altered by polyglutamine expansion in ataxin-7.

Discussion

Immunoprecipitation of endogenous ataxin-7 from HeLa cells specifically retrieved TFTC/STAGA subunits together with a characteristic GCN5 HAT activity.Similarly,antibodiesimmunoprecipatingTFTCsco-immunoprecipitated ataxin-7. Altogether, these results demonstrate that ataxin-7 is a novel *bona fide* subunit of TFTC-like complexes and is the orthologue of the yeast SAGA-associated factor Sgf73 (Table 1).

Ataxin-7 is the fourth polyglutamine disease gene product, in addition to the androgen receptor, TBP and atrophin-1 [23], whose normal function is involved in transcriptional regulation. Interestingly, previous findings indicate that transcriptional deregulation contributes to the pathogenesis of polyglutamine-induced neuronal dysfunction [24,25] and might be mediated by defects in histone acetylation. This hypothesis is based on the observations that CBP HAT activity is impaired by expanded polyglutamine *in vitro* and that histone deacetylase inhibitors are protective in mouse and *Drosophila* models of Huntington's disease [26–28]. Severe down-regulation of genes encoding components of the phototransduction pathway has been consistently reported in photoreceptors from three different SCA7 mouse models [29–31]. Thus it is tempting to speculate that polyglutamine expansion in ataxin-7 would impair TFTC/STAGA HAT function and/or recruitment, thereby contributing to the transcriptional alterations observed in SCA7 patients. Expanded ataxin-7 aggregation may decrease the amounts of TFTC/STAGA complexes containing ataxin-7 and/or sequester TFTC/STAGA subunits from their normal localization in neurons in SCA7 mouse models. However, several pieces of experimental evidence suggest that this may not be the case. First, Yoo et al. [30] demonstrated that SCA7 toxicity results from a gain-of-function mechanism using mice in which a 266 CAG repeat was targeted into the mouse *Sca7* locus. Compound heterozygous mice carrying an expanded *Sca7* allele and a *Sca7* null allele were not more severely affected than heterozygous mice carrying an expanded *Sca7* allele and a wild-type *Sca7* allele [30]. Secondly, overexpression of normal ataxin-7 did not modulate phenotype severity in our transgenic SCA7 mouse model [31].

We demonstrated that both normal and mutant ataxin-7 can be incorporated into TFTC/STAGA complexes immunopurified from SCA7 patient cell lines, and that the incorporation of mutant ataxin-7 did not alter the composition or the *in vitro* HAT activity of TFTC/STAGA complexes. Decreased HAT activity of mutant ataxin-7-containing complexes *in vivo* is unlikely to occur, as a partial loss of ataxin-7 function does not appear to contribute to SCA7 pathogenesis. Thus a polyglutamine expansion could instead lead to non-regulated transcriptional activity of the TFTC/STAGA complexes. Finally, SCA7 mice revealed selective stabilization of expanded ataxin-7 compared with normal ataxin-7 [30,32]. It would thus be worth testing whether polyglutamine expansion-containing TFTC-type HAT complexes alter the chromatin structure in the retina of SCA7 mice, and whether the observed alteration in the gene expression pattern in the rod photoreceptors is a consequence of aberrant TFTC function and/or recruitment.

We are grateful to G. Stevanin for SCA7 lymphoblastoid cell lines. This study was supported by grants from INSERM, the CNRS and the Hôpital Universitaire de Strasbourg to D.D. and L.T., from the European Union (EUROSCA integrated project; LSHM-CT-2004-503304) to D.D., and from the Association pour la Recherche sur le Cancer, the Fondation pour la Recherche Médicale, the Fonds National de la Science (FNS) ACI, the European Union (HPRN-CT-2000-00087, HPRN-CT-2000-00088 and HPRN-CT-005 04228), INTAS and AICR (03-084) to L.T.

References

1. David, G., Abbas, N., Stevanin, G., Durr, A., Yvert, G., Cancel, G., Weber, C., Imbert, G., Saudou, F., Antoniou, E. et al. (1997) Nat. Genet. **17**, 65–70
2. Michalik, A., Martin, J.J. and Van Broeckhoven, C. (2004) Eur. J. Hum. Genet. **12**, 2–15
3. Mushegian, A.R., Vishnivetskiy, S.A. and Gurevich, V.V. (2000) Biochemistry **39**, 6809–6813
4. Scheel, H., Tomiuk, S. and Hofmann, K. (2003) Hum. Mol. Genet. **12**, 2845–2852
5. Gavin, A.C., Bosche, M., Krause, R., Grandi, P., Marzioch, M., Bauer, A., Schultz, J., Rick, J.M., Michon, A.M., Cruciat, C.M. et al. (2002) Nature (London) **415**, 141–147
6. Sanders, S.L., Jennings, J., Canutescu, A., Link, A.J. and Weil, P.A. (2002) Mol. Cell. Biol. **22**, 4723–4738
7. Grant, P.A., Duggan, L., Cote, J., Roberts, S.M., Brownell, J.E., Candau, R., Ohba, R., Owen-Hughes, T., Allis, C.D., Winston, F. et al. (1997) Genes Dev. **11**, 1640–1650
8. Grant, P.A., Schieltz, D., Pray-Grant, M.G., Steger, D.J., Reese, J.C., Yates, III, J.R. and Workman, J.L. (1998) Cell **94**, 45–53
9. Yang, X.J., Ogryzko, V.V., Nishikawa, J., Howard, B.H. and Nakatani, Y. (1996) Nature (London) **382**, 319–324
10. Xu, W., Edmondson, D.G. and Roth, S.Y. (1998) Mol. Cell. Biol. **18**, 5659–5669
11. Wieczorek, E., Brand, M., Jacq, X. and Tora, L. (1998) Nature (London) **393**, 187–191
12. Brand, M., Yamamoto, K., Staub, A. and Tora, L. (1999) J. Biol. Chem. **274**, 18285–18289
13. Martinez, E., Kundu, T.K., Fu, J. and Roeder, R.G. (1998) J. Biol. Chem. **273**, 23781–23785
14. Ogryzko, V.V., Kotani, T., Zhang, X., Schiltz, R.L., Howard, T., Yang, X.J., Howard, B.H., Qin, J. and Nakatani, Y. (1998) Cell **94**, 35–44
15. Grant, P.A. and Berger, S.L. (1999) Semin. Cell Dev. Biol. **10**, 169–177
16. Carrozza, M.J., Utley, R.T., Workman, J.L. and Cote, J. (2003) Trends Genet. **19**, 321–329
17. Martinez, E., Palhan, V.B., Tjernberg, A., Lymar, E.S., Gamper, A.M., Kundu, T.K., Chait, B.T. and Roeder, R.G. (2001) Mol. Cell. Biol. **21**, 6782–6795
18. Hardy, S., Brand, M., Mittler, G., Yanagisawa, J., Kato, S., Meisterernst, M. and Tora, L. (2002) J. Biol. Chem. **277**, 32875–32882
19. Wu, P.Y., Ruhlmann, C., Winston, F. and Schultz, P. (2004) Mol. Cell **15**, 199–208
20. Timmers, H.T. and Tora, L. (2005) Trends Biochem. Sci. **30**, 7–10
21. Brand, M., Moggs, J.G., Oulad-Abdelghani, M., Lejeune, F., Dilworth, F.J., Stevenin, J., Almouzni, G. and Tora, L. (2001) EMBO J. **20**, 3187–3196
22. Yvert, G., Lindenberg, K.S., Picaud, S., Landwehrmeyer, G.B., Sahel, J.A. and Mandel, J.L. (2000) Hum. Mol. Genet. **9**, 2491–2506
23. Zhang, S., Xu, L., Lee, J. and Xu, T. (2002) Cell **108**, 45–56

24. Cha, J.H. (2000) Trends Neurosci. **23**, 387–392
25. Sugars, K.L. and Rubinsztein, D.C. (2003) Trends Genet. **19**, 233–238
26. Steffan, J.S., Bodai, L., Pallos, J., Poelman, M., McCampbell, A., Apostol, B.L., Kazantsev, A., Schmidt, E., Zhu, Y.Z., Greenwald, M. et al. (2001) Nature (London) **413**, 739–743
27. Hockly, E., Richon, V.M., Woodman, B., Smith, D.L., Zhou, X., Rosa, E., Sathasivam, K., Ghazi-Noori, S., Mahal, A., Lowden, P.A. et al. (2003) Proc. Natl. Acad. Sci. U.S.A. **100**, 2041–2046
28. Ferrante, R.J., Kubilus, J.K., Lee, J., Ryu, H., Beesen, A., Zucker, B., Smith, K., Kowall, N.W., Ratan, R.R., Luthi-Carter, R. and Hersch, S.M. (2003) J. Neurosci. **23**, 9418–9427
29. La Spada, A.R., Fu, Y.H., Sopher, B.L., Libby, R.T., Wang, X., Li, L.Y., Einum, D.D., Huang, J., Possin, D.E., Smith, A.C. et al. (2001) Neuron **31**, 913–927
30. Yoo, S.Y., Pennesi, M.E., Weeber, E.J., Xu, B., Atkinson, R., Chen, S., Armstrong, D.L., Wu, S.M., Sweatt, J.D. and Zoghbi, H.Y. (2003) Neuron **37**, 383–401
31. Helmlinger, D., Abou-Sleymane, G., Yvert, G., Rousseau, S., Weber, C., Trottier, Y., Mandel, J.L. and Devys, D. (2004) J. Neurosci. **24**, 1881–1887
32. Yvert, G., Lindenberg, K.S., Devys, D., Helmlinger, D., Landwehrmeyer, G.B. and Mandel, J.L. (2001) Hum. Mol. Genet. **10**, 1679–1692

Biochem. Soc. Symp. **73**, 165–180
(Printed in Great Britain)

16

Regulation of NF-κB function

Kirsteen J. Campbell and Neil D. Perkins[1]

School of Life Sciences, Division of Gene Regulation and Expression, MSI/WTB Complex, Dow Street, University of Dundee, Dundee DD1 5EH, Scotland, U.K.

Abstract

In the 20 years since its discovery, research into the NF-κB (nuclear factor-κB) family of transcription factors has revealed an amazing diversity of functions. NF-κB proteins are regulators of the immune, inflammatory, stress, proliferative and apoptotic responses of a cell to a very large number of different stimuli. NF-κB complexes can be found in all cell types, indicating that the number of different contexts in which NF-κB can become induced is enormous. Moreover, many reports suggest apparently opposing or contradictory functions for NF-κB. It is clear that it is not simply enough to understand the pathways leading to nuclear localization and DNA binding of NF-κB subunits. It is also important that we comprehend the regulation of NF-κB subunit functionality if we are to understand the NF-κB pathway as a whole. These issues include the mechanisms controlling the specificity and timing of genes regulated by NF-κB under particular circumstances. They also include the reasons why NF-κB subunits can sometimes repress rather than activate transcription and how the NF-κB response is integrated with other important transcription factor pathways in the cell, such as the induction of the p53 tumour suppressor following DNA damage or oncogene activation. Understanding the mechanisms that regulate NF-κB function has important implications for our understanding of the role that NF-κB subunits play in human inflammatory diseases and cancer, and could also impact on the use of future NF-κB-based clinical therapies.

Introduction

NF-κB (nuclear factor-κB) was first reported as a nuclear protein necessary for Igκ light chain transcription in B cells [1]. It is now known that NF-κB exists in most cell types, held in the cytoplasm as an inactive form directly bound to a member of the IκB (inhibitor of NF-κB) protein family [2]. NF-κB complexes are composed of homo- or hetero-dimers from the multigene family of RelA/

[1]To whom correspondence should be addressed (email n.d.perkins@dundee.ac.uk).

p65, c-Rel, RelB, NF-κB1 (p50/p105) and NF-κB2 (p52/p100) [2]. Each family member contains an N-terminal 300-amino-acid domain [the RHD (Rel homology domain)] that is responsible for DNA binding and dimerization. The RHD also contains a nuclear localization sequence and mediates non-covalent binding to IκB [2]. RelA, c-Rel and RelB contain a variable C-terminal transactivation domain [2]. p50 and p52 are synthesized as longer precursors (p105 and p100 respectively) that undergo endoproteolytic cleavage to form mature products [3]. p50 can also be produced by co-translational processing of p105 [4]. Distinct roles for these subunits have been characterized through the generation of mice and cell lines deficient in one or more NF-κB family member [5,6]. In addition to this complexity, there are a number of IκBs, including IκBα, IκBβ and IκBε [2]. IκBs are ankyrin-repeat-containing proteins that interact with the nuclear localization sequence in the RHD of NF-κB proteins to ensure sequestration in the cytoplasm [2]. p100 and p105 proteins also act as IκBs; prior to proteolytic cleavage, their ankyrin repeat domains ensure cytoplasmic localization [3]. Furthermore, as p50 and p52 lack transactivation domains, they can only activate transcription when complexed with RelA, RelB, c-Rel or Bcl-3 [3]. Bcl-3 is another IκB family member, but functions as a co-activator, interacting with p50 and p52, promoting their nuclear localization and activating transcription. Inappropriate IKK (IκB kinase) and NF-κB activity, where NF-κB becomes aberrantly and constitutively nuclear-localized, is associated with a range of disorders, including cancer, neurodegenerative disease, arthritis, asthma and inflammatory bowel disease [7–9]. Therefore the pharmaceutical industry is actively pursuing the development of IKK inhibitors as therapeutic compounds for the treatment of many human diseases [9,10].

Canonical and non-canonical activation of NF-κB

Two distinct pathways to NF-κB activation have emerged in recent years. The canonical pathway of NF-κB activation releases dimers composed of RelA, c-Rel and p50 from IκB sequestration [11]. Release from IκB sequestration in the cytoplasm allows translocation of NF-κB to the nucleus, where it can influence the transcription of a diverse set of target genes [12]. Upon stimulation with pro-inflammatory cytokines or microbial infection, the canonical pathway is activated through the IKK-mediated phosphorylation of NF-κB-bound IκBα at Ser^{32} and Ser^{36} (or equivalent residues in other IκB proteins) [11]. This results in polyubiquitination at Lys^{21} and Lys^{22} by a specific ubiquitin ligase of the SCF (Skp1/Cul1/F-box) family [11]. This triggers ubiquitin-dependent IκB degradation by the 26 S proteasome, leaving NF-κB dimers to translocate to the nucleus [11]. The IKK signalosome is a cytoplasmic complex of over 700 kDa, and is known to include kinase activity in the form of IKKα and IKKβ, and a non-kinase regulatory subunit, IKKγ [11]. Canonical activation depends on the IKKβ activity of the IKK complex. Although IKKγ has no intrinsic kinase activity, *in vivo* it is required for IKKα and IKKβ activity and is thought to function as a scaffold protein [11].

In addition to the canonical pathway, an alternative (non-canonical) pathway of NF-κB activation occurs in response to stimulation of the CD40 receptor and lymphotoxin-β receptor or by B-cell-activating factor and LMP-1 (latent membrane protein-1) of the Epstein–Barr virus [3,13,14]. The non-canonical pathway of NF-κB activation involves p100. Non-canonical inducing stimuli activate the NF-κB-inducing kinase, which in turn induces IKKα activity. IKKα then phosphorylates p100, inducing its polyubiquitination and proteolytic removal of the C-terminal, ankyrin-repeat-containing, domain, allowing p52-containing dimers to translocate to the nucleus [3,13,14]. This is different from signal-induced p105 processing, which requires IKKβ [3]. These distinct activation pathways therefore control the composition of nuclear NF-κB dimers.

Typical compared with atypical NF-κB activators

In addition to the canonical and non-canonical pathways, we believe that NF-κB function can be understood further by considering the inducers of NF-κB as belonging to either typical or atypical classes of activators. Among the better characterized inducers of NF-κB are inflammatory cytokines and bacterial products that culminate in activation of the IKK signalosome [11]. Upon stimulation by TNFα (tumour necrosis factor α), IL-1 (interleukin-1) or LPS (lipopolysaccharide), nuclear translocation of NF-κB is rapid and peaks by 1 h, which, by our criteria, classifies these as 'typical' activators of NF-κB. NF-κB is also activated by stimuli relevant to tumour initiation, promotion and therapy [12]. Such inducers include UV-C, ionizing radiation, phorbol esters, hypoxia and the chemotherapeutic agents daunorubicin, doxorubicin, etoposide, camptothecin, vincristine, vinblastine and cisplatin [12]. Activators such as UV-C and some chemotherapeutic compounds can be classified as 'atypical'. Here, induction of NF-κB DNA binding occurs with slower kinetics and through a pathway distinct from that utilized by inflammatory cytokines. For example, UV-C and doxorubicin have been shown to induce NF-κB DNA binding in an IKK-independent manner without phosphorylation of Ser^{32} and Ser^{36} of IκBα [15–18]. However, other reports have indicated that IKK activity is required for these activators [19,20], but may be utilized in a novel way, with a requirement for the zinc finger domain of IKKγ, which is not found with typical inducers such as TNFα and LPS [21,22]. In addition, other stimuli, such as hypoxic injury and oxidative stress, have been shown to activate NF-κB through induction of tyrosine phosphorylation of IκBα [23,24]. The situation appears complex, but it is reasonable to say that NF-κB activation by atypical activators occurs with delayed kinetics and utilizes IKK in a manner distinct from typical inducers such as TNFα, IL-1 and LPS.

NF-κB target genes

For simplicity, NF-κB target genes can be grouped into four subsets based on function. The first group is involved in the inflammatory response and immunoregulatory functions. This group is typified by pro-inflammatory cytokines, chemokines and their receptors, such as TNFα, IL-1, IL-6, MIP-1α (monocyte chemoattractant protein-1α) and RANTES (regulated upon activation, normal T-cell expressed and secreted), together with adhesion molecules such as VCAM-1 (vascular cell adhesion molecule-1), ICAM-1 (intercellular cell adhesion molecule-1) and ELAM-1 (endothelial–leucocyte adhesion molecule-1) [12]. Amplification of the NF-κB response therefore occurs, as binding of TNFα and IL-1 to their receptors strongly activates NF-κB [12]. Chronic inflammatory diseases such as rheumatoid arthritis can develop if this production of inflammatory cytokines goes unchecked [9]. Therefore the second subset of NF-κB target genes are negative regulators of NF-κB. Expression of these proteins provides a feedback mechanism to curtail the NF-κB response. This is exemplified by IκBα, the anti-apoptotic protein A20, and the tumour suppressor CYLD [12,25]. The third subset are genes with an apoptotic function. These include anti-apoptotic gene products such as IAP (inhibitor of apoptosis) proteins, TRAFs (TNFα-receptor-associated factors) and Bcl-xL, as well as pro-apoptotic genes such as those encoding Fas and Fas Ligand (see below) [26]. The final subset of NF-κB target genes encode positive regulators of the cell cycle. For example, cyclin D1 and Myc promote cell growth, and their expression is regulated by NF-κB [27,28]. NF-κB target genes from all four of these groups can contribute towards tumorigenesis [7].

NF-κB activity is subject to regulation at multiple levels

Numerous mechanisms exist to ensure appropriate control of transcription factor activity. In particular, regulation of NF-κB activity occurs at multiple levels to generate exquisite control of transcriptional activity (Figure 1) [29]. NF-κB activity is firstly regulated at the level of IκB proteins. Upon stimulation by TNFα or IL-1, activated NF-κB induces IκBα expression. This IκBα moves to the nucleus, where it binds to RelA and p50, preventing DNA binding, before shuttling back to the cytoplasm [30]. This rapidly turns off the NF-κB response and can result in an oscillatory activation profile. Computational models have revealed that IκBβ and IκBε, whose expression is not regulated by NF-κB, stabilize the NF-κB response during sustained stimulation, dampening the oscillatory potential [31]. This bimodal NF-κB response manifests physiologically as an ability to generate substantial responses to short stimulations while also responding to stable long-term stimulation. Hoffmann and colleagues [31] further identified two classes of TNFα-inducible NF-κB target genes, illustrated by the chemokines IP-10 (interferon γ inducible protein 10; activated by short stimulation with TNFα) and RANTES (requires sustained NF-κB activation for expression). Thus IκBα, IκBβ and IκBε can control the intensity, duration and (to a certain extent) specificity of the NF-κB response.

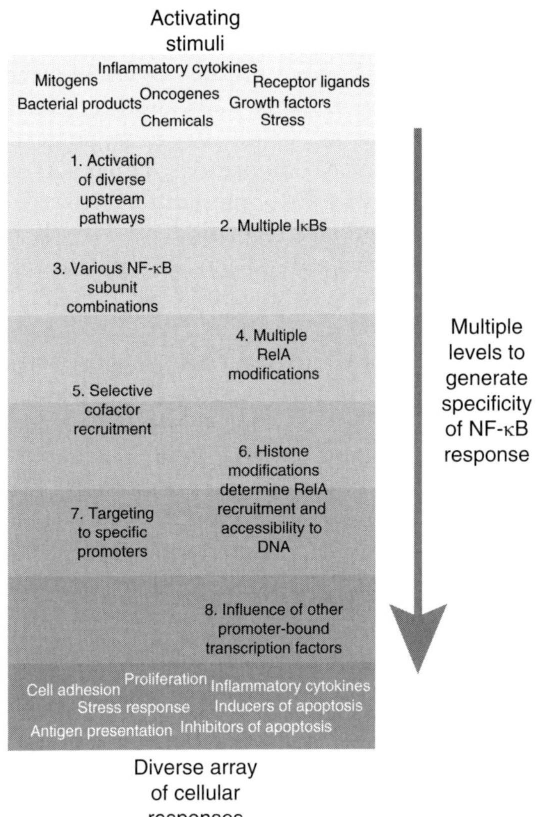

Figure 1 Specificity of NF-κB function is determined at multiple levels. Model illustrating the multiple levels at which NF-κB activation can be regulated to generate diverse responses, dependent on stimulus and cell context. Signal divergence at these levels allows NF-κB to elicit a complex range of responses.

Specific kinase cascades have been implicated upstream of IKK and these can determine the nature of the NF-κB response. MEKK3 (mitogen-activated protein kinase kinase kinase 3) is essential for the rapid activation of NF-κB by TNFα and the formation of an IκBα–NF-κB–IKK complex [32,33]. Using alternative kinase pathways upstream of IKK, the cell can regulate the kinetics of NF-κB activation. This is illustrated by the activation of MEKK2, which unlike MEKK3 was found to be necessary for the second phase of TNFα- or IL-1α-induced NF-κB activity and the assembly of an IκBβ–NF-κB–IKK complex [32]. Although TNFα-induced NF-κB DNA binding in MEKK3 null cells is impaired, UV-induced DNA binding is unaffected [33]. Therefore different kinase pathways upstream of IKK could explain why atypical activators trigger NF-κB DNA binding with much slower kinetics than typical activators.

Several NF-κB dimer combinations are possible, and the composition of the NF-κB complex can change with time following cell stimulation [34]. Since different NF-κB subunits have distinct DNA sequence affinities and transactivation properties, this diversity adds another level of specificity to the NF-κB pathway. However, as NF-κB sites in target genes are frequently bound by multiple NF-κB dimer combinations, differences in DNA-binding specificity cannot account completely for most differential regulation of NF-κB target genes [6,34]. Transcriptional regulation by NF-κB can be achieved through differential interactions with co-activator and co-repressor proteins. Association with co-activators such as p300/CBP [CREB (cAMP response element-binding protein) binding protein], P/CAF (p300/CBP-associated factor) and the steroid receptor co-activator Src1 enhance activation of NF-κB through recruitment of histone acetylase activity [35–37]. Similarly, association with HDACs (histone deacetylases) can repress NF-κB through hypoacetylation of promoters where RelA is bound [38–41]. Promoter-bound NF-κB can also be influenced by the presence of heterologous transcription factors. Examples of this include an interaction between Sp1 and NF-κB, which induces HIV-1 gene transcription [42]. Similarly, IRFs (interferon response factors) and ATF2 (activating transcription factor 2)/c-Jun co-operate with NF-κB to induce expression of interferon-β [43]. The recent discovery that κB site sequence can determine both the DNA-binding affinity of NF-κB complexes as well as the ability of bound NF-κB dimers to interact productively with cofactors provides a new dimension to promoter-specific NF-κB activity [44]. Therefore NF-κB activity at specific promoters may be determined by the presence or absence of other transcription factors and cofactors.

Post-translational modification of the NF-κB subunits provides another level at which NF-κB activity can be regulated, and has emerged as a dynamic mechanism controlling RelA transactivation, DNA-binding capabilities and stability. These modifications have recently been reviewed elsewhere [45,46], and for the RelA subunit alone include phosphorylation, dephosphorylation, acetylation, deacetylation, ubiquitination and prolyl isomerization. In particular, phosphorylation of the RelA transactivation domain appears to determine whether RelA activates or represses transcription [47] (and see below), while phosphorylation within the RHD can determine targeting to different promoters [48]. Much remains to be discovered about the importance of post-translational modifications of NF-κB, however. In particular, virtually nothing is known about how such events regulate the function of subunits other than RelA.

A final level of control of NF-κB activity is exercised by the chromatin environment of target genes. This was first illustrated by LPS stimulation of macrophages, which results in biphasic recruitment of NF-κB to target gene promoters, determined by histone acetylation [49]. At constitutively acetylated promoters, NF-κB DNA binding is rapid, whereas hypoacetylated promoters require additional events leading to their hyperacetylation prior to NF-κB recruitment and activation of transcription. This demonstrates the important point that total NF-κB DNA binding, as determined by electrophoretic mobility shift assay, is not adequate evidence of activity at individual promoters. Histone phosphorylation can also mark some promoters for RelA recruitment

[50]. This could be a reciprocal relationship, as RelA-associated IKKα was recently identified as a putative histone kinase [51,52]. Therefore the chromatin environment of NF-κB target genes may contribute to the specificity of the NF-κB response. Together, these levels of control must account for the ability of NF-κB to respond to over 150 distinct activators [12], with defined patterns of gene expression triggering diverse cellular outcomes [12]. Delineating the determinants of NF-κB functionality should lead to a better understanding of the role of NF-κB in disease.

RelA and apoptosis

A large number of reports describe NF-κB-dependent regulation of apoptosis [26]. The majority of these reports describe NF-κB, and the RelA subunit in particular, as being an anti-apoptotic transcription factor. This view first arose from early work describing the phenotype of *rela* null mice. These mice die during embryogenesis as a result of massive liver apoptosis arising from increased sensitivity to TNFα [53]. This embryonic lethality was completely suppressed in the absence of TNFα or TNFα receptor 1 [54,55]. Consistent with this hypothesis, NF-κB has been shown to be required for suppression of TNFα-, etoposide- and daunorubicin-induced apoptosis [26]. Mechanistically, this can be explained by the fact that NF-κB regulates anti-apoptotic genes, such as those encoding A20, IAP1, IAP2, TRAF1, TRAF2, Bfl-1/A1 and Bcl-xL [26]. However, NF-κB target genes also include those encoding pro-apoptotic proteins, such as Fas, FasL, DR4 (death receptor 4) and DR5 [26]. Therefore, whether NF-κB prevents or facilitates apoptosis is stimulus-, cell type- and context-dependent.

Table 1 highlights examples of cases where NF-κB, and specifically the RelA subunit, has been reported to be anti-apoptotic in contexts relevant to tumour initiation, development or therapy. The majority of reports of NF-κB function in these contexts suggest an anti-apoptotic function for RelA. Other studies concluded that there is no role for NF-κB in the regulation of chemotherapy-induced apoptosis. For example, Bentires-Alj and co-workers [56] studied the effects of stable expression of IκBαSR (super-repressor IκB) in lymphoid T cells, and colorectal adenocarcinoma, breast cancer and ovarian carcinoma cells. Surprisingly, they found that survival following treatment with TNFα, mitomycin C, camptothecin, daunorubicin, taxol or vinblastine was not significantly different between parental cells and those expressing IκBαSR. Another study utilized IκBαSR in breast and colon carcinoma cells, but identified no role for NF-κB in the control of daunorubicin-induced apoptosis or cell cycle progression [57]. On the other hand, additional reports revealed both pro- and anti-apoptotic functions of NF-κB in HeLa cells, dependent on stimulation with TNFα (anti-apoptotic), hydrogen peroxide and pervanadate (pro-apoptotic) [58]. This was echoed in a more recent study where it was shown that, in response to camptothecin, NF-κB inhibition protected against p53-mediated neuronal cell death [59]. However, a dual role for RelA in these cells was revealed,

Table 1 Inhibition of apoptosis by NF-κB/RelA. Listed are some examples of studies where RelA has been described as having an anti-apoptotic function in contexts relevant to cancer initiation, progression and therapy. Abbreviation: PDTC, pyrrolidine dithiocarbamate.

Stimulus	Cell type	Evidence for NF-κB/RelA-dependence of effect	Refs
Ionizing radiation and daunorubicin	HT1080 fibrosarcoma cell line	Stable expression of IκBSR	[89]
TNF and etoposide	HT1080 fibrosarcoma cell line	Stable expression of IκBSR	[90–92]
TNF and camptothecin	HT1080 and Lovo colorectal tumour cells in mouse xenograft model	Stable expression or adenoviral delivery of IκBαSR enhanced apoptosis	[91]
Camptothecin	Neurons	κB decoy oligonucleotides and PDTC abolished neuroprotective effect of pifithrin α	[93]
Doxorubicin	Mouse fibroblasts and p53 null cells	Decreased viability of *rela* null cells compared with wild-type	[20]
Doxorubicin and SN-38	HeLa human cervical carcinoma cells	Stable expression of IκBαSR increased apoptotic response	[19]
Doxorubicin	Mouse fibroblasts	Increased apoptosis of *ikkα/β* double null cells or cells stably expressing IκBαSR compared with wild type	[94]
Doxorubicin	Mouse fibroblasts	Increased doxorubicin-induced apoptosis of *ikkα/β* double null cells upon NF-κB inhibition by blocking the proteosome	[17]
Camptothecin	Mouse pre-B cells	Twice as many apoptotic cells when IκBαSR was expressed	[95]
Sodium butyrate/trichostatin A	Human non-small cell lung cancer cells	Stable expression of IκBαSR increased apoptosis	[96]
Etoposide and doxorubicin	Human pancreatic carcinoma cells	Transient expression of IκBαSR or a proteosome inhibitor enhanced apoptosis	[97]
Doxorubicin	Mouse liver model	Expression of liver-specific IκBαSR increased cell death	[98]
Daunorubicin and TNF	Colon cancer cells	Expression of IκBαSR increased toxicity of these agents	[98]

as long-term NF-κB inhibition also led to neuronal death. Further examples suggesting that RelA can facilitate apoptosis are shown in Table 2.

It is possible that the situation has been confused by some of the methodology used to determine dependence on NF-κB. For example, relatively non-specific compounds such as proteasome inhibitors are frequently used to infer NF-κB-specific effects without formally showing NF-κB dependence (Tables 1 and 2). Similarly, a dominant negative IκB super-repressor (IκBSR) is commonly utilized to reveal NF-κB-dependent effects (Tables 1 and 2). IκBSR is unable to be phosphorylated and degraded by classical activation pathways, as it has an N-terminal truncation, or point mutations removing Ser^{32} and Ser^{36} phosphorylation sites or Lys^{22} and Lys^{23} ubiquitination sites [11]. However, IκBSR can affect non-NF-κB pathways. Overexpression of IκBSR may affect nuclear import/export pathways [60], and overexpressed IκBα can interact with other proteins, such as CDK4 (cyclin-dependent kinase 4) [61] and p53 [62,63]. Indeed, IκBSR has been reported to prevent p53-dependent apoptosis through interaction with p53 [63]. Therefore, uncertainty still exists about the actual function of RelA and NF-κB in these contexts.

We have recently investigated the role of RelA as a regulator of apoptosis following stimulation with atypical inducers of NF-κB. We had noticed that, in many reports where atypical activators induced NF-κB that primed cells for apoptosis, little was reported of the transcriptional targets and activity of NF-κB. Furthermore, many studies of atypical inducers rely on NF-κB DNA binding as an indicator of transcriptional activity. Surprisingly, we found that following cell stimulation with UV light or the chemotherapeutic drugs daunorubicin or doxorubicin (adriamycin), RelA repressed rather than induced the expression of anti-apoptotic genes such as those encoding Bcl-xL and XIAP [39]. We have observed similar results with the ARF tumour suppressor: although ARF does not induce NF-κB DNA binding, it modulates existing NF-κB activity, converting RelA into a repressor of Bcl-xL expression, thus sensitizing cells to apoptotic stimuli [47,64]. In both cases transcriptional repression by RelA is associated with its differential phosphorylation and increased association with HDACs, although UV and daunorubicin/doxorubicin appear to work through distinct signalling pathways to ARF [39,47]. While these results are probably not the whole story concerning the pro-apoptotic effects of RelA and NF-κB, they do illustrate an important point, namely that NF-κB functionality can be determined by post-translational modification of the NF-κB subunits.

NF-κB has both oncogenic and tumour suppressor properties

Considerable evidence suggests a link between NF-κB and cancer. However, this relationship is not straightforward. The RelA subunit of NF-κB in particular can display both oncogene- and tumour suppressor-like capabilities. The functional differences described above, particularly differential regulation of apoptosis, will probably contribute to the nature of NF-κB regulation of tumorigenesis.

Table 2 Facilitation or induction of apoptosis by NF-κB/RelA. Listed are examples of studies where RelA has been described as facilitating apoptosis in contexts relevant to cancer initiation, progression and therapy. Abbreviations: TRAIL, TNF-related apoptosis-inducing ligand; MEF, mouse embryonic fibroblast; RNAi, RNA interference; PDTC, pyrrolidine dithiocarbamate; siRNA, small interfering RNA.

Stimulus	Cell type	Evidence for NF-κB RelA-dependence of effect	Refs
UV-B, etoposide and tenoposide	T-lymphocytes	Stable expression of IκBαSR inhibited apoptosis; mutation of κB site in FasL promoter prevented up-regulation	[99]
Inducible expression of p53, actinomycin D and doxorubicin	Osteosarcoma, RKO and mouse fibroblast cells	Expression of IκBαSR prevented apoptosis; *rela* null cells were less sensitive to apoptosis	[78]
Camptothecin	Neurons	Stable expression of IκBαSR, RNAi to RelA or pharmacological inhibition of NF-κB protected against cell death	[59]
H_2O_2 and pervanadate, but not TNF	Cervical carcinoma cells	Stable expression of IκBαSR; pharmalogical inhibition of NF-κB	[58]
Doxorubicin	Endothelial cells and cardiomyocytes	SN50 peptide inhibition of NF-κB nuclear translocation; IκBαSR	[100]
Paclitaxel	Breast cancer cells	Stable expression of IκBαSR or IκBα antisense mRNA	[101]
Serum starvation	Embryonic kidney cells	Dominant negative RelA	[71]
UV-C	Melanoma cells	IκBαSR abolished Fas-mediated death	[102]
Etoposide and TRAIL	Epithelial cells	Dominant negative IκB	[103]
Doxorubicin	Neuroblastoma cells	Stable expression of IκBαSR or PDTC inhibition of NF-κB	[104]
Doxorubicin and analogues	Myeloid, lymphoid and non-small-cell lung carcinoma cells (H1299)	Receptor-interacting protein (RIP) deficiency; PDTC or NEMO-binding domain peptide suppressed NF-κB activation and apoptosis	[105]
UV-C, daunorubicin, doxorubicin	Osteosarcoma (U-2 OS and SAOS2), non-small-cell lung carcinoma cells (H1299)	RelA siRNA; *rela* null cells were less sensitive to apoptosis and MEFs	[39]

NF-κB is required for the transforming properties of oncogenes such as activated Ras and Bcr-Abl [65]. Viral oncoproteins have also been shown to utilize NF-κB for transformation [66]. Examples of this are the TAX oncoprotein of HTLV-1 (human T-cell leukaemia virus-1), which interacts directly with IKKγ to cause constitutive activation of IKK, and hence NF-κB. Similarly, the Epstein–Barr virus protein LMP-1 increases NF-κB DNA binding and transcriptional activity. Furthermore, activation of NF-κB and IKK was recently found to make a significant contribution to inflammation-associated tumorigenesis in intestinal and liver tumours [67]. Interestingly, these studies also revealed a different role for NF-κB in different cell types. For example, in an ulcerative colitis model for colon cancer, regulation of apoptosis by IKK/NF-κB was found to be a critical at early stages in the development of enterocytes into viable tumours [68]. A critical role for NF-κB/IKK in non-transformed myeloid cells was also identified, but here the production of cytokines and chemokines stimulated cancer cell proliferation and growth at later stages of tumour growth [68]. Therefore, although NF-κB and IKK contribute to tumour development in both cell types, this occurs through at least two distinct mechanisms, presumably reflecting cell type-specific differences in gene regulation.

There are also numerous examples where failure to negatively regulate NF-κB results in cancer. Constitutive NF-κB activity, through altered activity of upstream proteins, has been identified in haematopoietic and solid tumours [65]. For example, analysis of Hodgkin's disease cell lines and patient biopsies revealed mutated IκBα that was incapable of interacting with NF-κB [69]. Similarly, the CYLD tumour suppressor functions as a negative regulator of NF-κB signalling, through de-ubiquitination of IKKγ, TRAF2 and TRAF6 [70]. Therefore NF-κB activity can contribute towards tumorigenesis in many contexts.

Considerable evidence also exists that RelA is required to prevent tumorigenesis. A tumour-suppressing role for RelA first emerged from studies showing that, upon serum starvation, RelA was actually required for apoptosis [71]. This was corroborated by studies showing that expression of RelA could drive pro-B cells and breast cancer cells to apoptosis [72,73]. These studies were taken into an *in vivo* setting when Khavari and colleagues showed that overexpression of RelA or p50 NF-κB subunits in the basal layer of mouse epidermis led to epidermal hypoplasia [74]. In the same setting, expression of IκBαSR caused hyperplasia [74]. Khavari and colleagues also found that human sporadic small-cell carcinoma may involve nuclear exclusion of the RelA subunit [75]. Indeed, co-expression of the IκBαSR and activated Ras resulted in neoplasia that resembled invasive small-cell carcinoma in human keratinocytes transplanted on to mice. van Hogerlinden and co-workers [76] also observed spontaneous development of squamous cell carcinomas in murine skin targeted with overexpressed IκBα, revealing that IκBα can be oncogenic. The hyperplasia observed by Khavari and colleagues has subsequently been attributed to enhanced JNK (c-Jun N-terminal kinase) activation induced by *ikk*β or *rela* gene deficiency in the epidermis, resulting in a failure to activate apoptosis and an increase in cell proliferation [77].

RelA has been reported to be required for p53-mediated cell death [78]. Furthermore, while the phenotype of *rela* null mice suggested an anti-apoptotic role for RelA, Gapuzan and co-workers [79] demonstrated that at least some derivatives of immortalized *rela* null mouse fibroblasts displayed a weakly transformed phenotype. These cells also formed tumours in nude mice which then regressed, indicating a requirement for additional mutations. These effects were reversed by re-expression of RelA, providing strong genetic evidence for a tumour suppressor role for RelA [79]. Furthermore, *in vivo* tumour models revealed that loss of RelA enhanced Ras- and E1A-induced tumorigenesis [80]. In agreement with these studies, it has emerged that, in colon- and breast-cancer cells, NF-κB target gene expression is repressed by β-catenin, found to be elevated in these tumours and in a complex with RelA and p50 [81]. Similarly, HSCO (hepatoma substracted cDNA library clone one) protein, found to be overexpressed in hepatocellular carcinomas, binds the RHD of RelA, sequestering it in the cytoplasm; it also prevents doxorubicin- and etoposide-induced, but not TNFα-induced, apoptosis [82]. Therefore considerable evidence has also revealed a tumour suppressor role for RelA.

Consistent with these studies, results from our laboratory and others have revealed mechanisms through which tumour suppressors can regulate the function of RelA and other NF-κB subunits and neutralize their oncogenic capabilities. In particular, our laboratory recently characterized regulation of RelA by the ARF tumour suppressor [47,64]. ARF expression is induced upon aberrant oncogene activation, whereupon it inactivates Hdm2/Mdm2 (human/murine double minute 2), the inhibitor of p53. Since one of the functions of p53 is to induce the expression of pro-apoptotic genes, the anti-apoptotic functions of RelA would reduce the effectiveness of the p53 response if left unchecked. We found that ARF activates the ATR (ataxia telangiectasia-related)/Chk1 (checkpoint kinase 1) kinases, in a manner functionally distinct from that seen upon DNA damage. Chk1 then phosphorylates Thr^{505} of RelA, a residue within its transactivation domain. This phosphorylation event converts RelA into a repressor of Bcl-xL expression, thus sensitizing cells to apoptotic stimuli [47,64]. At the early stages in the development of potential tumour cells, this mechanism will help to prevent the oncogenic function of RelA and can be thought of as recruiting RelA to the ARF tumour suppressor programme. If this mechanism is lost during tumour cell development, then the anti-apoptotic RelA activity of RelA is no longer suppressed and can contribute to tumour cell growth and survival [7].

There are many other reports of tumour suppressors regulating NF-κB function. For example, multiple levels of cross-talk exist between p53 and NF-κB, with both antagonistic and co-operative functions being described [7]. Interestingly, we demonstrated that p53 can also regulate the activity of the p52 NF-κB subunit [83]. p52 NF-κB is a regulator of the cyclin D1 promoter [84]. Upon p53 expression, we observed loss of p52–Bcl-3 activator complexes and an increase in p52–HDAC1 complexes, resulting in repression of cyclin D1 expression [83]. Similar to regulation of RelA by ARF, this provides an example where NF-κB subunit activity is not merely inhibited by a pathway, but subverted and recruited to the tumour suppressor programme. Further evidence for this emerging theme of tumour suppressor modulation of NF-κB arises from the finding that the putative tumour suppressor protein ING4 (inhibitor of growth 4) interacts with RelA and represses transcription of NF-κB target genes, to inhibit brain tumour growth and angiogenesis [85]. Moreover, the PTEN (phosphatase and tensin homologue deleted on chromosome 10) tumour suppressor protein (the natural antagonist for phosphoinostide 3-kinase) also inhibits NF-κB transcriptional activity through inhibition of the RelA transactivation domain [86]. In addition, the p16^{INK4a} CDK4/6 inhibitor, which shares a genetic locus with ARF, has also been reported to interact physically with RelA, and p16 overexpression inhibits RelA transactivation [87]. Indeed, a study of primary melanomas and metastases revealed an inverse correlation between p16 expression and RelA nuclear localization [88]. It is conceivable that, similar to ARF, these other pathways do not merely inhibit RelA activity, but induce RelA's ability to repress transcription, thus converting RelA to its tumour suppressor form.

The devil is in the details

The complexity of the NF-κB response demands precise regulatory mechanisms to ensure that the specificity and selectivity of this pathway is utilized appropriately by the cell. Although many of these mechanisms often appear as relatively minor components of the NF-κB response, when compared with the more obvious regulatory events such as IκB degradation and NF-κB nuclear translocation, they are nonetheless critical in determining NF-κB function. Since NF-κB subunit functionality is ultimately key to determining the role that NF-κB plays in normal processes and disease, these regulatory mechanisms should not be ignored or sidelined. This could be particularly true when implementing NF-κB-based therapies, such as IKK inhibitors, or more general treatments, such as chemotherapeutic drugs that are known to activate or modulate NF-κB. The effectiveness of such treatments could well be determined by the NF-κB functional status in the target tissue or cell type. Furthermore, potential side effects might be influenced by the beneficial effects of NF-κB elsewhere. This is not an argument for refraining from the pursuit of NF-κB-based therapies, which have the potential to treat a wide range of diseases. Instead, understanding the mechanisms that regulate NF-κB functionality will lead to more effective use of existing and future therapies.

We thank all the members of N.D.P.s laboratory and our colleagues in the Division of Gene Regulation and Expression for their help and assistance. N.D.P. is funded by a Royal Society University Fellowship, and K.J.C. is funded by a grant from Active Motif.

References

1. Sen, R. and Baltimore, D. (1986) Cell **46**, 705–716
2. Hayden, M.S. and Ghosh, S. (2004) Genes Dev. **18**, 2195–2224
3. Beinke, S. and Ley, S.C. (2004) Biochem. J. **382**, 393–409
4. Lin, L., DeMartino, G.N. and Greene, W.C. (1998) Cell **92**, 819–828
5. Gerondakis, S., Grossmann, M., Nakamura, Y., Pohl, T. and Grumont, R. (1999) Oncogene

18, 6888–6895
6. Hoffmann, A., Leung, T.H. and Baltimore, D. (2003) EMBO J. **22**, 5530–5539
7. Perkins, N.D. (2004) Trends Cell Biol. **14**, 64–69
8. Grilli, M. and Memo, M. (1999) Biochem. Pharmacol. **57**, 1–7
9. D'Acquisto, F., May, M.J. and Ghosh, S. (2002) Mol. Interventions **2**, 22–35
10. Haefner, B. (2002) Drug Discovery Today **7**, 653–663
11. Karin, M. and Ben-Neriah, Y. (2000) Annu. Rev. Immunol. **18**, 621–623
12. Pahl, H.L. (1999) Oncogene **18**, 6853–6866
13. Perkins, N.D. (2003) Oncogene **22**, 7553–7556
14. Senftleben, U., Cao, Y.X., Xiao, G.T., Greten, F.R., Krahn, G., Bonizzi, G., Chen, Y., Hu, Y.L., Fong, A., Sun, S.C. and Karin, M. (2001) Science **293**, 1495–1499
15. Bender, K., Gottlicher, M., Whiteside, S., Rahmsdorf, H.J. and Herrlich, P. (1998) EMBO. J. **17**, 5170–5181
16. Kato, T., Delhase, M., Hoffmann, A. and Karin, M. (2003) Mol. Cell **12**, 829–839
17. Tergaonkar, V., Bottero, V., Ikawa, M., Li, Q.T. and Verma, I.M. (2003) Mol. Cell. Biol. **23**, 8070–8083
18. Li, N.X. and Karin, M. (1998) Proc. Natl. Acad. Sci. U.S.A. **95**, 13012–13017
19. Bottero, V., Busuttil, V., Loubat, A., Magne, N., Fischel, J.L., Milano, G. and Peyron, J.F. (2001) Cancer Res. **61**, 7785–7791
20. Panta, G.R., Kaur, S., Cavin, L.G., Cortes, M.L., Mercurio, F., Lothstein, L., Sweatman, T.W., Israel, M. and Arsura, M. (2004) Mol. Cell. Biol. **24**, 1823–1835
21. Huang, T.T., Wuerzbrger-Davis, S.M., Wu, Z.H. and Miyamoto, S. (2003) Cell **115**, 565–576
22. Huang, T.T., Feinberg, S.L., Suryanarayanan, S. and Miyamoto, S. (2002) Mol. Cell. Biol. **22**, 5813–5825
23. Imbert, V., Rupec, R.A., Livolsi, A., Pahl, H.L., Traenckner, E.B., Mueller-Dieckmann, C., Farahifar, D., Rossi, B., Auberger, P., Baeuerle, P.A. and Peyron, J. (1996) Cell **86**, 787–798
24. Koong, A.C., Chen, E.Y. and Giaccia, A.J. (1994) Cancer Res. **54**, 1425–1430
25. Jono, H., Lim, J.H., Chen, L.F., Xu, H., Trompouki, E., Pan, Z.K., Mosialos, G. and Li, J.D. (2004) J. Biol. Chem. **279**, 36171–36174
26. Kucharczak, J., Simmons, M.J., Fan, Y.J. and Gelinas, C. (2003) Oncogene **22**, 8961–8982
27. Guttridge, D.C., Albanese, C., Reuther, J.Y., Pestell, R.G. and Baldwin, A.S. (1999) Mol. Cell. Biol. **19**, 5785–5799
28. Larosa, F.A., Pierce, J.W. and Sonenshein, G.E. (1994) Mol. Cell. Biol. **14**, 1039–1044
29. Perkins, N.D. (1997) Int. J. Biochem. Cell Biol. **29**, 1433–1448
30. Arenzana-Seisdedos, F., Thompson, J., Rodriguez, M.S., Bachelerie, F., Thomas, D. and Hay, R.T. (1995) Mol. Cell. Biol. **15**, 2689–2696
31. Hoffmann, A., Levchenko, A., Scott, M.L. and Baltimore, D. (2002) Science **298**, 1241–1245
32. Schmidt, C., Peng, B., Li, Z., Sclabas, G.M., Fujioka, S., Niu, J., Schmidt-Supprian, M., Evans, D.B., Abbruzzese, J.L. and Chiao, P.J. (2003) Mol. Cell **12**, 1287–1300
33. Yang, J., Lin, Y., Guo, Z., Cheng, J., Huang, J., Deng, L., Liao, W., Chen, Z., Liu, Z. and Su, B. (2001) Nat. Immunol. **2**, 620–624
34. Saccani, S., Pantano, S. and Natoli, G. (2003) Mol. Cell **11**, 1563–1574
35. Baek, S.H., Ohgi, K.A., Rose, D.W., Koo, E.H., Glass, C.K. and Rosenfeld, M.G. (2002) Cell **110**, 55–67
36. Perkins, N.D., Felzien, L.K., Betts, J.C., Leung, K.Y., Beach, D.H. and Nabel, G.J. (1997) Science **275**, 523–527
37. Sheppard, K.A., Rose, D.W., Haque, Z.K., Kurokawa, R., McInerney, E., Westin, S., Thanos, D., Rosenfeld, M.G., Glass, C.K. and Collins, T. (1999) Mol. Cell. Biol. **19**, 6367–6378
38. Ashburner, B.P., Westerheide, S.D. and Baldwin, A.S. (2001) Mol. Cell. Biol. **21**, 7065–7077
39. Campbell, K.J., Rocha, S. and Perkins, N.D. (2004) Mol. Cell **13**, 853–865
40. Yeung, F., Hoberg, J.E., Ramsey, C.S., Keller, M.D., Jones, D.R., Frye, R.A. and Mayo, M.W. (2004) EMBO J. **23**, 2369–2380
41. Lee, S.K., Kim, J.H., Lee, Y.C., Cheong, J. and Lee, J.W. (2000) J. Biol. Chem. **275**, 12470–12474
42. Perkins, N.D., Agranoff, A.B., Pascal, E. and Nabel, G.J. (1994) Mol. Cell. Biol. **14**, 6570–6583
43. Thanos, D. and Maniatis, T. (1995) Cell **83**, 1091–1100

44. Leung, T.H., Hoffmann, A. and Baltimore, D. (2004) Cell **118**, 453–464
45. Viatour, P., Merville, M.P., Bours, V. and Chariot, A. (2005) Trends Biochem. Sci. **30**, 43–52
46. Campbell, K.J. and Perkins, N.D. (2004) Biochem. Soc. Trans. **32**, 1087–1089
47. Rocha, S., Garrett, M.D., Campbell, K.J., Schumm, K. and Perkins, N.D. (2005) EMBO J. **24**, 1157–1169
48. Anrather, J., Racchumi, G. and Iadecola, C. (2005) J. Biol. Chem. **280**, 244–252
49. Saccani, S., Pantano, S. and Natoli, G. (2001) J. Exp. Med. **193**, 1351–1359
50. Saccani, S., Pantano, S. and Natoli, G. (2002) Nat. Immunol. **3**, 69–75
51. Anest, V., Hanson, J.L., Cogswell, P.C., Steinbrecher, K.A., Strahl, B.D. and Baldwin, A.S. (2003) Nature (London) **423**, 659–663
52. Yamamoto, Y., Verma, U.N., Prajapati, S., Kwak, Y.T. and Gaynor, R.B. (2003) Nature (London) **423**, 655–659
53. Beg, A.A. and Baltimore, D. (1996) Science **274**, 782–784
54. Doi, T.S., Marino, M.W., Takahashi, T., Yoshida, T., Sakakura, T., Old, L.J. and Obata, Y. (1999) Proc. Natl. Acad. Sci. U.S.A. **96**, 2994–2999
55. Alcamo, E., Mizgerd, J.P., Horwitz, B.H., Bronson, R., Beg, A.A., Scott, M., Doerschuk, C.M., Hynes, R.O. and Baltimore, D. (2001) J. Immunol. **167**, 1592–1600
56. Bentires-Alj, M., Hellin, A.C., Ameyar, M., Chouaib, S., Merville, M.P. and Bours, V. (1999) Cancer Res. **59**, 811–815
57. Hellin, A.C., Bentires-Alj, M., Verlaet, M., Benoit, V., Gielen, J., Bours, V. and Merville, M.P. (2000) J. Pharmacol. Exp. Ther. **295**, 870–878
58. Kaltschmidt, B., Kaltschmidt, C., Hofmann, T.G., Hehner, S.P., Droge, W. and Schmitz, M.L. (2000) Eur. J. Biochem. **267**, 3828–3835
59. Aleyasin, H., Cregan, S.P., Iyirhiaro, G., O'Hare, M.J., Callaghan, S.M., Slack, R.S. and Park, D.S. (2004) J. Neurosci. **24**, 2963–2973
60. Wu, B.Y., Woffendin, C., MacLachlan, I. and Nabel, G.J. (1997) J. Virol. **71**, 3161–3167
61. Li, J., Joo, S.H. and Tsai, M.D. (2003) Biochemistry **42**, 13476–13483
62. Chang, N.S. (2002) J. Biol. Chem. **277**, 10323–10331
63. Zhou, M., Gu, L., Zhu, N., Woods, W.G. and Findley, H.W. (2003) Oncogene **22**, 8137–8144
64. Rocha, S., Campbell, K.J. and Perkins, N.D. (2003) Mol. Cell **12**, 15–25
65. Baldwin, A.S. (2001) J. Clin. Invest. **107**, 241–246
66. Hiscott, J., Kwon, H. and Genin, P. (2001) J. Clin. Invest. **107**, 143–151
67. Clevers, H. (2004) Cell **118**, 671–674
68. Greten, F.R., Eckmann, L., Greten, T.F., Park, J.M., Li, Z.W., Egan, L.J., Kagnoff, M.F. and Karin, M. (2004) Cell **118**, 285–296
69. Cabannes, E., Khan, G., Aillet, F., Jarrett, R.F. and Hay, R.T. (1999) Oncogene **18**, 3063–3070
70. Wilkinson, K.D. (2003) Nature (London) **424**, 738–739
71. Grimm, S., Bauer, M.K. A., Baeuerle, P.A. and Schulze-Osthoff, K. (1996) J. Cell Biol. **134**, 13–23
72. Ricca, A., Biroccio, A., Trisciuoglio, D., Cippitelli, M., Zupi, G. and Del Bufalo, D. (2001) Br. J. Cancer **85**, 1914–1921
73. Sheehy, A.M. and Schlissel, M.S. (1999) J. Biol. Chem. **274**, 8708–8716
74. Seitz, C.S., Lin, Q., Deng, H. and Khavari, P.A. (1998) Proc. Natl. Acad. Sci. U.S.A. **95**, 2307–2312
75. Dajee, M., Lazarov, M., Zhang, J.Y., Cai, T., Green, C.L., Russell, A.J., Marinkovich, M.P., Tao, S.Y., Lin, Q., Kubo, Y. and Khavari, P.A. (2003) Nature (London) **421**, 639–643
76. van Hogerlinden, M., Rozell, B.L., Ahrlund-Richter, L. and Toftgard, R. (1999) Cancer Res. **59**, 3299–3303
77. Zhang, J.Y., Green, C.L., Tao, S. and Khavari, P.A. (2004) Genes Dev. **18**, 17–22
78. Ryan, K.M., Ernst, M.K., Rice, N.R. and Vousden, K.H. (2000) Nature (London) **404**, 892–897
79. Gapuzan, M.E.R., Yufit, P.V. and Gilmore, T.D. (2002) Oncogene **21**, 2484–2492
80. Ryan, K.M., O'Prey, J. and Vousden, K.H. (2004) Cancer Res. **64**, 4415–4418
81. Deng, J.O., Miller, S.A., Wang, H.Y., Xia, W.Y., Wen, Y., Zhou, B.H.P., Li, Y., Lin, S.Y. and Hung, M.C. (2002) Cancer Cell **2**, 323–334
82. Higashitsuji, H., Nagao, T., Nonoguchi, K., Fujii, S., Itoh, K. and Fujita, J. (2002) Cancer

Cell **2**, 335–346
83. Rocha, S., Martin, A.M., Meek, D.W. and Perkins, N.D. (2003) Mol. Cell. Biol. **23**, 4713–4727
84. Westerheide, S.D., Mayo, M.W., Anest, V., Hanson, J.L. and Baldwin, A.S. (2001) Mol. Cell. Biol. **21**, 8428–8436
85. Garkavtsev, I., Kozin, S.V., Chernova, O., Xu, L., Winkler, F., Brown, E., Barnett, G.H. and Jain, R.K. (2004) Nature (London) **428**, 328–332
86. Mayo, M.W., Madrid, L.V., Westerheide, S.D., Jones, D.R., Yuan, X.J., Baldwin, A.S. and Whang, Y.E. (2002) J. Biol. Chem. **277**, 11116–11125
87. Wolff, B. and Naumann, M. (1999) Oncogene **18**, 2663–2666
88. Ghiorzo, P., Mantelli, M., Gargiulo, S., Gramigni, C., Pastorino, L., Banelli, B., Villaggio, B., Coccia, M.C., Sementa, A.R., Garre, C. and Bianchi-Scarra, G. (2004) Hum. Pathol. **35**, 1029–1037
89. Wang, C.Y., Mayo, M.W. and Baldwin, Jr, A.S. (1996) Science **274**, 784–787
90. Wang, C.Y., Mayo, M.W., Korneluk, R.G., Goeddel, D.V. and Baldwin, Jr, A.S. (1998) Science **281**, 1680–1683
91. Wang, C.Y., Cusack, Jr, J.C., Liu, R. and Baldwin, Jr, A.S. (1999) Nat. Med. **5**, 412–417
92. Wang, C.Y., Guttridge, D.C., Mayo, M.W. and Baldwin, Jr, A.S. (1999) Mol. Cell. Biol. **19**, 5923–5929
93. Culmsee, C., Siewe, J., Junker, V., Retiounskaia, M., Schwarz, S., Camandola, S., El-Metainy, S., Behnke, H., Mattson, M.P. and Krieglstein, J. (2003) J. Neurosci. **23**, 8586–8595
94. Tergaonkar, V., Pando, M., Vafa, O., Wahl, G., and Verma, I. (2002) Cancer Cell **1**, 493–503
95. Huang, T.T., Wuerzberger-Davis, S.M., Seufzer, B.J., Shumway, S.D., Kurama, T., Boothman, D.A. and Miyamoto, S. (2000) J. Biol. Chem. **275**, 9501–9509
96. Mayo, M.W., Denlinger, C.E., Broad, R.M., Yeung, F., Reilly, E.T., Shi, Y. and Jones, D.R. (2003) J. Biol. Chem. **278**, 18980–18989
97. Arlt, A. (2001) Oncogene **20**, 859–868
98. Lavon, I., Pikarsky, E., Gutkovich, E., Goldberg, I., Bar, J., Oren, M. and Ben-Neriah, Y. (2003) Cancer Res. **63**, 25–30
99. Kasibhatla, S., Genestier, L., Echeverri, F., Mahboubi, A and Green, D.R. (1998) Mol. Cell **1**, 543–551
100. Wang, S., Kotamraju, S., Konorev, E., Kalivendi, S., Joseph, J. and Kalyanaraman, B. (2002) Biochem. J. **367**, 729–740
101. Huang, Y., Johnson, K.R., Norris, J.S. and Fan, W. (2000) Cancer Res. **60**, 4426–4432
102. Ivanov, V.N. and Ronai, Z. (2000) Oncogene **19**, 3003–3012
103. Gibson, S.B., Oyer, R., Spalding, A.C., Anderson, S.M. and Johnson, G.L. (2000) Mol. Cell. Biol. **20**, 205–212
104. Bian, X., McAllister-Lucas, L.M., Shao, F., Schumacher, K.R., Feng, Z., Porter, A.G., Castle, V.P. and Opipari, Jr, A.W. (2001) J. Biol. Chem. **276**, 48921–48929
105. Ashikawa, K., Shishodia, S., Fokt, I., Priebe, W. and Aggarwal, B.B. (2004) Biochem. Pharmacol. **67**, 353–364

Biochem. Soc. Symp. **73**, 181–189
(Printed in Great Britain)

17

The p53 response during DNA damage: impact of transcriptional cofactors

Amanda S. Coutts and Nicholas La Thangue[1]

Laboratory of Cancer Biology, Division of Medical Sciences, University of Oxford, Oxford OX3 9DU, U.K.

Abstract

Defects in the DNA damage response pathways can lead to tumour development. The tumour suppressor p53 is a key player in the DNA damage response, and the precise regulation of p53 is critical for the suppression of tumorigenesis. DNA damage induces the activity of p53, via damage sensors such as ATM (ataxia telangiectasia mutated) and ATR (ataxia telangiectasia-related), which leads to the transcriptional regulation of a variety of genes involved in cell cycle control and apoptosis. p53 is therefore tightly controlled, and its activity is regulated at a multiplicity of levels. An increasing array of cofactors are now known to influence p53 activity. Here we will discuss several of the cofactors that impact on p53 activity, specifically those involved in the function of the two novel p53 cofactors JMY (junction-mediating and regulatory protein) and Strap (serine/threonine-kinase-receptor-associated protein).

Introduction

The *p53* gene is one of the most commonly mutated genes in human cancer; approx. 50% of human cancers contain *p53* mutations, while the other half are thought to contain alterations in components of the p53 pathway [1]. p53 is a critical human tumour suppressor that plays a key role in the checkpoint response to DNA damage, and the precise regulation of p53 is necessary for both the checkpoint response and the suppression of tumorigenesis. The checkpoint response to DNA damage involves a myriad of cellular activities, including cell cycle arrest, DNA repair and apoptosis, and defects in the DNA damage response pathway lead to tumour development [2].

[1]To whom correspondence should be addressed (email nick.lathangue@ndcls.ox.ac.uk).

p53 is a nuclear transcription factor that affects cellular functions which include transcription, DNA synthesis and repair, cell cycle arrest, senescence and apoptosis [3]. Under normal conditions, p53 is held in a latent inactive state, but undergoes a significant increase in protein stability upon exposure to DNA damage. DNA damage stabilizes p53 in part via the DNA damage signalling pathway that involves the sensor kinases, including ATM (ataxia telangiectasia mutated) and ATR (ataxia telangiectasia-related), and effector kinases, such as Chk1 (checkpoint kinase 1) and Chk2, which leads to the transcriptional regulation of a variety of genes involved in cell cycle control and apoptosis [2,4]. p53 is therefore tightly controlled, and its activity is regulated at a multiplicity of levels. While post-translational modification plays an important role in p53 regulation [5], an increasing array of cofactors are now known to influence p53 activity. Here we will discuss several of the cofactors that impact on p53 activity, specifically those involved in the function of the two novel p53 cofactors JMY (junction-mediating and regulatory protein) and Strap (serine/threonine-kinase-receptor-associated protein).

The Mdm2 (murine double minute 2) oncoprotein

Mdm2 (or Hdm2 in humans; hereafter referred to as Mdm2) is an important human oncogene and a key regulator of the p53 response, in both the presence and the absence of DNA damage, where it plays a role in negatively regulating p53 activity [6,7]. Mdm2 binds to the N-terminal region of p53 and represses p53 activity via two main mechanisms: by promoting degradation and by blocking p53 transcriptional activation [8–10]. The importance of Mdm2 in regulating p53 activity is underscored by the finding that *mdm2*$^{-/-}$ mouse embryos die early in development due to massive p53-dependent apoptosis, which is rescued in the *mdm2*$^{-/-}$/*p53*$^{-/-}$ mouse [6,7,11]. Furthermore, because *mdm2* transcription is regulated by p53, transcriptional activation by p53 provides an autoregulatory feedback loop [12].

p53 is tightly regulated, and is normally maintained at low levels. After DNA damage, p53 is stabilized and its half-life increases. Because Mdm2 contains within it RING-finger domain E3 ligase activity responsible for p53 ubiquitination and proteasomal degradation, the negative regulation by Mdm2 must be relieved in order for stabilization and nuclear accumulation of p53 to occur. The mechanism of action of the Arf protein, which binds to Mdm2 causing the release of p53, is a paradigm in the regulation of p53, and we refer the reader to several recent reviews on this topic [13,14]. Similarly, post-translational control of p53 is well documented to influence p53 activity, and again we refer the reader to some excellent reviews [5,15].

However, the interaction of Mdm2 with p53 can be influenced by several other groups of proteins (Figure 1). For example, Mdm2 can, under certain circumstances, localize to the nucleolus [16,17], and contains within its C-terminal region an NoLS (nucleolar localization) signal [18]. A growing body of evidence supports the fact that nucleolar sequestration of Mdm2 influences p53 activity [19–22]. More recently, a role for ribosomal proteins has been suggested

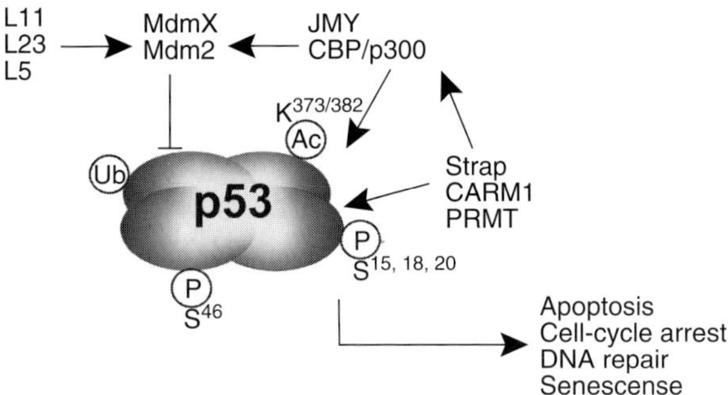

Figure 1 p53 cofactors in the DNA damage response. Schematic diagram representing p53 cofactors and their potential inter-relationships. DNA damage results in the stabilization and activation of p53. The outcome of p53 activation is dependent on the type of damage, cellular context and repertoire of cofactors. P, phosphorylation; Ac, acetylation; Ub, ubiquitination; S, serine; K, lysine.

in regulating the nucleolar localization of Mdm2. Mdm2 has been shown to interact with the ribosomal protein L11 [23], and this interaction is increased in response to the DNA-damaging agent actinomycin D [24]. L11 prevents Mdm2 ubiquitinating and degrading p53, most probably by inhibiting the E3 ubiquitin ligase activity of Mdm2 [24]. The ribosomal protein L23 also interacts with Mdm2, and L23 and L11 can interact simultaneously with Mdm2 [25,26]. Reducing the levels of L23 concomitantly increases p53 transcriptional activity [26], and the overexpression of L23 has been found to inhibit Mdm2-mediated p53 ubiquitination [25]. The ribosomal protein L5 can also interact with Mdm2 [27], although the functional significance of this interaction has not been determined. Further, the interaction of Mdm2 with L23, L11 and L5 can occur in the cytoplasm [25]. The fact that multiple ribosomal proteins interact with Mdm2 suggests an important link between ribosomal biogenesis and p53 activity, perhaps highlighting a pathway that integrates the p53 response with protein synthesis.

It is known that binding of Mdm2 conceals the N-terminal activation domain of p53 [10]. MdmX, which is a structural homologue of Mdm2, also binds to p53 and inhibits its transcriptional activity [28]. Furthermore, the interaction between Mdm2 and MdmX results in an increased half-life of Mdm2 [29]. Despite this, MdmX blocks Mdm2-mediated p53 degradation [30]. Although the role of MdmX in DNA damage-mediated control of p53 activity remains unclear, Mdm2 targets MdmX for degradation after DNA damage, which appears to be dependent on ATM-directed MdmX phosphorylation [31,32]. In addition, a role for MdmX in p53-dependent tumour suppressor activity has recently been demonstrated [33].

CBP/p300 family of transcriptional co-activators

CBP [CREB (cAMP response element-binding protein) binding protein] and p300 are ubiquitous transcriptional co-activators that interact with a plethora of sequence-specific transcription factors, and a variety of studies support a role for CBP/p300 in regulating p53-dependent transcription. CBP/p300 proteins bind to the p53 activation domain and stimulate p53 transcriptional activity [34–38]. Further, CBP/p300 proteins are acetyltransferases that acetylate not only chromatin, but also a variety of transcription factors, including p53 [39]. The C-terminal region of p53 contains two sites for acetylation by CBP/p300 and one for the CBP/p300-associated histone acetyltransferase P/CAF (p300/CBP-associated factor) [40]. It appears that, after DNA damage, p53 is phosphorylated at Ser^{15}, which increases the association between CBP/p300 and p53, and thereafter the acetylation of p53 [41]. Although the influence that acetylation has on p53 activity remains unclear, several reports suggest that acetylation increases p53 stability and promotes co-activator recruitment, leading to increased transcriptional activity [35,42–44]. Mdm2 prevents the interaction between p53 and p300, and it has also been shown to reduce p53 acetylation by recruiting a histone deacetylase 1 complex that binds Mdm2 and acts synergistically with Mdm2 to reduce the acetylation of p53 [45]. The regulatory events that may impact upon p300 activity are further underscored by the more recent findings that Mdm2 also influences the stability of P/CAF [46].

In addition, p300 binds to Mdm2, and a role for p300 in the Mdm2-dependent regulation of p53 levels has been suggested [47,48]. Kobet and co-workers [49] purified a p300–Mdm2–p53 protein complex from HeLa cell nuclear extracts, supporting the existence of a tripartite complex in cells. In addition, Mdm2 is acetylated *in vitro* by CBP/p300, within its C-terminal RING-finger domain, which may influence p53 degradation by impairing E3 ligase activity [50].

The JMY cofactor

JMY is a CBP/p300 cofactor that augments the p53 response [51]. The JMY transcript is expressed in a variety of tissues and encodes a protein of 110 kDa with no significant sequence similarity to other known proteins. JMY and p300 interact under physiological conditions, probably as a result of a direct interaction, and two regions in JMY are capable of interacting with p300. The fact that JMY increases p53-dependent transcription and apoptosis provides evidence that JMY plays a functional role in the p53 response. Importantly, ectopic expression of JMY did not alter p53 protein levels, which is consistent with the idea that it functions as a transcriptional cofactor [51]. Furthermore, actinomycin D treatment causes the formation of a complex containing p300, JMY and p53 (Figure 2), demonstrating the likely importance of JMY in the p53 stress response [51]. Most interestingly, JMY

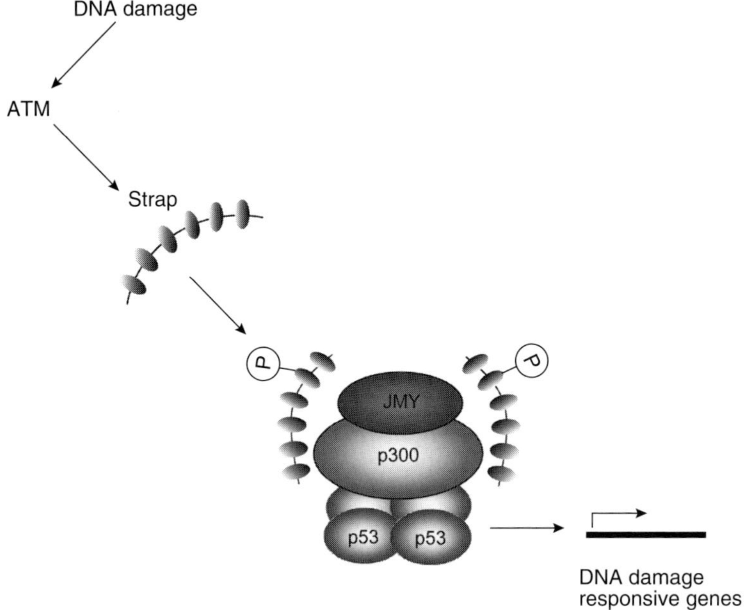

Figure 2 Role of JMY and Strap in the p53 response. Model summarizing the role of Strap in linking ATM kinase with the DNA damage response. It is envisaged that DNA damage activates ATM kinase, which thereafter phosphorylates Strap at Ser^{203}. Phosphorylated Strap is stabilized and undergoes accumulation in the nucleus, where it assembles into a co-activator complex that includes p300 and cofactors such as JMY. A part of the role of Strap is to augment p53 activity through increased acetylation and transcriptional activity, resulting in the activation of DNA damage response genes.

contains a C-terminal proline-rich domain which is absent in certain protein isoforms that display reduced activation of p53-dependent transcription and apoptosis, but co-operate with p53 to promote cell cycle arrest more efficiently than wild-type JMY [51]. It is possible, but as yet unproven, that the JMY expression level and isoform composition influences the properties of the p53 response.

Cofactor Strap

Strap was originally identified as a novel JMY-interacting protein [52]. Notably, Strap contains a tandem series of six TPR (tetratricopeptide) repeats, a motif thought to function in protein–protein interactions [52]. As Strap is composed almost entirely of TPR motifs, it is likely that this protein is involved in regulating and mediating the assembly of multi-protein complexes. In support of this, Strap contains distinct regions that interact with either JMY or p300, and which facilitate the interaction between JMY and p300 [52]. Strap is also involved in the p53 response, and ectopic expression of Strap increases the levels and half-life of endogenous p53. Furthermore, Strap prevents Mdm2-mediated down-regulation of p53 levels and increases p53-dependent apoptosis, providing further evidence for an important role for Strap in regulating the p53 response [52].

More recently, the DNA-damage-responsive ATM kinase was found to phosphorylate Strap, causing increased nuclear accumulation and association with p300 [53]. Specifically, ATM phosphorylates Strap at Ser^{203}, and this is important in the control of Strap stability and subcellular localization (Figure 2). Consistent with this, in AT (ataxia telangiectasia) cells, which have missing or inactive ATM kinase, Strap remains cytoplasmic and cannot locate to the nucleus during DNA damage. In normal cells, the DNA damage-dependent increase in Strap prompts p53 acetylation through recruitment of p300/JMY, thereafter enhancing p53-dependent apoptosis [53]. This requires ATM-dependent phosphorylation, since a mutant derivative of Strap lacking the ATM phosphorylation site has a reduced effect on p53 [53]. These studies have elucidated a new DNA damage-responsive co-activator complex that integrates upstream signals derived from the ATM family of sensor kinases with the transcription apparatus.

Chromatin-modulating proteins

Because p53 acts as a transcription factor, chromatin needs to be modified in order to facilitate changes in transcriptional activity. As discussed above, the p300/CBP co-activators, upon recruitment to p53, acetylate both the local chromatin environment and p53 itself [39]. In addition to p300, other distinct co-activator complexes have been shown to function in p53-dependent transcription, including the SAGA (Spt/Ada/Gcn5 acetyltransferase) and NuA4–Tip60 complexes (reviewed in [54]), whose common subunit, the ATM-related protein TRRAP (transformation/transcription domain-associated protein), binds directly to p53 and is involved in p53-dependent transcription [55]. However, in a similar fashion to the role of acetylation in gene activation, histone lysine and arginine methylation are enzymatic events that influence gene expression [56]. The arginine methyltransferases CARM1 (co-activator-associated arginine methyltransferase 1) and PRMT1 (protein-arginine N-methyltransferase 1) have been implicated in p53 transcription, where they co-operate with p300 in activating the *GADD45* (growth arrest and DNA damage-inducible protein) gene and assemble with the *GADD45* promoter during the p53 response [57]. Interestingly, PRMT1 and p300 both interact with the N-terminal region of p53, whereas CARM1 binds preferentially to the C-terminal region of p53 [57].

An important question relates to the interplay between chromatin modulators such as PRMT1 and CARM1, and cofactors such as JMY and Strap. Is there, for example, any sequential control and temporal order in the assembly of co-activator complexes at p53-responsive promoters, or do different factors act in a gene-specific fashion? We await the answers to these interesting questions.

Conclusions

It has become increasingly clear that numerous p53 cofactors exist that exert positive and negative effects on p53 activity. Although a more complete understanding of the factors themselves and signals that control their activity is required in order to gain a greater understanding of the multiple regulatory levels that p53 is subjected to, we can, nevertheless, reach some general conclusions. For example, subnuclear localization of both p53 and its various cofactors clearly plays an important role in the DNA damage response. Significantly, the nucleolus and nucleolar proteins are important players in the response of p53 after DNA damage, and the list of nucleolar proteins capable of affecting the p53 response continues to expand. Future research will no doubt shed light on the signals that govern these interactions, and the spatial and temporal influences on p53 activity.

An important area of research that needs to be explored more fully is the role of tissue-specific differences in the levels of p53 cofactors in tumour development and therapeutic responses. Similarly, differences in the tissue-specific distribution of the p53 cofactors JMY and Strap may contribute to the differing response of p53 to damaging agents. We suspect that a more complete understanding of the regulation of the factors that can control p53 activity will be of significance in developing a better understanding of therapeutic strategies to treat cancer.

Further, the interplay between cofactors in the regulation of p53-dependent genes, particularly chromatin control, is likely to yield important insights into the DNA damage response. We have only discussed a few of the numerous cofactors that assemble with and influence p300/CBP activity. A detailed understanding of the interplay between these factors and their role in the p53 response will undoubtedly uncover new complexities in p53 control. We await with interest information on the pathways and physiological cues that control the expanding group of cofactors implicated in p53 regulation.

We thank Marie Caldwell for help in preparing the manuscript. We apologize to those whose work could not be cited due to space constraints. Work in our laboratory is supported by the Medical Research Council, Cancer Research UK, the Leukaemia Research Fund, the Association for International Cancer Research and the EU.

References

1. Soussi, T. and Beroud, C. (2001) Nat. Rev. Cancer **1**, 233–240
2. Ljungman, M. and Lane, D.P. (2004) Nat. Rev. Cancer **4**, 727–737
3. Zhou, B.B. and Elledge, S.J. (2000) Nature (London) **408**, 433–439
4. Meek, D.W. (2004) DNA Repair **3**, 1049–1056
5. Bode, A.M. and Dong, Z. (2004) Nat. Rev. Cancer **4**, 793–805
6. Jones, S.N., Roe, A.E., Donehower, L.A. and Bradley, A. (1995) Nature (London) **378**, 206–208
7. Montes de Oca Luna, R., Wagner, D.S. and Lozano, G. (1995) Nature (London) **378**, 203–206
8. Haupt, Y., Maya, R., Kazaz, A. and Oren, M. (1997) Nature (London) **387**, 296–299
9. Kubbutat, M.H., Jones, S.N. and Vousden, K.H. (1997) Nature (London) **387**, 299–303
10. Momand, J., Zambetti, G.P., Olson, D.C., George, D. and Levine, A.J. (1992) Cell **69**, 1237–1245

11. de Rozieres, S., Maya, R., Oren, M. and Lozano, G. (2000) Oncogene **19**, 1691–1697
12. Barak, Y., Juven, T., Haffner, R. and Oren, M. (1993) EMBO J. **12**, 461–468
13. Lowe, S.W. and Sherr, C.J. (2003) Curr. Opin. Genet. Dev. **13**, 77–83
14. Sherr, C.J. (2001) Nat. Rev. Mol. Cell Biol. **2**, 731–737
15. Xu, Y. (2003) Cell Death Differ. **10**, 400–403
16. Ashcroft, M., Taya, Y. and Vousden, K.H. (2000) Mol. Cell. Biol. **20**, 3224–3233
17. Poyurovsky, M.V., Jacq, X., Ma, C., Karni-Schmidt, O., Parker, P.J., Chalfie, M., Manley, J.L. and Prives, C. (2003) Mol. Cell **12**, 875–887
18. Lohrum, M.A., Ashcroft, M., Kubbutat, M.H. and Vousden, K.H. (2000) Nat. Cell Biol. **2**, 179–181
19. Bernardi, R., Scaglioni, P.P., Bergmann, S., Horn, H.F., Vousden, K.H. and Pandolfi, P.P. (2004) Nat. Cell Biol. **6**, 665–672
20. Kurki, S., Latonen, L. and Laiho, M. (2003) J. Cell Sci. **116**, 3917–3925
21. Kurki, S., Peltonen, K., Latonen, L., Kiviharju, T.M., Ojala, P.M., Meek, D. and Laiho, M. (2004) Cancer Cell **5**, 465–475
22. Weber, J.D., Taylor, L.J., Roussel, M.F., Sherr, C.J. and Bar-Sagi, D. (1999) Nat. Cell Biol. **1**, 20–26
23. Lohrum, M.A., Ludwig, R.L., Kubbutat, M.H., Hanlon, M. and Vousden, K.H. (2003) Cancer Cell **3**, 577–587
24. Zhang, Y., Wolf, G.W., Bhat, K., Jin, A., Allio, T., Burkhart, W.A. and Xiong, Y. (2003) Mol. Cell. Biol. **23**, 8902–8912
25. Dai, M.S., Zeng, S.X., Jin, Y., Sun, X.X., David, L. and Lu, H. (2004) Mol. Cell. Biol. **24**, 7654–7668
26. Jin, A., Itahana, K., O'Keefe, K. and Zhang, Y. (2004) Mol. Cell. Biol. **24**, 7669–7680
27. Marechal, V., Elenbaas, B., Piette, J., Nicolas, J.C. and Levine, A.J. (1994) Mol. Cell. Biol. **14**, 7414–7420
28. Shvarts, A., Steegenga, W.T., Riteco, N., van Laar, T., Dekker, P., Bazuine, M., van Ham, R.C., van der Houven van Oordt, W., Hateboer, G., van der Eb, A.J. and Jochemsen, A.G. (1996) EMBO J. **15**, 5349–5357
29. Tanimura, S., Ohtsuka, S., Mitsui, K., Shirouzu, K., Yoshimura, A. and Ohtsubo, M. (1999) FEBS Lett. **447**, 5–9
30. Jackson, M.W. and Berberich, S.J. (2000) Mol. Cell. Biol. **20**, 1001–1007
31. Pereg, Y., Shkedy, D., de Graaf, P., Meulmeester, E., Edelson-Averbukh, M., Salek, M., Biton, S., Teunisse, A.F., Lehmann, W.D., Jochemsen, A.G. and Shiloh, Y. (2005) Proc. Natl. Acad. Sci. U.S.A. **102**, 5056–5061
32. Kawaim H., Wiederschain, D., Kitao, H., Stuart, J., Tsai, K.K. and Yuan, Z.M. (2003) J. Biol. Chem. **278**, 45946–45953
33. Danovi, D., Meulmeester, E., Pasini, D., Migliorini, D., Capra, M., Frenk, R., de Graaf, P., Francoz, S., Gasparini, P., Gobbi, A. et al. (2004) Mol. Cell. Biol. **24**, 5835–5843
34. Avantaggiati, M.L., Ogryzko, V., Gardner, K., Giordano, A., Levine, A.S. and Kelly, K. (1997) Cell **89**, 1175–1184
35. Gu, W. and Roeder, R.G. (1997) Cell **90**, 595–606
36. Lill, N.L., Grossman, S.R., Ginsberg, D., DeCaprio, J. and Livingston, D.M. (1997) Nature (London) **387**, 823–827
37. Scolnick, D.M., Chehab, N.H., Stavridi, E.S., Lien, M.C., Caruso, L., Moran, E., Berger, S.L. and Halazonetis, T.D. (1997) Cancer Res. **57**, 3693–3696
38. Lee, C.W., Sorensen, T.S., Shikama, N. and La Thangue, N.B. (1998) Oncogene **16**, 2695–2710
39. Chan, H.M. and La Thangue, N.B. (2001) J. Cell Sci. **114**, 2363–2373
40. Liu, L., Scolnick, D.M., Trievel, R.C., Zhang, H.B., Marmorstein, R., Halazonetis, T.D. and Berger, S.L. (1999) Mol. Cell. Biol. **19**, 1202–1209
41. Lambert, P.F., Kashanchi, F., Radonovich, M.F., Shiekhattar, R. and Brady, J.N. (1998) J. Biol. Chem. **273**, 33048–33053
42. Barlev, N.A., Liu, L., Chehab, N.H., Mansfield, K., Harris, K.G., Halazonetis, T.D. and Berger, S.L. (2001) Mol. Cell **8**, 1243–1254
43. Ito, A., Lai, C.H., Zhao, X., Saito, S., Hamilton, M.H., Appella, E. and Yao, T.P. (2001) EMBO J. **20**, 1331–1340
44. Yuan, Z.M., Huang, Y., Ishiko, T., Nakada, S., Utsugisawa, T., Shioya, H., Utsugisawa, Y., Yokoyama, K., Weichselbaum, R., Shi, Y. and Kufe, D. (1999) J. Biol. Chem. **274**, 1883–

1886
45. Ito, A., Kawaguchi, Y., Lai, C.H., Kovacs, J.J., Higashimoto, Y., Appella, E. and Yao, T.P. (2002) EMBO J. **21**, 6236–6245
46. Jin, Y., Zeng, S.X., Lee, H. and Lu, H. (2004) J. Biol. Chem. **279**, 20035–20043
47. Grossman, S.R., Perez, M., Kung, A.L., Joseph, M., Mansur, C., Xiao, Z.X., Kumar, S., Howley, P.M. and Livingston, D.M. (1998) Mol. Cell **2**, 405–415
48. Wadgaonkar, R. and Collins, T. (1999) J. Biol. Chem. **274**, 13760–13767
49. Kobet, E., Zeng, X., Zhu, Y., Keller, D. and Lu, H. (2000) Proc. Natl. Acad. Sci. U.S.A. **97**, 12547–12552
50. Wang, X., Taplick, J., Geva, N. and Oren, M. (2004) FEBS Lett. **561**, 195–201
51. Shikama, N., Lee, C.W., France, S., Delavaine, L., Lyon, J., Krstic-Demonacos, M. and La Thangue, N.B. (1999) Mol. Cell **4**, 365–376
52. Demonacos, C., Krstic-Demonacos, M. and La Thangue, N.B. (2001) Mol. Cell **8**, 71–84
53. Demonacos, C., Krstic-Demonacos, M., Smith, L., Xu, D., O'Connor, D.P., Jansson, M. and La Thangue, N.B. (2004) Nat. Cell Biol. **6**, 968–976
54. Berger, S L. (2002) Curr. Opin. Genet. Dev. **12**, 142–148
55. Ard, P.G., Chatterjee, C., Kunjibettu, S., Adside, L.R., Gralinski, L.E. and McMahon, S.B. (2002) Mol. Cell. Biol. **22**, 5650–5661
56. Stallcup, M.R. (2001) Oncogene **20**, 3014–3020
57. An, W., Kim, J. and Roeder, R.G. (2004) Cell **117**, 735–748

Biochem. Soc. Symp. **73**, 191–201
(Printed in Great Britain)

18

The modulation of WT1 transcription function by cofactors

Stefan G.E. Roberts[1]

Faculty of Life Sciences, The Michael Smith Building, University of Manchester, Oxford Road, Manchester M13 9PT, U.K.

Abstract

Wilms' tumour is a paediatric malignancy of the kidneys that affects one in every 10 000 live births, making it the most common solid tumour in the young. This cancer arises due to a failure of the metanephric mesenchyme to differentiate and form the kidney filtration units and tubules, which instead undergo uncontrolled proliferation. WT1 (Wilms' tumour 1) was identified as a factor that is frequently mutated in Wilms' tumours. WT1 plays a central role in the development of the genito-urinary organs and also other regions of the embryo. A major function of WT1 is to act as a regulator of transcription, controlling the expression of genes that are involved in proliferation and differentiation. WT1 can either activate or repress transcription of its target genes. Thus the transcription function of WT1 is highly context-specific, and can be modulated by a number of cofactors. Here, the known interaction partners of WT1 and the mechanisms by which they modulate WT1 transcription function will be discussed.

Introduction

Wilms' tumour is a paediatric malignancy of the kidneys that affects 1 in 10 000 children (reviewed in [1,2]). The tumour arises from abnormal development of the kidneys that leads to increased cellular proliferation instead of differentiation. The Wilms' tumour suppressor protein WT1 (Wilms' tumour 1) was identified on the basis of its frequent mutation in Wilms' tumours. However, WT1 mutations are present in only 10–15% of Wilms' tumours, and while some also show elevated expression of WT1, it is clear that other factors can also contribute to Wilms tumorigenesis.

[1]Email Stefan.Roberts@manchester.ac.uk

WT1 null mice die *in utero*, displaying complete agenesis of the kidneys and also several other organs, including gonad, spleen and adrenal glands [3,4]. Thus WT1 is a critical regulator of the development of the genito-urinary system and other organs. In addition, areas of the nervous system, including the retinal ganglia and olfactory epithelium, fail to develop in WT1 knockout mice [5,6]. The mice die, however, due to defects in heart development that most probably arise from a requirement for WT1 in the formation of the mesothelium that lines the internal organs [3].

The initial analysis of the primary structure of WT1 revealed the presence of the four zinc fingers similar to those found in the transcription factor Egr1 (Figure 1) [7]. The N-terminus of WT1 contains a proline-rich region that is typical of transcriptional regulators. It was thus proposed that WT1 might be a transcription factor, and subsequent transient transfection experiments demonstrated that WT1 was able to repress transcription at reporter constructs containing the promoters of several genes, including those encoding insulin-like growth factor II and platelet-derived growth factor A [8–10]. These gene products, among others, are elevated in Wilms' tumours, and the finding that WT1 could repress their transcription provided a working hypothesis for how loss of WT1 function might lead to Wilms' tumorigenesis.

WT1 is subject to alternative splicing at two sites (Figure 1). One of the alternative forms of WT1 contains an additional three amino acids (lysine, threonine, serine; KTS) between zinc fingers 3 and 4. This alternative splice form is conserved in WT1 from all species for which sequence information is available [11]. The other alternative splice site inserts an additional 17 amino acids at the C-terminus of the transcriptional regulatory region and is found only in mammalian WT1 (17aa form). The number of potential WT1 isoforms is increased further by both upstream and downstream alternative translation events (reviewed in [1]). Moreover, RNA editing of WT1 has been reported that results in either leucine or proline at position 280 [12]. Finally, a recent

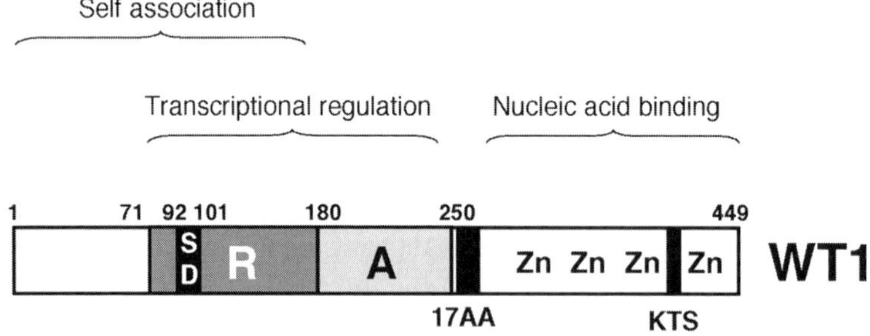

Figure 1 Schematic diagram of WT1. Shown are the repression domain (R), the activation domain (A) and zinc fingers (Zn). The 10-amino-acid suppression domain (SD) is incorporated, and the alternative splice sites of WT1 (KTS and 17AA) are indicated by solid bars. The N-terminal self-association region is also indicated. The amino acid numbers refer to the isoform of WT1 that contains both the KTS and 17-amino-acid splice insertions.

report has described an alternative promoter of the *WT1* gene that resides within the first intron [13]. This promoter results in the generation of a novel, smaller form of WT1 that lacks the N-terminal transcriptional repression region. This alternative promoter is paternally imprinted, a property that appears to be lost in some Wilms' tumours.

Analysis of the +KTS and −KTS forms of WT1 suggests that they may have different functions in gene expression. Although both forms of WT1 can bind to both DNA and RNA, the −KTS form shows a preference for the former and the +KTS form for the latter (for review see [1]). In addition, immunofluorescence studies have demonstrated that the +KTS and −KTS forms associate with different regions of the nucleus [14]. WT1 −KTS associates with transcription factors within the nucleus, while WT1 +KTS associates with coiled bodies and splicing factors [15,16].

Mice have been generated that express exclusively either the +KTS or the −KTS form of WT1 ([17]; reviewed in [18]). Unlike the completely WT1 null mice, both the +KTS and −KTS mice survive to birth. This suggests that there is functional overlap between the two isoforms. Indeed, both mice show development of the adrenal glands and spleen, suggesting that the +KTS and −KTS forms can perform the same functions in these organs. However, in the gonads and kidney, neither the +KTS or the −KTS form of WT1 was able to provide complete function by itself. Thus mice that lack the +KTS form of WT1 develop as females even when they are XY status. On the other hand, although both the +KTS and −KTS forms of WT1 are required for complete development of the kidney, the −KTS form plays a more important role, and its ablation causes a severe reduction in nephrogenesis.

Mice expressing a truncated form of WT1 that lacks nucleic-acid-binding activity have also been generated [19–21]. These mice mimic the effects of WT1 mutations found in the human condition DDS (Denys Drash syndrome), which is typified by gonad abnormalities, nephropathy and a high incidence of Wilms' tumour. Because WT1 self-associates via its N-terminus, it is believed that these forms of WT1, through heterodimerization, act as dominant negative inhibitors of wild-type WT1 [22–24]. The 'Denys Drash mice' exhibit genito-urinary abnormalities similar to those observed in human DDS patients.

In addition to the above, mice have been generated that lack the 17aa isoform of WT1 [25]. These mice do not exhibit any defects in genito-urinary development or reproduction. However, mice that express truncated forms of WT1 (the DDS mutations described above) show differences when the truncated WT1 contains the 17 amino acids [21]. Thus, although the 17-amino-acid insert does not appear to be critical in mouse development, its presence can exert an effect in a defective WT1 background. Mice have also been generated that are unable to produce WT1 that is derived from the upstream translation event [26]. Again, these mice do not show any obvious defects in development of the genito-urinary system. Thus it is possible that the 17aa and N-terminally extended forms of WT1 have redundant functions, or that they are not needed in mouse development.

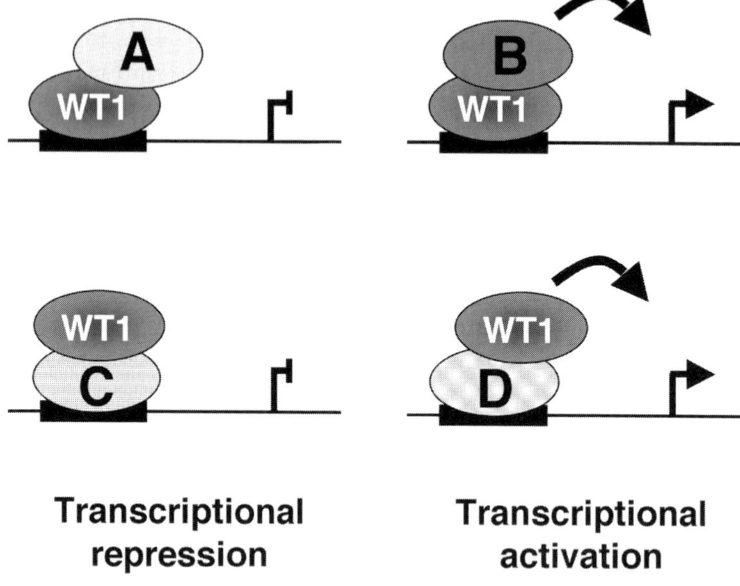

Figure 2 Mechanisms of action of WT1 cofactors. Factors A, B, C and D are shown as potential WT1-interacting proteins. Transcriptional repression can result from factor A binding to DNA-bound WT1 or from WT1 interacting with DNA-bound factor C. Transcriptional activation can occur when co-activator B interacts with DNA-bound WT1 or when WT1 interacts with DNA-bound factor D.

Transcriptional regulation by WT1

As stated above, WT1 was initially characterized as a transcriptional regulator, and specifically as a repressor of transcription. However, further work suggested that this may not be the full story. The ability of WT1 to repress transcription was highly dependent on the cell line that was used and also the precise promoter construct that was employed (reviewed in [1]). In addition, the transcription effects mediated by WT1 were dependent upon the vector that was used to drive WT1 expression. At this time, reports were emerging that WT1 could activate transcription at some promoters (e.g. syndecan 1 [27] and p21 [28]). DNA microarray analyses were in agreement with this notion, finding that amphiregulin and podocalyxin were targets of WT1, and that the effect mediated was transcriptional activation [29,30]. Amphiregulin is a growth factor of the epidermal growth factor family and is expressed in the kidneys during nephrogenesis. In the kidney, podocalyxin is expressed specifically in the podocytes, where it is important for the elaborate morphology of these cells, which form a crucial part of the filtration barrier. Indeed, WT1 expression is restricted to the podocytes of the adult kidney, where it is required to maintain their differentiated state. More recently, nephrin, another gene product that is critical in podocyte morphology, was described as a target gene that is transactivated by WT1 [31,32]. The gene encoding the anti-apoptotic factor Bcl-2 has also been reported as a target of WT1 [33]. Several labs have reported the effects of WT1 on apoptosis (reviewed in [1]). Significantly, Bcl-2 knockout mice show defects in kidney development typified by a significant decrease in the number of nephrons.

It is now clear that transcriptional regulation by WT1 is highly context-dependent. The prevailing view is that the disparate effects of WT1 are mediated by factors that associate with WT1 (Figure 2). Indeed, gel filtration chromatography studies suggest that WT1 is contained within large complexes [34]. Thus the last few years have seen a major effort to identify interaction partners for WT1. These WT1-binding proteins may provide clues to the cell-type- and promoter-dependent activities of WT1. Here I will discuss our current understanding of the function of the WT1 interaction partners. This will be restricted to the WT1 interaction partners that have been implicated in the transcription function of WT1. However, it is important to appreciate that, as transcription and RNA processing are integrated activities, WT1-binding proteins that have been proposed to play a role in RNA processing may well ultimately be found to affect transcriptional regulation by WT1.

DNA-binding proteins that interact with WT1

The p53 family

WT1 has been found to interact with members of the p53 family. In baby rat kidney cells, WT1 and p53 co-immunoprecipitate, and WT1 can stabilize p53 and thereby enhance DNA binding by p53 and transcriptional activation [35]. The interaction involves the zinc finger region of WT1. p53 can also augment transcriptional repression by WT1 in reporter assays. Analysis of p53/WT1 compound transgenic mice adds further weight to the notion of a functional interplay between p53 and WT1 [36]. WT1/p53 null mouse embryos reach a later stage in gestation before resorption than WT1 null mice. However, the underlying physiological mechanisms that are involved remain to be determined. WT1 can also interact with p73 via its zinc fingers [37]. This interaction modulates DNA binding by WT1, and reduces transcriptional activation mediated by both WT1 and p73.

Cholesterol biosynthesis pathway

A microarray screen was used to find genes that are regulated following the ablation of WT1 expression in M15 cells using an antisense approach [38]. In addition, a microarray was performed following ectopic expression of WT1 in HEK 293 cells. A common finding using both approaches was a cluster of genes involved in cholesterol biosynthesis that were found to be repressed by WT1. The transcription of these genes is normally activated by SREBPs (sterol responsive element-binding proteins), which bind to their promoter regions. WT1 can interact directly with SREBP, and it was suggested that WT1 acts as a co-repressor of SREBPs. In the developing kidney, WT1 expression is correlated with a reduction in the expression of the proteins involved in cholesterol biosynthesis.

WT1 partners in the gonad

The WT1 –KTS isoform can interact with SF-1 (steroidogenic factor-1) [39]. Together, they synergistically activate the MIS (Mullerian inhibiting substance) promoter. WT1 mutations that result in male pseudohermaphroditism are unable to co-operate with SF-1, suggesting a molecular basis for the WT1-associated syndromes that result in gonadal defects. Dax-1 can antagonize the co-operative transcriptional activation mediated by WT1/SF-1 by binding to SF-1. Interestingly, the co-activator FHL2 (four-and-a-half LIM domain protein) interacts with WT1 and potentiates WT1-mediated transcriptional activation of the Dax-1 promoter [40]. FHL2 and WT1 are co-expressed in developing testes and ovary. It was proposed that, during differentiation of the ovary, WT1 and FHL2 co-operate to activate the Dax-1 gene, which in turn causes repression of the MIS gene. Thus WT1, SF-1 and Dax-1 integrate to form a molecular switch in gonad differentiation.

'Classical' transcription targets of WT1

The transcriptional activation region of WT1 resides within residues 180–250 [41]. As yet we know very little about how the WT1 transcriptional activation domain functions. A glutathione S-transferase fusion protein containing the WT1 activation domain can bind to the general transcription factor TFIIB, which is an interaction partner for many activation domains [42]. However, there are no functional data regarding the mechanism of action of the WT1 transcriptional activation domain.

WT1 has been demonstrated to interact with the co-activators CBP [CREB (cAMP response element-binding protein)-binding protein] and p300 [43]. These interactions enhance transcriptional activation by WT1. CBP interacts with the zinc finger DNA-binding region of WT1. The zinc finger region of WT1 is not in itself a transcriptional activation domain, but it is possible that it co-operates with the transcriptional activation domain in the context of the intact protein.

It has been reported that WT1 can interact with the adenovirus E1b 55 kDa protein [44]. This interaction also occurs via the zinc fingers. The interaction with E1b 55K sequesters WT1 along with p53 into cytoplasmic bodies. Although it is not clear how this regulates gene expression, E1b 55K is able to abrogate WT1-mediated induction of apoptosis.

Non DNA-binding proteins

WT1 has been shown to interact with a number of proteins that do not directly contact DNA, but are implicated in transcription through known pathways or via protein sequence motifs. Immunoprecipitation of WT1 from cell extracts identified Hsp70 (heat-shock protein 70) as a WT1-interacting protein [45]. Hsp70 interacts with the N-terminus of WT1, a region known to be involved in WT1 self-association. It was also found that expression of WT1 leads to transactivation of the Hsp70 promoter. The promoter element responsible for this effect is the HSE (heat shock element). A model is proposed in which WT1 binds to Hsp70 via its N-terminus, and in so doing releases HSF (heat-shock factor) from Hsp70, which then transactivates the Hsp70 promoter. Hsp70 does not appear to affect transcriptional regulation mediated by WT1.

BMZF2 (bone marrow zinc finger 2) was isolated as a WT1-interacting protein by affinity chromatography using a column containing the WT1 zinc fingers [46]. BMZF2 inhibits transcriptional activation mediated by WT1 in reporter assays. BMZF2 is expressed primarily in the embryo, where expression is observed in the developing kidney. In addition, as its name suggests, BMZF2 is expressed in the haematopoietic lineage, in which WT1 is also present.

Ciao1 was isolated in a yeast two-hybrid screen to identify factors that interact with the full-length WT1 –KTS isoform [47]. Ciao1 contains seven WD40 repeats, a motif known to be involved in the formation of protein complexes. The zinc finger region of WT1 appears to be involved in the interaction with Ciao1. In transfection assays Ciao1 inhibits WT1-mediated transcriptional activation at a synthetic reporter.

Par-4 (prostate apoptosis response-4)

WT1 has also been shown to interact with several proteins that have not previously been associated with gene regulation. One of these, Par-4, is a protein that is involved in apoptosis. Par-4 was isolated as a WT1-interacting protein in a yeast two-hybrid screen using intact WT1 as bait [48]. Par-4 was shown to augment transcriptional repression by WT1, which was mediated by a direct interaction with the WT1 zinc fingers. It has become apparent that Par-4 can contact a second region in WT1 that is formed by the insertion of the extra 17 amino acids by alternative splicing [49]. In this context, the region encompassing the 17 amino acids forms a second transcriptional activation domain in WT1 that is dependent upon Par-4 as a transcriptional co-activator. Thus Par-4 can act as either a transcriptional co-activator or as a co-repressor for WT1, and this is dependent upon the splice isoform of WT1. The interaction of the 17aa isoform of WT1 with Par-4 appears to confer a cell-survival function and is able to rescue cells from UV-mediated apoptosis [49].

WTIP (WT1-interacting protein)

A yeast two-hybrid screen using the WT1 –KTS isoform as bait isolated a novel LIM-domain-containing protein, WTIP [50]. WTIP is co-expressed with WT1 in the forming glomeruli of the developing kidney, and was also expressed in a podocyte cell line. In the latter case WTIP was localized to the cytoplasm. However, inhibition of the CRM1 (chromosome region maintenance 1) nuclear export pathway results in nuclear accumulation of WTIP and subsequent interaction with WT1. This leads to inhibition of transcriptional activation by

WT1. These results suggested a context-dependent function for the WT1–WTIP interaction. Treatment of cultured podocytes with puromycin aminonucleoside causes disruption of cell–cell contacts and retraction of the podocyte foot processes. This leads to nuclear translocation of WTIP and association with WT1 [51]. Thus it is possible that the WT1–WTIP association is regulated by damage to the podocytes, and that this then elicits a changed programme of gene expression. This scenario is in agreement with the notion that WT1 is needed in the podocyte cells to help maintain their differentiated state.

BASP1 (brain acid-soluble protein 1)

Using an *in vitro* transcription assay, we functionally identified a WT1 transcriptional co-suppressor protein [52]. The N-terminus of WT1 contains a region (suppression domain) that inhibits the transcriptional activation domain (see Figure 1). This suppression domain is a potent inhibitor of transcriptional activation *in vitro* at a synthetic reporter. Using a functional transcription assay system to monitor WT1 transcriptional co-suppressor activity, BASP1 was identified as an interaction partner for the WT1 suppression domain. BASP1 is expressed in the developing nephrons of the embryonic kidney and, like WT1, becomes restricted to the podocyte cells of the adult kidney. BASP1 elicits

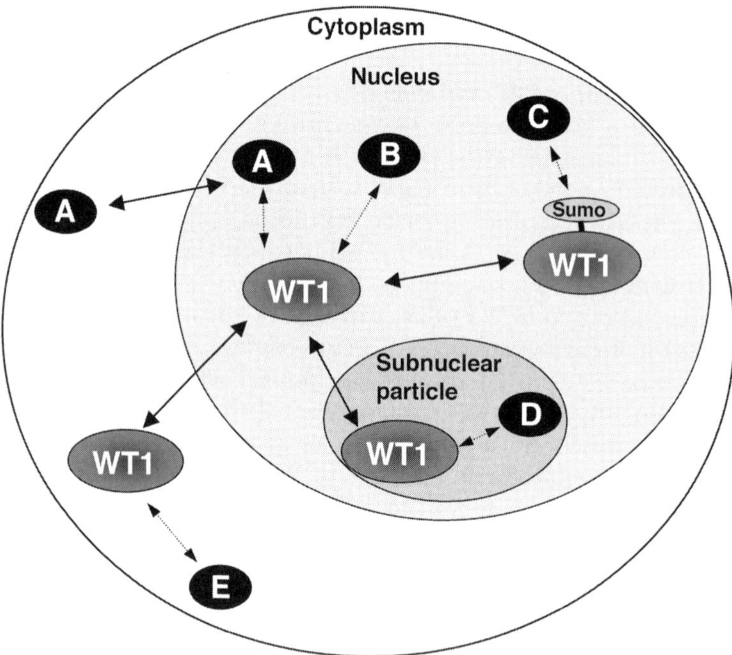

Figure 3 Regulation of WT1–cofactor interactions. A schematic of a mammalian cell is shown, with cytoplasm, nucleus and subnuclear particle indicated. WT1 and cofactors A–E are shown, with solid arrows indicating movement between compartments or modification (e.g. sumoylation). Dotted arrows indicate points of interaction between cofactor and WT1.

WT1-dependent transcriptional co-suppressor activity in transient transfection assays. In addition, ablation of BASP1 expression using siRNA (small interfering RNA) leads to transcriptional activation of the amphiregulin gene by WT1.

Modulation of WT1 transcription function

Our understanding of WT1 function so far suggests that a large number of factors can modulate its activities. This raises the question of how these different interactions are integrated into a coherent functional outcome (Figure 3). What are the circumstances under which these different interactions with WT1 occur? This could be achieved by regulation of the expression of the WT1 cofactor. For example, Par-4 levels are elevated in response to pro-apoptotic conditions. Also, the WT1–WTIP interaction is modulated by translocation of WTIP to the nucleus following damage to podocytes. In both of these cases, the interaction of WT1 with the cofactor is context-specific, either by modulation of the level of the cofactor or by relocalization of the cofactor.

It is also possible that the ability of WT1 to interact with its cofactors is regulated by direct modification of WT1 itself. For example, WT1 has recently been shown to be sumoylated [53], and this may modulate its ability to interact with target factors. Phosphorylation of WT1 [54] might also regulate the ability of WT1 to interact with other proteins. In addition, WT1 has been shown to undergo nucleocytoplasmic shuttling, which could affect its availability to interact with other factors [55,56]. Some of the interactions with WT1 have been found to be WT1-isoform-specific. In addition, the subnuclear localization of WT1 shows isoform specificity. These properties of WT1 are likely to lead to the modulation of WT1 interaction events, with resulting changes in WT1-mediated transcriptional events. Future studies will focus on how WT1-associated proteins modify WT1 function in development and disease.

I thank the AICR, Cancer Research UK and the Wellcome Trust for their support. S.G.E.R. is a Wellcome Trust Senior Research Fellow.

References

1. Discenza, M.T. and Pelletier, J. (2004) Physiol. Genomics **16**, 287–300
2. Scholz, H. and Kirschner, K.M. (2005) Physiology **20**, 54–59
3. Kreidberg, J.A., Sariola, H., Loring, J.M., Maeda, M., Pelletier, J., Housman, D. and Jaenisch, R. (1993) Cell **74**, 679–691
4. Herzer, U., Crocoll, A., Barton, D., Howells, N. and Englert, C. (1999) Curr. Biol. **9**, 837–840
5. Wagner, K.D., Wagner, N., Vidal, V.P., Schley, G., Wilhelm, D., Schedl, A., Englert, C. and Scholz, H. (2002) EMBO J. **21**, 1398–1405
6. Wagner, N., Wagner, K.D., Hammes, A., Kirschner, K.M., Vidal, V.P., Schedl, A. and Scholz, H. (2005) Development **132**, 1327–1336
7. Madden, S.L., Cook, D.M., Morris, J.F., Gashler, A., Sukhatme, V.P. and Rauscher, III, F.J. (1991) Science **253**, 1550–1553
8. Gashler, A.L., Bonthron, D.T., Madden, S.L., Rauscher, F.J., Collins, T. and Sukhatme, V.P. (1992) Proc. Natl. Acad. Sci. U.S.A. **89**, 10984–10988
9. Wang, Z.Y., Madden, S.L., Deuel, T.F. and Rauscher, F.J. (1992) J. Biol. Chem. **267**, 21999–22002

10. Drummond, I.A., Madden, S.L., Rohwer-Nutter, P., Bell, G.I., Sukhatme, V.P. and Rauscher, F.J. (1992) Science **257**, 674–678
11. Kent, J., Coriat, A.M., Sharpe, P.T., Hastie, N.D. and van Heyningen, V. (1995) Oncogene **11**, 1781–1792
12. Sharma, P.M., Bowman, M., Madden, S.L., Rauscher, F.J. and Sukumar, S. (1994) Genes Dev. **8**, 720–731
13. Dallosso, A.R., Hancock, A.L., Brown, K.W., William, A.C., Jackson, S. and Malik, K. (2004) Hum. Mol. Genet. **13**, 405–415
14. Larsson, S.H., Charlieu, J.P., Miyagawa, K., Engelkamp, D., Rassoulzadegan, M., Ross, A., Cuzin, F., van Heyningen, V. and Hastie, N.D. (1995) Cell **81**, 391–401
15. Davies, R.C., Calvio, C., Bratt, E., Larsson, S.H., Lamond, A.I. and Hastie, N.D. (1998) Genes Dev. **12**, 3217–3225
16. Little, N.A., Hastie, N.D. and Davies, R.C. (2000) Hum. Mol. Genet. **9**, 2231–2239
17. Hammes, A., Guo, J.K., Lutsch, G., Leheste, J.R., Landrock, D., Ziegler, U., Gubler, M.C. and Schedl, A. (2001) Cell **106**, 319–329
18. Hastie, N.D. (2001) Cell **106**, 391–394
19. Gao, F., Maiti, S., Sun, G., Ordonez, N.G., Udtha, M., Deng, J.M., Behringer, R.R. and Huff, V. (2004) Mol. Cell. Biol. **24**, 9899–9910
20. Patek, C.E., Fleming, S., Miles, C.G., Bellamy, C.O., Ladomery, M., Spraggon L., Mullins, J., Hastie, N.D. and Hooper, M.L. (2003) Hum. Mol. Genet. **12**, 2379–2394
21. Natoli, T.A., Liu, J., Eremina, V., Hodgens, K., Li, C., Hamano, Y., Mundel, P., Kalluri, R., Miner. J.H., Quaggin, S.E. and Kreidberg, J.A. (2002) J. Am. Soc. Nephrol. **13**, 2058–2067
22. Holmes, G., Boterashvili, S., English, M., Wainwright, B., Licht, J. and Little, M. (1997) Biochem. Biophys. Res. Commun. **233**, 723–728
23. Moffett, P., Bruening, W., Nakagama, H., Bardeesy, N., Housman, D., Housman, D.E. and Pelletier, J. (1995) Proc. Natl. Acad. Sci. U.S.A. **92**, 11105–11109
24. Reddy, J.C., Morris, J.C., Wang, J., English, M.A., Haber, D.A., Shi, Y. and Licht, J.D. (1995) J. Biol. Chem. **270**, 10878–10884
25. Natoli, T.A., McDonald, A., Alberta, J.A., Taglienti, M.E., Housman, D.E. and Kreidberg, J.A. (2002) Mol. Cell. Biol. **22**, 4433–4438
26. Miles, C.G., Slight, J., Spraggon, L., O'Sullivan, M., Patek, C. and Hastie, N.D. (2003) Mol. Cell. Biol. **23**, 2608–2613
27. Cook, D.M., Hinkes, M.T., Bernfield, M. and Rauscher, F.J. (1996) Oncogene **13**, 1789–1799
28. Englert, C., Maheswaran, S., Garvin, A.J., Kreidberg, J. and Haber, D.A. (1997) Cancer Res. **57**, 1429–1434
29. Lee, S.B., Huang, K., Palmer, R., Truong, V.B., Herzlinger, D., Kolquist, K.A., Wong, J., Paulding, C., Yoon, S.K., Gerald, W. et al. (1999) Cell **98**, 663–673
30. Palmer, R.E., Kotsianti, A., Cadman, B., Boyd, T., Gerald, W. and Haber, D.A. (2001) Curr. Biol. **11**, 1805–1809
31. Guo, G., Morrison, D.J., Licht, J.D. and Quaggin, S.E. (2004) J. Am. Soc. Nephrol. **15**, 2851–2856
32. Wagner, N., Wagner, K.D., Xing, Y., Scholz, H. and Schedl, A. (2004) J. Am. Soc. Nephrol. **15**, 3044–3051
33. Mayo, M.W., Wang, C.Y., Drouin, S.S., Madrid, L.V., Marshall, A.F., Reed, J.C., Weissman, B.E. and Baldwin, A.S. (1999) EMBO J. **18**, 3990–4003
34. Iben, S., and Royer-Pokora, B. (1999) Oncogene **18**, 2533–2536
35. Maheswaran, S., Englert, C., Bennett, P., Heinrich, G. and Haber, D.A. (1995) Genes Dev. **9**, 2143–2156
36. Menke, A.L., Clarke, A.R., Leitch, A., Ijpenberg, A., Williamson, K.A., Spraggon, L., Harrison, D.J. and Hastie, N.D. (2002) Cancer Res. **62**, 6615–6620
37. Scharnhorst, V., Dekker, P., van der Eb, A.J. and Jochemsen, A.G. (2000) J. Biol. Chem. **275**, 10202–10211
38. Rae, F.K., Martinez, G., Gillinder, K.R., Smith, A., Shooter, G., Forrest, A.R. Grimmond, S.M. and Little, M.H. (2004) Oncogene **23**, 3067–3079
39. Nachtigal, M.W., Hirokawa, Y., Enyeart-VanHouten, D.L., Flanagan, J.N., Hammer, G.D. and Ingraham, H.A. (1998) Cell **93**, 445–454
40. Du, X., Hublitz, P., Gunther, T., Wilhelm, D., Englert, C. and Schule, R. (2002) Biochim. Biophys. Acta **1577**, 93–101

41. Madden, S.L., Cook, D.M. and Rauscher, F.J. (1993) Oncogene **8**, 1713–1720
42. McKay, L.M., Carpenter, B. and Roberts, S.G.E. (1999) Oncogene **18**, 6546–6554
43. Wang, W., Lee, S.B., Palmer, R., Ellisen, L.W. and Haber, D.A. (2001) J. Biol. Chem. **276**, 16810–16816
44. Maheswaran, S., Englert, C., Lee, S.B., Ezzel, R.M., Settleman, J. and Haber, D.A. (1998) Oncogene **16**, 2041–2050
45. Maheswaran, S., Englert, C., Zheng, G., Lee, S.B., Wong, J., Harkin, D.P., Bean, J., Ezzell, R., Garvin, A.J., McCluskey, R.T. et al. (1998) Genes Dev. **12**, 1108–1120
46. Lee, T.H., Lwu, S., Kim, J. and Pelletier, J. (2002) J. Biol. Chem. **277**, 44826–44837
47. Johnstone, R.W., Wang, J., Tommerup, N., Vissing, H., Roberts, T. and Shi, Y. (1998) J. Biol. Chem. **273**, 10880–10887
48. Johnstone, R.W., See, R.H., Sells, S.F., Wang, J., Muthukkumar, S., Englert, C., Haber, D.A., Licht, J.D., Sugrue, S.P., Roberts, T. et al. (1996) Mol. Cell. Biol. **16**, 6945–6956
49. Richard, D.J., Schumacher, V., Royer-Pokora, B. and Roberts, S.G.E. (2001) Genes Dev. **15**, 328–339
50. Srichai, M.B., Konieczkowski, M., Padiyar, A., Konieczkowski, D.J., Mukherjee, A., Hayden, P.S., Kamat, S., El-Meanawy, M.A., Khan, S., Mundel, P. et al. (2004) J. Biol. Chem. **279**, 14398–14408
51. Rico, M., Mukherjee, A., Konieczkowski, M., Bruggeman, L.A., Miller, R.T., Khan, S., Schelling, J.R. and Sedor, J.R. (2005) Am. J. Physiol. Renal Physiol. **289**, F431–F441
52. Carpenter, B., Hill, K.J., Charalambous, M., Wagner, K. J., Lahiri, D., James, D.I., Anderson, J.S., Schumacher, V., Royer-Pokora, B., Mann, M. et al. (2004) Mol. Cell. Biol. **24**, 537–549
53. Smolen, G.A., Vassileva, M.T., Wells, J., Matunis, M.J. and Haber, D.A. (2004) Cancer Res. **64**, 7846–7851
54. Ye, Y., Raychaudhuri, B., Gurney, A., Campbell, C.E. and Williams, B.R. (1996) EMBO J. **15**, 5606–5615
55. Niksic, M., Slight, J., Sanford, J.R., Caceres, J.F. and Hastie, N.D. (2004) Hum. Mol. Genet. **13**, 463–471
56. Vajjhala, P.R. Macmillan, E., Gonda, T. and Little, M. (2003) FEBS Lett. **554**, 143–148

Biochem. Soc. Symp. **73**, 203–216
(Printed in Great Britain)

19

The RNA polymerase I transcription machinery

Jackie Russell and Joost C.B.M. Zomerdijk[1]

Division of Gene Regulation and Expression, School of Life Sciences, University of Dundee, Wellcome Trust Biocentre, Dundee DD1 5EH, U.K.

Abstract

The rRNAs constitute the catalytic and structural components of the ribosome, the protein synthesis machinery of cells. The level of rRNA synthesis, mediated by Pol I (RNA polymerase I), therefore has a major impact on the life and destiny of a cell. In order to elucidate how cells achieve the stringent control of Pol I transcription, matching the supply of rRNA to demand under different cellular growth conditions, it is essential to understand the components and mechanics of the Pol I transcription machinery. In this review, we discuss: (i) the molecular composition and functions of the Pol I enzyme complex and the two main Pol I transcription factors, SL1 (selectivity factor 1) and UBF (upstream binding factor); (ii) the interplay between these factors during pre-initiation complex formation at the rDNA promoter in mammalian cells; and (iii) the cellular control of the Pol I transcription machinery.

Introduction

The Pol I (RNA polymerase I) transcription machinery comprises three main components dedicated to the transcription of the rRNA genes: the Pol I enzyme, the TBP (TATA-binding protein)–TAF (TBP-associated factor) complex SL1 (selectivity factor 1)/TIF-IB (transcription initiation factor-IB) and the transactivator protein UBF (upstream binding factor) (reviewed in [1–6]). Confined to the nucleolar subcompartment of the nucleus, this machinery directs transcription from the rDNA promoter at some 200–400 rRNA (ribosomal RNA) genes in the nucleolar organizer regions. The Pol I transcription machinery churns out 50% of the nascent RNA in the cell in the form of pre-rRNA (47 S precursor rRNA), which is processed into three mature rRNAs (18 S, 5.8 S and 28 S), and these, together with the 5 S rRNA synthesized

[1]To whom correspondence should be addressed (email j.zomerdijk@dundee.ac.uk).

by Pol III, comprise the enzymatic and structural scaffold of the ribosome [7]. Transcription of the rRNA genes drives the biogenesis of the several millions of ribosomes needed in an actively growing cell, thereby dictating the overall level of cellular protein synthesis. Thus transcription by Pol I is essential for, and inextricably linked to, cell growth and, hence, to normal cell division (reviewed in [6,8]). Predictably, therefore, Pol I transcription is at the receiving end of many of the signalling pathways implicated in the control of cell growth, nutrient sensing and proliferation [1,6,9–17].

The Pol I enzyme complexes

DNA-dependent RNA polymerases are highly conserved multisubunit enzyme complexes in eukaryotes. The composition of the Pol I enzyme complex has been derived primarily from yeast genetic and biochemical studies [18]. Our understanding of the Pol I enzyme complex in mammalian cells has advanced rapidly in recent years following identification of mammalian subunits both by comparison with yeast Pol I and by MS analysis of purified Pol I complexes. Pol I enzymes from yeast to humans show conservation with Pols II and III and with the $\alpha_2\beta\beta'\omega$ subunit composition of prokaryotic core RNA polymerase (reviewed in [19]) (Table 1). The largest and second largest subunits of Pol I share substantial similarity with those of Pol II and Pol III and with prokaryotic β' and β subunits respectively, possessing most of the enzymatic functions (substrate binding, active centre, template movement). The heterodimer AC40–AC19 of Pol I is shared with Pol III, and is homologous with the RPB3–RPB11 heterodimer of Pol II and functionally analogous to the prokaryotic α_2 homodimer. The RPB6 subunit of Pol I, shared with Pols II and III, is a structural and functional homologue of the bacterial ω subunit. There are four additional subunits that are required to synthesize RNA from a non-specific DNA template, shared between Pols I, II and III (RPB5, RPB8, RPB10 and RPB12). Subunit A12.2 is unique to Pol I (Table 1). In addition to these 10 subunits, collectively referred to as the 'core' complex, Pol I heterodimer A14–A43 shares genetic, biochemical and structural characteristics with heterodimers RPB4–RPB7 and C17–C25 of Pols II and III respectively, and there is sequence similarity between subunits RPB7, A43 and C25 [20–22]. There are two additional subunits, A49 and A34.5, that are specific to Pol I [18,23,24] and hence likely targets of the Pol I cognate transcription factors.

The ~600 kDa yeast Pol I complex is therefore composed of 14 subunits, and homologues of all of the yeast Pol I subunits are found in Metazoa, with the exception of the Pol I-specific subunit A14, for which a mammalian homologue has yet to be identified. Mammalian homologues of the Pol I-specific subunit A34.5 were discovered only recently: the mouse homologue is PAF49 (polymerase I-associated factor of 49 kDa) [25]. Human Pol I proteomics led us to identify a 72 kDa subunit of Pol I with homology to the yeast Pol I subunit A34.5 (K.I. Panov, T.B. Panova and J.C.B.M. Zomerdijk, unpublished work), previously named ASE-1 [antisense for ERCC-1 (excision repair cross-complementing-1)] [26] or CAST (CD3ε-associated signal transducer) [27] and, curiously, also

Table 1 Comparison of Pol I subunit composition in yeast and humans. Most Pol I subunits are essential for viability in *Saccharomyces cerevisiae* (as determined by null mutant analysis); those subunits not essential are marked viable (^{v}) or conditional mutants (^{c}) (adapted from [18]). 'Unique or shared' refers to subunits unique in Pol I, or shared with Pol III, or shared with Pols II and III. Human Pol I subunits are named primarily following yeast nomenclature

S. cerevisiae Pol I subunits	Unique or shared	Human Pol I subunits	Homologues*	Interactions in mammalian PIC
RPA190	I	hRPA190 (A194)	β'; RPB1 (B220)	
RPA135	I	hRPA127	β; RPB2 (B140)	
RPA49^{c}	I	PAF53	None	UBF [29]
RPA43	I	hRPA43	RPB7	hRRN3 [41]
RPA40 (AC40)	I, III	hRPA40	α; RPB3 (B45)	
RPA34.5^{v}	I	CAST (PAF49)	None	UBF and SL1 [25,26]†
RPB5 (ABC27)	I, II, III	hRPB5	–	
RPB6 (ABC23)	I, II, III	hRPB6	ω	
RPA19 (AC19)	I, III	hRPA19	α; RPB11(B12.5)	
RPB8 (ABC14.5)	I, II, III	RPB8	–	
RPA14^{v}	I	?	RPB4	
RPA12^{c}	I	hRPA12.2	RPB9	
RPB10 (ABC10β)	I, II, III	RPB10	–	
RPB12 (ABC10α)	I, II, III	RPB12	–	

*Structural and/or functional homologues with bacterial RNA polymerase (β', β, α_2, ω) and yeast Pol II (B).
†K.I. Panov, T.B. Panova and J.C.B.M. Zomerdijk, unpublished work.

implicated in T-cell activation (Table 1). The A34.5 and A49 subunits are absent from yeast Pol I isolated under high-salt conditions (referred to as A*), which displays a reduced specific activity in RNA synthesis from calf thymus DNA and a higher sensitivity to α-amanitin, suggesting a role for these subunits in elongation or processivity [28]. PAF49 and PAF53, the mammalian homologues of yeast A34.5 and A49, also dissociate from Pol I under certain conditions, most notably upon serum starvation of cells [25,29].

The structures of yeast Pol I and Pol II complexes have been studied extensively by electron microscopy and crystallography [30–33]. The cryo-electron-microscopic structure of yeast Pol I displays an overall similarity with the crystal structure of yeast Pol II (lacking RPB4 and RPB7), except at the positions of the Pol I-specific subunits A34.5, A49, A43 and A14 [33]. The A34.5 subunit of Pol I is positioned adjacent to the second largest subunit A135, at the entry to the cleft which binds the DNA template [31], where it is proposed to stabilize the interaction of the DNA template with the 'core' enzyme. The A49 subunit is located on the head region of the clamp, formed by the largest subunit of Pol I [33,34] and constituting one side of the cleft. A49 could potentially affect the conformation of the clamp [33], moving it inwards to hold downstream DNA more firmly, or interact directly with the DNA, thereby increasing the processivity of the enzyme [30,31]. Another structure specific to Pol I is the stalk, formed by subunits A43 and A14 and postulated to function as an interface for the assembly of initiation factors through its binding to the essential Pol I-associated factor Rrn3p. Interestingly, the stalk protrudes from the Pol I structure at the same site as the C-terminal domain of the largest subunit in Pol II, which has a role in initiation via its interaction with the basal transcription factors [33]. There are different conformational subpopulations of the yeast Pol I enzyme [34], consistent with conformational flexibility in other RNA polymerases, where the clamp maintains an open conformation in the initiation-competent enzyme, being converted into a closed conformation during elongation [19,34].

We have shown previously that the Pol I enzyme in human cells is found in at least two functionally distinct complexes (Pol Iα and Pol Iβ). These complexes are considerably larger than the yeast Pol I complex, at >1 MDa, and contain Pol I-associated factors in addition to the core subunits [35] (K.I. Panov and J.C.B.M. Zomerdijk, unpublished work). Pol Iβ, comprising approx. 10% of the total Pol I extractable from the cell nucleus, is the initiation-competent form of the enzyme, able to direct accurate initiation of transcription from the rDNA promoter. By contrast, Pol Iα, representing the bulk of the Pol I, is unable to direct rDNA promoter-specific transcription, although it does possess Pol activity, catalysing the random synthesis of RNA. Given these properties and its relative abundance compared with Pol Iβ, the Pol Iα complex might represent the elongating Pol I complex, or polymerase with the capacity to be converted readily into initiation-competent Pol Iβ, or spent polymerase complexes following transcription. The Pol Iβ enzyme complex is distinguishable by its associated factors, likely to be involved specifically in initiation. hRRN3 (human RRN3; equivalent to mouse TIF-IA/yeast Rrn3p) [35–39] is one such Pol Iβ-specific factor. RRN3 is tethered to the A43 subunit of Pol I [40,41] and links Pol I to the promoter-bound essential transcription factor SL1. The role of RRN3 in bridging the polymerase and promoter recognition components of the Pol I transcription machinery (SL1 or core factor) is conserved from yeast to humans, and is required for PIC (pre-initiation complex) assembly [35,37,38].

The association of hRRN3/TIF-IA with Pol I, to generate Pol Iβ [35,36], is an important point of control in pre-rRNA synthesis. The hRRN3/TIF-IA–Pol I interaction is impaired and cellular rRNA synthesis is down-regulated in stationary-phase, nutrient-limited or cycloheximide-treated mammalian cells [41,42]. There is evidence that rDNA transcription is regulated by the phosphorylation state of RRN3, which determines the steady-state concentration of the RRN3–Pol I complex Pol Iβ [41]. Perhaps as a result of a reversal of phosphorylation events, Rrn3 is inactivated during the process of rDNA transcription [43]. The phosphorylation of TIF-IA by JNK2 (c-Jun N-terminal kinase 2), which can occur upon cellular stress, leads to inactivation of TIF-IA and down-regulation of rRNA synthesis [44]. RRN3 activity and the RRN3–Pol I interaction are also regulated by the ERK (extracellular-signal-regulated kinase)/MAPK (mitogen-activated protein kinase) and mTOR (mammalian target of rapamycin) signalling pathways, which are involved in growth factor signalling and nutrient sensing. Increased RRN3/TIF-IA activity, rRNA gene transcription and cell proliferation correlate with phosphorylation of RRN3/TIF-IA by ERK and RSK (p90 ribosomal S6 kinase) in growth factor-stimulated mouse NIH 3T3 cells [10]. In separate studies, RRN3 activity has been reported to be down-regulated [12] or unaffected [9] following inactivation of mTOR by rapamycin treatment of NIH 3T3 cells. Perhaps, therefore, the specific physiology of a cell determines the involvement of mTOR in the regulation of RRN3 activity. The mTOR pathway has been implicated in PIC formation, as the down-regulation of RRN3 by rapamycin was correlated with changes in the phosphorylation status of RRN3, an inability of RRN3 to interact with Pol I and SL1, and translocation of TIF-IA/RRN3 to the cytoplasm [12]. Furthermore, inactivation of yeast TOR by rapamycin is accompanied by release of Pol I from the nucleolus and inhibition of rDNA transcription [45], and the TOR signalling pathway has been shown to regulate the Rrn3p–Pol I interaction and the Rrn3-dependent recruitment of Pol I to the promoter [14]. Curiously, in yeast, phosphorylation of Pol I, not Rrn3p, is necessary for the formation of a stable Pol I–Rrn3p complex for efficient transcription initiation *in vitro*, and a change in the phosphorylation state of Pol I is correlated with the association of Pol I with Rrn3p *in vivo* [46].

TBP–TAF complex SL1

RNA polymerase enzymes themselves have no intrinsic ability to recognize and bind specifically to promoter DNA sequences. Therefore PIC formation in transcription calls for the recruitment of the polymerases to the promoter via transcription factors. Sigma factors fulfil this role in bacteria, and basal or general transcription factors co-operate in this function in eukaryotes, where the three classes of highly related enzymes, Pol I, Pol II and Pol III, catalyse the transcription of specific sets of genes. A complex of TBP and TAFs is required for the accurate initiation of transcription by all three eukaryotic RNA polymerases [47], with few exceptions. The complement of TAFs in each complex, although variable in Pol II transcription, is specific to each class of genes. Binding to TBP of TAF subunits specific to one class of polymerases can preclude the binding of TAFs from a different class [48]. The TBP–TAF complexes specific to Pols I, II and III perform distinct roles in mediating a specific interaction between the polymerases and their respective promoters.

In transcription by mammalian Pol II, the TBP–TAF complex TFIID (transcription factor IID) can bind the promoter through interaction of the TBP protein with TATA boxes and of the TAF proteins with other promoter sequence elements. Pol II and other factors are recruited subsequently to form the PIC [49]. In transcription by Pol III, the initial phases converge on the recruitment of TBP–TAF complex TFIIIB and the Pol III enzyme in formation of the PIC [23,50], although the type of Pol III promoter determines whether loading of promoter-type specific TFIIIB at the promoter DNA is facilitated by the specific DNA-binding abilities of TFIIIA and TFIIIC, TFIIIC alone, or the multisubunit complex PBP–PTF–$SNAP_C$.

In transcription by mammalian Pol I (reviewed in [6]), the TBP–TAF complex SL1 (murine TIF-IB) is essential [48,51–53]. This ~300 kDa complex is composed of TBP and at least three TAFs [48,51,52], TAF_I48, TAF_I63 and TAF_I110 (mouse TAF_I48, TAF_I68 and TAF_I95 respectively). Reconstitution of transcription *in vitro* using recombinant SL1 comprising only these four components is inefficient, suggesting that more components and/or specific modifications might be required to reproduce the full activity of SL1 [54]. We have recent evidence for the existence of another TAF_I in SL1 (J.J. Gorski, S. Pathak and J.C.B.M. Zomerdijk, unpublished work). SL1 functions through the essential core element of the rDNA promoter, overlapping the start site of transcription [55]. SL1 alone does not produce a DNase I footprint, but a combination of SL1 and UBF extends the footprint of UBF alone, demonstrating co-operativity between these proteins at the promoter [56,57].

SL1 directs promoter- and species-specific Pol I transcription. The SL1 complexes of humans and mice are not interchangeable between these systems, in contrast with Pol I and the activator UBF. These findings imply a specific interaction between SL1 and its cognate rDNA promoter (reviewed in [58]). Indeed, we have shown recently that human SL1 can bind the human rDNA promoter independently and direct the initiation of transcription specifically from the human rDNA promoter in the absence of UBF [59], consistent with the findings that TIF-IB and rat SL1 can bind and direct transcription from their cognate rDNA promoters independently of UBF [60,61].

In addition to promoter recognition and binding, SL1 has an essential role in the recruitment of Pol I to the start site of transcription [35,59]. SL1 subunits TAF_I63 and TAF_I110 interact with the Pol I-associated RRN3 [35–37], which in turn is attached to Pol I through subunit A43 in the 'stalk' structure of the enzyme complex and, perhaps, PAF67 [41,42].

Another function for SL1 in PIC formation has been reported recently: we found that SL1 can stabilize UBF binding at the rDNA promoter [59]. In the

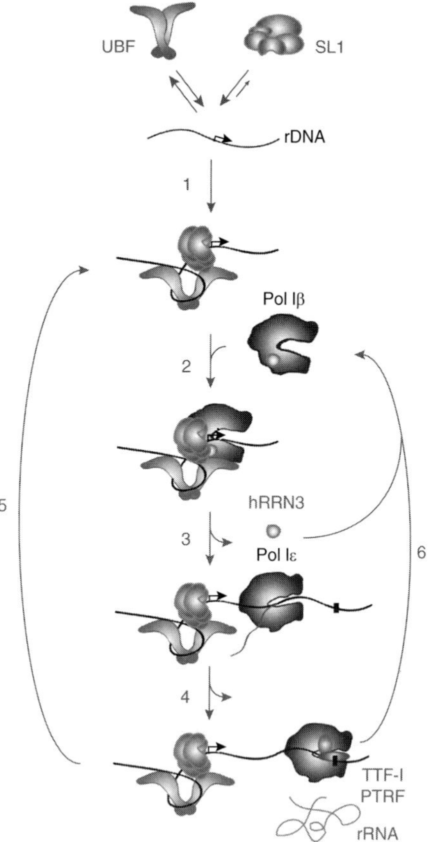

Figure 1 The Pol I transcription cycle. The TBP–TAF complex SL1 interacts stably with the rDNA promoter, near the start site of transcription (black arrow). Dimers of UBF interact dynamically with rDNA promoter elements (grey arrows), such that UBF–rDNA complexes might be relatively short-lived. UBF and SL1 interact co-operatively at the promoter elements, and SL1 reduces the dissociation of UBF from the rDNA, thereby increasing the residence time of UBF at the promoter (1). rDNA promoter topology is altered dramatically by UBF binding. SL1 in the UBF–SL1–rDNA complex recruits the initiation-competent Pol I–hRRN3 complex Pol Iβ through the interactions of TAF_I subunits with hRRN3. This culminates in formation of the complete PIC, poised for initiation of transcription (2). UBF stimulates the promoter escape/clearance of Pol I. In the transition from initiation to elongation, hRRN3 dissociates from Pol I, and the Pol I enzyme is converted into a transcript-elongating enzyme complex, Pol Iε (3). SL1 and UBF remain promoter-bound following promoter clearance by Pol I (4) and, thus, can function as a re-initiation scaffold (5), to allow multiple rounds of transcription from the same rDNA promoter. Pol Iε transcription terminates and rRNA is released at termination sites (black rectangle on rDNA); this process is directed by TTF-I (transcription termination factor-I) and PTRF (polymerase I and transcript release factor). Following termination of transcription, it is possible that the spent Pol I complex is converted back into initiation-competent Pol Iβ and thence recruited to the rDNA promoter by SL1 to initiate a new round of transcription (6).

absence of SL1, UBF associates with and dissociates from the rDNA rapidly, whereas the presence of SL1 at the rDNA promoter alters significantly the binding equilibrium of UBF, thereby increasing the residence time of UBF at the promoter. It is possible that the increased lifetime of the rDNA–UBF complex is essential to facilitate the UBF-mediated activation of Pol I transcription.

Collectively, these results support a unified model for SL1 function in the mammalian system, where SL1 drives or nucleates PIC formation specifically at the rDNA promoter, leading ultimately to the productive initiation of transcription by Pol I (Figure 1). SL1 remains stably bound to the rDNA promoter following initiation of transcription and escape by Pol I, and the SL1 complex, together with UBF at the promoter, might serve as a re-initiation scaffold to support multiple rounds of transcription [62].

PIC assembly could perhaps be regulated through SL1, influencing the level of transcription by Pol I. Indeed, an increase in SL1 occupancy of the rDNA promoter correlates with increased Pol I transcription in IGF-1 (insulin-like growth factor-1)-stimulated HEK293 cells [13]. Although phosphorylation of SL1 in response to growth stimulatory factors has not yet been reported, some of the SL1 subunits are known to be phosphoproteins. SL1 activity can be controlled by acetylation, as acetylation of the mouse SL1 (TIF-IB) subunit TAF_I68 by P/CAF [p300/CBP (CREB-binding protein)-associated factor] stimulates binding of SL1 to the rDNA promoter and Pol I-mediated transcription [63], and it is possible that this acetylation is a regulated, rather than constitutive, event in cells. Furthermore, SL1 could be a limiting component of rRNA gene expression in cells, and therefore an increase in its abundance would stimulate Pol I transcription. Although tissue-specific or differential expression of the TAFs of SL1 has not been reported as yet, regulated expression of these Pol I-specific TAF_I subunits could impact on transcription by Pol I. Up-regulation of TBP expression, following activation of the MAPK signalling pathway, can lead to stimulation of transcription from Pol I and Pol III promoters in certain cell types, implying that overexpression of this single subunit can drive the biogenesis of the separate TBP–TAF complexes required in the co-ordinated transcription of rRNA and tRNA genes [64,65].

Transcription activator UBF

Reconstituted transcription assays have revealed the components minimally required for transcription initiation. 'Basal' levels of transcription from the rDNA promoter can be supported by SL1 and initiation-competent Pol Iβ alone [60,61]. UBF efficiently activates rDNA transcription above this basal level. Stimulated transcription by UBF in the reconstituted system constitutes 'true' or 'net' activation, as distinct from stimulation of transcription through the potential anti-repression function of UBF in the context of chromatin transcription *in vivo* [66–68].

UBF is proposed to function as an architectural protein at the rDNA promoter [69,70], where it binds as a dimer to the UCE (upstream control element), which has a modulatory role in Pol I transcription, and to the essential core element. UBF (97 kDa) is composed of several distinct domains [69,70]. The N-terminal dimerization domain of UBF is essential for its transactivation function [69,71]. The large central portion contains six HMG (high-mobility group) boxes [72,73] which exhibit a relaxed sequence specificity of DNA binding and display the ability to bend DNA [69,74–79]. In UBF, these HMG boxes do not contribute equally to DNA binding; rather, it is thought that HMG boxes 1–3 interact with the DNA minor groove in the interaction of UBF with the rDNA promoter [74,79]. Binding of a UBF dimer induces bending of the DNA, to create a loop of some 140 bp of DNA in a single 360° turn, generating a structure termed the enhancesome [78]. It has been proposed that tandem enhancesomes at the rDNA promoter would juxtapose the UCE and core promoter elements [78,79]. There is co-operativity of binding of UBF and SL1 to the UCE and the core elements of the rDNA promoter [56,57], although the precise arrangement and stoichiometry of UBF and SL1 at the rDNA promoter remain unknown. Interaction between UBF and SL1 [69,80,81] requires the highly acidic C-terminal domain of UBF [69,82,83] and may involve subunits TAF_I48 and TBP of SL1 [84,85]. UBF also interacts with Pol I [29,86] through Pol I subunits ASE1/CAST and PAF53 [29,87] (K.I. Panov, T.B. Panova and J.C.B.M. Zomerdijk, unpublished work) (Table 1).

These interactions and the co-operativity demonstrated between UBF and SL1 at the rDNA promoter led to the proposal that UBF recruits SL1 and Pol I to the rDNA promoter, activating transcription by facilitating PIC assembly. This is the principal mechanism used by activators of Pol II transcription [88]. However, in contrast with the sequence-specific Pol II transcriptional activators, binding of UBF occurs rather indiscriminately [75,76] and extensively throughout the rDNA repeats *in vivo* [89,90], inconsistent with a role for UBF in nucleating PIC assembly specifically at the rDNA promoter. Moreover, as discussed, SL1 alone can nucleate PIC formation [59–61], stabilizing UBF binding at the rDNA promoter [59] and recruiting initiation-competent Pol I to the core promoter [35].

Recently, using a highly purified and reconstituted transcription system, combined with an immobilized ribosomal promoter DNA template, we have defined a specific and novel function for UBF in activating the rate of RNA synthesis at promoter escape or clearance by Pol I (K.I. Panov, J.K. Friedrich, J. Russell and J.C.B.M. Zomerdijk, unpublished work), a step we had identified previously as the rate-limiting step for Pol I transcription *in vitro* [62]. This mechanism enables UBF to activate transcription both from previously inactive promoters following PIC assembly and from SL1-engaged promoters at each successive round of transcription following re-initiation (Figure 1). Stimulation of promoter escape/clearance by activators might also make an important contribution to efficient transcription by Pols II and III.

Given its role in activating transcription, any variation in UBF expression levels or activity would be anticipated to have a dramatic effect on the level of transcription by Pol I. Indeed, UBF, limiting in some cells [91], is expressed at relatively high levels in cardiac hypertrophy and liver cancer cells, which support high levels of rRNA synthesis [92,93]. In mitogen-stimulated lymphocytes, UBF overexpression is apparent prior to increased rRNA synthesis, and a decrease in

UBF expression occurs upon induced differentiation [94]. UBF expression can be up-regulated by serum induction [9], operating through the mTOR pathway, and also by the *myc* proto-oncogene [95].

How is the activity of the UBF protein regulated? Phosphorylation of UBF can affect its interaction with other PIC components or DNA. C-terminal phosphorylation of UBF increases its ability to interact with SL1 [69,82,83]. The protein kinase CK2 phosphorylates UBF at serine residues within the C-terminal acidic domain, and this contributes to transcriptional activation [96]. Phosphoinositide 3-kinase phosphorylates UBF predominantly in the C-terminus upon IGF-1 (insulin-like growth factor 1) treatment of serum-starved mouse cells, where IRS-1 (insulin receptor substrate-1) targets phosphoinositide 3-kinase to UBF following translocation of IRS-1 to the nucleolus [15], and this correlates with increased rRNA synthesis. C-terminal phosphorylation of UBF can also be mediated by the mTOR signalling pathway and requires S6 kinase 1. This occurs upon serum induction of rDNA transcription in NIH 3T3 cells and promotes the interaction between UBF and SL1 [9]. The ERK/MAPK pathway targets UBF upon stimulation of Pol I transcription by epidermal growth factor in human neuroepithelioma cells. ERK1/2-mediated phosphorylation in HMG boxes 1 and 2 alters the interaction of UBF with DNA [11]. UBF activity is also regulated by competitive acetylation/deacetylation. In this case, the ability of UBF to bind DNA is not affected [68]. Rather, binding of CBP acetyltransferase to UBF might up-regulate UBF by preventing the binding of the tumour suppressor Rb (retinoblastoma) (and associated histone deacetylases) to UBF, precluding the negative effect of Rb on the binding of UBF to SL1 [97].

Concluding remarks

From the evidence obtained so far, we conclude that the Pol I TBP–TAF complex SL1 performs three important functions that, together, drive PIC formation at the rDNA promoter. SL1 recognizes and binds the core promoter elements through its TAF_I proteins. The SL1 complex, positioned accurately at the rDNA promoter, recruits the initiation-competent form of Pol I, the Pol I–RRN3 complex Pol Iβ, to the start site of transcription, through the interactions of TAF_I110 and TAF_I63 with RRN3. SL1 also stabilizes UBF at the rDNA promoter, and the consequential increase in residence time of UBF at the promoter is predicted to enhance the efficacy of UBF in its role as an activator of Pol I transcription. The dramatic alterations in DNA architecture created by UBF at the rDNA promoter are likely to be important for its activation function, although the significance of these has yet to be realized. We consider that the interaction of UBF with subunits of Pol I is crucial for the ability of UBF to stimulate the escape of Pol I from the promoter in the transition state between initiation and elongation of transcription.

Pol I transcription can be regulated at the levels of PIC formation and UBF activity. These are influenced by the downstream effectors of growth factor-responsive signalling pathways (including the MAPK, mTOR, and ERK pathways), the availability of nutrients (mTOR) and/or the cellular stress response pathways (JNK2). The level of rRNA transcription activity has a direct effect on the overall level of cellular protein synthesis, thereby influencing the growth of the cell and cell division (reviewed in [6]). There is a correlation between the level of cellular rRNA transcription and the growth status of cells, in that slow-growing or differentiated cells display low levels of rRNA transcription, whereas rapidly growing and cancerous cells maintain high levels of rRNA synthesis. Down-regulation of Pol I transcription can lead to the activation of the p53 cellular response pathway, and therefore directly influence the cell's decision to arrest or exit from the cell cycle in response to DNA damage, stress or malfunction [98]. Up-regulation of rRNA transcription, with the growth and proliferative advantage it bestows on cells, could represent an early event in oncogenesis; significantly, the oncoprotein Myc has been shown recently to control cellular rRNA synthesis [99–101].

We thank present and past members of our laboratory for their research and interest. We apologize to our colleagues whose work could not be cited due to space limitations. The Wellcome Trust funds research in the laboratory of J.C.B.M.Z., and J.C.B.M.Z. is a Wellcome Trust Senior Research Fellow in the Basic Biomedical Sciences.

References

1. Grummt, I. (2003) Genes Dev. **17**, 1691–1702
2. Moss, T. and Stefanovsky, V.Y. (2002) Cell **109**, 545–548
3. Reeder, R.H. (1999) Prog. Nucleic Acid Res. Mol. Biol. **62**, 293–327
4. Hannan, K.M., Hannan, R.D. and Rothblum, L.I. (1998) Front. Biosci. **3**, d376–d398
5. Comai, L. (2004) Adv. Protein Chem. **67**, 123–155
6. Russell, J. and Zomerdijk, J.C. (2005) Trends Biochem. Sci. **30**, 87–96
7. Moore, P.B. and Steitz, T.A. (2002) Nature (London) **418**, 229–235
8. Warner, J.R. (1999) Trends Biochem. Sci. **24**, 437–440
9. Hannan, K.M., Brandenburger, Y., Jenkins, A., Sharkey, K., Cavanaugh, A., Rothblum, L., Moss, T., Poortinga, G., McArthur, G.A., Pearson, R.B. et al. (2003) Mol. Cell. Biol. **23**, 8862–8877
10. Zhao, J., Yuan, X., Frodin, M. and Grummt, I. (2003) Mol. Cell **11**, 405–413
11. Stefanovsky, V.Y., Pelletier, G., Hannan, R., Gagnon-Kugler, T., Rothblum, L.I. and Moss, T. (2001) Mol. Cell **8**, 1063–1073
12. Mayer, C., Zhao, J., Yuan, X. and Grummt, I. (2004) Genes Dev. **18**, 423–434
13. James, M.J. and Zomerdijk, J.C. (2004) J. Biol. Chem. **279**, 8911–8918
14. Claypool, J.A., French, S.L., Johzuka, K., Eliason, K., Vu, L., Dodd, J.A., Beyer, A.L. and Nomura, M. (2004) Mol. Biol. Cell **15**, 946–956
15. Drakas, R., Tu, X. and Baserga, R. (2004) Proc. Natl. Acad. Sci. U.S.A. **101**, 9272–9276
16. Sun, H., Tu, X., Prisco, M., Wu, A., Casiburi, I. and Baserga, R. (2003) Mol. Endocrinol. **17**, 472–486
17. Moss, T. (2004) Curr. Opin. Genet. Dev. **14**, 210–217
18. Carles, C. and Riva, M. (1998) in Transcription of Ribosomal RNA Genes by Eukaryotic RNA Polymerase I (Paule, M.R., ed.), pp. 9–38, Springer-Verlag, Berlin, Heidelberg, New York
19. Cramer, P. (2002) Curr. Opin. Struct. Biol. **12**, 89–97
20. Peyroche, G., Levillain, E., Siaut, M., Callebaut, I., Schultz, P., Sentenac, A., Riva, M. and Carles, C. (2002) Proc. Natl. Acad. Sci. U.S.A. **99**, 14670–14675
21. Meka, H., Daoust, G., Arnvig, K.B., Werner, F., Brick, P. and Onesti, S. (2003) Nucleic Acids Res. **31**, 4391–4400
22. Siaut, M., Zaros, C., Levivier, E., Ferri, M.L., Court, M., Werner, M., Callebaut, I., Thuriaux, P., Sentenac, A. and Conesa, C. (2003) Mol. Cell. Biol. **23**, 195–205
23. Geiduschek, E.P. and Kassavetis, G.A. (2001) J. Mol. Biol. **310**, 1–26

24. Chedin, S., Ferri, M.L., Peyroche, G., Andrau, J.C., Jourdain, S., Lefebvre, O., Werner, M., Carles, C. and Sentenac, A. (1998) Cold Spring Harbor Symp. Quant. Biol. **63**, 381–389
25. Yamamoto, K., Yamamoto, M., Hanada, K., Nogi, Y., Matsuyama, T. and Muramatsu, M. (2004) Mol. Cell. Biol. **24**, 6338–6349
26. Whitehead, C.M., Winkfein, R.J., Fritzler, M.J. and Rattner, J.B. (1997) Chromosoma, **106**, 493–502
27. Yamazaki, T., Hamano, Y., Tashiro, H., Itoh, K., Nakano, H., Miyatake, S. and Saito, T. (1999) J. Biol. Chem. **274**, 18173–18180
28. Huet, J., Buhler, J.M., Sentenac, A. and Fromageot, P. (1975) Proc. Natl. Acad. Sci. U.S.A. **72**, 3034–3038
29. Hanada, K., Song, C.Z., Yamamoto, K., Yano, K., Maeda, Y., Yamaguchi, K. and Muramatsu, M. (1996) EMBO J. **15**, 2217–2226
30. Cramer, P., Bushnell, D.A. and Kornberg, R.D. (2001) Science, 292, 1863–1876
31. Gnatt, A.L., Cramer, P., Fu, J., Bushnell, D.A. and Kornberg, R.D. (2001) Science, **292**, 1876–1882
32. Kettenberger, H., Armache, K.J. and Cramer, P. (2004) Mol. Cell **16**, 955–965
33. Bischler, N., Brino, L., Carles, C., Riva, M., Tschochner, H., Mallouh, V. and Schultz, P. (2002) EMBO J. **21**, 4136–4144
34. De Carlo, S., Carles, C., Riva, M. and Schultz, P. (2003) J. Mol. Biol. **329**, 891–902
35. Miller, G., Panov, K.I., Friedrich, J.K., Trinkle-Mulcahy, L., Lamond, A.I. and Zomerdijk, J.C. (2001) EMBO J. **20**, 1373–1382
36. Bodem, J., Dobreva, G., Hoffmann-Rohrer, U., Iben, S., Zentgraf, H., Delius, H., Vingron, M. and Grummt, I. (2000) EMBO Rep. **1**, 171–175
37. Moorefield, B., Greene, E.A. and Reeder, R.H. (2000) Proc. Natl. Acad. Sci. U.S.A. **97**, 4724–4729
38. Peyroche, G., Milkereit, P., Bischler, N., Tschochner, H., Schultz, P., Sentenac, A., Carles, C. and Riva, M. (2000) EMBO J. **19**, 5473–5482
39. Yamamoto, R.T., Nogi, Y., Dodd, J.A. and Nomura, M. (1996) EMBO J. **15**, 3964–3973
40. Thuriaux, P., Mariotte, S., Buhler, J.M., Sentenac, A., Vu, L., Lee, B.S. and Nomura, M. (1995) J. Biol. Chem. **270**, 24252–24257
41. Cavanaugh, A.H., Hirschler-Laszkiewicz, I., Hu, Q., Dundr, M., Smink, T., Misteli, T. and Rothblum, L.I. (2002) J. Biol. Chem. **277**, 27423–27432
42. Yuan, X., Zhao, J., Zentgraf, H., Hoffmann-Rohrer, U. and Grummt, I. (2002) EMBO Rep. **3**, 1082–1087
43. Hirschler-Laszkiewicz, I., Cavanaugh, A.H., Mirza, A., Lun, M., Hu, Q., Smink, T. and Rothblum, L.I. (2003) J. Biol. Chem. **278**, 18953–18959
44. Mayer, C., Bierhoff, H. and Grummt, I. (2005) Genes Dev. **19**, 933–941
45. Tsang, C.K., Bertram, P.G., Ai, W., Drenan, R. and Zheng, X.F. (2003) EMBO J. **22**, 6045–6056
46. Fath, S., Milkereit, P., Peyroche, G., Riva, M., Carles, C. and Tschochner, H. (2001) Proc. Natl. Acad. Sci. U.S.A. **98**, 14334–14339
47. Hernandez, N. (1993) Genes Dev. **7**, 1291–1308
48. Comai, L., Zomerdijk, J.C., Beckmann, H., Zhou, S., Admon, A. and Tjian, R. (1994) Science **266**, 1966–1972
49. Verrijzer, C.P. and Tjian, R. (1996) Trends Biochem. Sci. **21**, 338–342
50. Schramm, L. and Hernandez, N. (2002) Genes Dev. **16**, 2593–2620
51. Comai, L., Tanese, N. and Tjian, R. (1992) Cell **68**, 965–976
52. Zomerdijk, J.C.B.M., Beckmann, H., Comai, L. and Tjian, R. (1994) Science **266**, 2015–2018
53. Zomerdijk, J.C.B.M. and Tjian, R. (1998) in Transcription of Eukaryotic Ribosomal RNA Genes by RNA Polymerase I (Paule, M.R., ed.), pp 67–73, Springer-Verlag, Berlin, Heidelberg, New York
54. Heix, J., Zomerdijk, J., Ravanpay, A., Tjian, R. and Grummt, I. (1997) Proc. Natl. Acad. Sci. U.S.A. **94**, 1733–1738
55. Zomerdijk, J.C.B.M. and Tjian, R. (1998) in Transcription of Ribosomal RNA Genes by Eukaryotic RNA Polymerase I (Paule, M.R., ed.), pp. 121–134, Springer-Verlag, Berlin, Heidelberg, New York
56. Learned, R.M., Learned, T.K., Haltiner, M.M. and Tjian, R.T. (1986) Cell **45**, 847–857
57. Bell, S.P., Learned, R.M., Jantzen, H.M. and Tjian, R. (1988) Science **241**, 1192–1197

58. Heix, J. and Grummt, I. (1995) Curr. Opin. Genet. Dev. **5**, 652–656
59. Friedrich, J.K., Panov, K.I., Cabart, P., Russell, J. and Zomerdijk, J.C. (2005) J. Biol. Chem., **280**, 29551–29558
60. Schnapp, A. and Grummt, I. (1991) J. Biol. Chem. **266**, 24588–24595
61. Smith, S.D., O'Mahony, D.J., Kinsella, B.T. and Rothblum, L.I. (1993) Gene Expression **3**, 229–236
62. Panov, K.I., Friedrich, J.K. and Zomerdijk, J.C. (2001) Mol. Cell. Biol. **21**, 2641–2649
63. Muth, V., Nadaud, S., Grummt, I. and Voit, R. (2001) EMBO J. **20**, 1353–1362
64. Zhong, S., Zhang, C. and Johnson, D.L. (2004) Mol. Cell. Biol. **24**, 5119–5129
65. White, R.J. (2004) Oncogene **23**, 3208–3216
66. Kuhn, A. and Grummt, I. (1992) Proc. Natl. Acad. Sci. U.S.A. **89**, 7340–7344
67. Brou, C., Kuhn, A., Staub, A., Chaudhary, S., Grummt, I., Davidson, I. and Tora, L. (1993) Nucleic Acids Res. **21**, 4011–4018
68. Pelletier, G., Stefanovsky, V.Y., Faubladier, M., Hirschler-Laszkiewicz, I., Savard, J., Rothblum, L.I., Cote, J. and Moss, T. (2000) Mol. Cell **6**, 1059–1066
69. Jantzen, H.M., Chow, A.M., King, D.S. and Tjian, R. (1992) Genes Dev. **6**, 1950–1963
70. Reeder, R.H., Pikaard, C.S. and McStay, B. (1995) in Nucleic Acids and Molecular Biology (Eckstein, F. and Lilley, D.M.J., eds.), vol. 9, pp. 251–263, Springer-Verlag, Berlin, Heidelberg, New York
71. McStay, B., Frazier, M.W. and Reeder, R.H. (1991) Genes Dev. **5**, 1957–1968
72. Bachvarov, D. and Moss, T. (1991) Nucleic Acids Res. **19**, 2331–2335
73. Jantzen, H.M., Admon, A., Bell, S.P. and Tjian, R. (1990) Nature (London) **344**, 830–836
74. Leblanc, B., Read, C. and Moss, T. (1993) EMBO J. **12**, 513–525
75. Copenhaver, G.P., Putnam, C.D., Denton, M.L. and Pikaard, C.S. (1994) Nucleic Acids Res. **22**, 2651–2657
76. Hu, C.H., McStay, B., Jeong, S.W. and Reeder, R.H. (1994) Mol. Cell. Biol. **14**, 2871–2882
77. Putnam, C.D., Copenhaver, G.P., Denton, M.L. and Pikaard, C.S. (1994) Mol. Cell. Biol. **14**, 6476–6488
78. Bazett Jones, D.P., Leblanc, B., Herfort, M. and Moss, T. (1994) Science, **264**, 1134–1137
79. Stefanovsky, V.Y., Pelletier, G., Bazett-Jones, D.P., Crane-Robinson, C. and Moss, T. (2001) Nucleic Acids Res. **29**, 3241–3247
80. Hempel, W.M., Cavanaugh, A.H., Hannan, R.D., Taylor, L. and Rothblum, L.I. (1996) Mol. Cell. Biol. **16**, 557–563
81. Bodeker, M., Cairns, C. and McStay, B. (1996) Mol. Cell. Biol. **16**, 5572–5578
82. Kihm, A.J., Hershey, J.C., Haystead, T.A., Madsen, C.S. and Owens, G.K. (1998) Proc. Natl. Acad. Sci. U.S.A. **95**, 14816–14820
83. Tuan, J.C., Zhai, W. and Comai, L. (1999) Mol. Cell. Biol. **19**, 2872–2879
84. Beckmann, H., Chen, J.L., O'Brien, T. and Tjian, R. (1995) Science **270**, 1506–1509
85. Kwon, H. and Green, M.R. (1994) J. Biol. Chem. **269**, 30140–30146
86. Schnapp, G., Santori, F., Carles, C., Riva, M. and Grummt, I. (1994) EMBO J. **13**, 190–199
87. Seither, P., Zatsepina, O., Hoffmann, M. and Grummt, I. (1997) Chromosoma **106**, 216–225
88. Ptashne, M. and Gann, A. (1997) Nature (London) **386**, 569–577
89. O'Sullivan, A.C., Sullivan, G.J. and McStay, B. (2002) Mol. Cell. Biol. **22**, 657–668
90. Mais, C., Wright, J.E., Prieto, J.L., Raggett, S.L. and McStay, B. (2005) Genes Dev. **19**, 50–64
91. Hannan, R., Stefanovsky, V., Arino, T., Rothblum, L. and Moss, T. (1999) Nucleic Acids Res. **27**, 1205–1213
92. Brandenburger, Y., Jenkins, A., Autelitano, D.J. and Hannan, R.D. (2001) FASEB J. **15**, 2051–2053
93. Huang, R., Wu, T., Xu, L., Liu, A., Ji, Y. and Hu, G. (2002) FASEB J. **16**, 293–301
94. Cabart, P. and Kalousek, I. (1998) Cell Mol. Biol. (Noisy-Le-Grand) **44**, 343–350
95. Poortinga, G., Hannan, K.M., Snelling, H., Walkley, C.R., Jenkins, A., Sharkey, K., Wall, M., Brandenburger, Y., Palatsides, M., Pearson, R.B. et al. (2004) EMBO J. **23**, 3325–3335
96. Voit, R., Kuhn, A., Sander, E.E. and Grummt, I. (1995) Nucleic Acids Res. **23**, 2593–2599
97. Hannan, K.M., Kennedy, B.K., Cavanaugh, A.H., Hannan, R.D., Hirschler-Laszkiewicz, I., Jefferson, L.S. and Rothblum, L.I. (2000) Oncogene **19**, 3487–3497
98. Yuan, X., Zhou, Y., Casanova, E., Chai, M., Kiss, E., Grone, H.-J., Schutz, G. and Grummt, I. (2005) Mol. Cell **19**, 77–87

99. Grandori, C., Gomez-Roman, N., Felton-Edkins, Z.A., Ngouenet, C., Galloway, D.A., Eisenman, R.N. and White, R.J. (2005) Nat. Cell Biol. 7, 311–318
100. Arabi, A., Wu, S., Ridderstrale, K., Bierhoff, H., Shiue, C., Fatyol, K., Fahlen, S., Hydbring, P., Soderberg, O., Grummt, I. et al. (2005) Nat. Cell Biol. 7, 303–310
101. Grewal, S.S., Li, L., Orian, A., Eisenman, R.N. and Edgar, B.A. (2005) Nat. Cell Biol. 7, 295–302

Biochem. Soc. Symp. **73**, 217–224
(Printed in Great Britain)

20

Fluorescence resonance energy transfer as a method for dissecting *in vivo* mechanisms of transcriptional activation

Sara K. Evans, David P. Aiello and Michael R. Green[1]

Howard Hughes Medical Institute, Programs in Gene Function and Expression and Molecular Medicine, University of Massachusetts Medical School, 364 Plantation Street, Worcester, MA 01605, U.S.A.

Abstract

The first step in transcriptional activation of protein-coding genes involves the assembly on the promoter of a large PIC (pre-initiation complex) comprising RNA polymerase II and a suite of general transcription factors. Transcription is greatly enhanced by the action of promoter-specific activator proteins (activators) that function, at least in part, by increasing PIC formation. Activator-mediated stimulation of PIC assembly is thought to result from a direct interaction between the activator and one or more components of the transcription machinery, termed the 'target'. The unambiguous identification of direct, physiologically relevant *in vivo* targets of activators has been a considerable challenge in the transcription field. The major obstacle has been the lack appropriate experimental methods to measure direct interactions with activators *in vivo*. The development of spectral variants of green fluorescent protein has made it possible to perform FRET (fluorescence resonance energy transfer) analysis in living cells, thereby allowing the detection of direct protein–protein interactions *in vivo*. Here we discuss how FRET can be used to identify activator targets and to dissect *in vivo* mechanisms of transcriptional activation.

Introduction

The expression of eukaryotic protein-coding genes is regulated at several steps; the majority of regulation, however, occurs at the level of transcription

[1]To whom correspondence should be addressed (email michael.green@umassmed.edu)

initiation. In eukaryotes, transcription of protein-coding genes requires RNA polymerase II and the ordered assembly of numerous GTFs (general transcription factors) on the core promoter to form a PIC (pre-initiation complex). The first step of PIC assembly is binding of the GTF TFIID (transcription factor IID) to the TATA box located in the core promoter. TFIID is a multi-subunit complex comprising TBP (TATA-binding protein) and at least 14 TAFs (TBP-associated factors) [1–3]. Following TFIID recruitment, other components of the transcription initiation machinery, including RNA polymerase II, TFIIA, TFIIB, TFIIE, TFIIF and TFIIH, are co-ordinately recruited to the core promoter and enter the PIC through a complex network of protein–protein interactions (reviewed in [4]).

Although GTFs can be sufficient for accurate transcription initiation *in vitro*, transcription is greatly stimulated by the action of a second class of factors termed promoter-specific activator proteins (activators) (reviewed in [5]). Activators are minimally composed of two distinct domains: a sequence-specific DNA-binding domain, which targets the activator to a specific *cis*-acting promoter element [in yeast, this element is called the UAS (upstream activating sequence)], and a separate activation domain, which engages in protein–protein interactions. Activation domains are typically categorized according to their amino acid content, and include acidic, glutamine-rich or proline-rich domains (reviewed in [6]).

A variety of studies indicate that activators function, at least in part, by increasing PIC formation [4,5,7,8]. Activator-mediated stimulation of PIC assembly is thought to result from a direct interaction between the activation domain and one or more components of the transcription machinery, termed the ‘target(s)’. In several instances, multiple putative targets have been identified for a single activator; for example, the acidic activator VP16 has been shown to interact with numerous factors, including the GTFs RNA polymerase II, TFIIA, TFIIB and TFIIH, as well as several classes of cofactors [9–14]. The unambiguous identification of direct, physiologically relevant *in vivo* targets of activators has been a considerable challenge in the transcription field. The major obstacle has been the lack appropriate experimental methods with which to measure direct interactions with activation domains *in vivo*.

Gal4: a model yeast transcriptional activator

The induction of genes involved in galactose utilization in budding yeast (*GAL* genes) represents a model experimental system for studying eukaryotic transcriptional activation mechanisms. Transcriptional stimulation of *GAL* genes requires Gal4, a well characterized acidic activator. Several components of the transcriptional machinery have been proposed to be a target of Gal4, including TBP [15,16], TAFs [16a], TFIIB [15], the Swi/Snf complex [17], mediator subunits Srb4 [18,19] and Gal11 [20,21], and the SAGA (Spt/Ada/Gcn5 acetyltransferase) complex [22–24]. The majority of these putative targets have been identified either by *in vitro* protein interaction studies or through inferences from various indirect *in vivo* experiments. Thus, for many of these factors, the question remains whether they interact directly with Gal4 *in vivo*.

FRET (fluorescence resonance energy transfer): a method for measuring direct protein–protein interactions *in vivo*

Originally devised as a technique for structure elucidation, FRET has been used to measure spatial proximity relationships in biological macromolecules for over three decades. Within the past few years, however, the development of spectral variants of GFP (green fluorescent protein) as reporters has made it possible to perform FRET analysis in living cells, thereby allowing the detection of protein–protein interactions *in vivo* (reviewed in [25]).

The general experimental strategy is shown in Figure 1. The technique utilizes a pair of fluorophores, termed a donor and an acceptor, which have overlapping emission/adsorption spectra. The most common pair of GFP-derived fluorophores for biological use are ECFP (enhanced cyan fluorescent protein) and EYFP (enhanced yellow fluorescent protein). Each fluorophore is fused to one of a protein pair of interest. The donor fluorophore is excited at its specific fluorescence excitation wavelength; if the donor and acceptor are not in close proximity, only the donor emission is detected (Figure 1A). However, if the two fluorophores are within a critical distance of each other, the excited state of the donor is transferred non-radiatively to the acceptor, and the acceptor emission is predominantly observed (a FRET signal; Figure 1B). In practice, following donor excitation, the emission intensity for both fluorophores has been measured using a variety of FRET imaging tools, including confocal laser

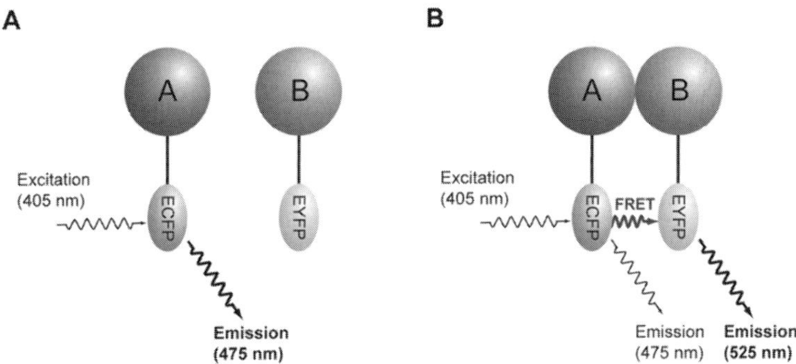

Figure 1 Experimental strategy for using FRET to detect direct protein–protein interactions *in vivo*. (**A**) If the donor (ECFP) and acceptor (EYFP) are not in close proximity, excitation of the donor fluorophore at its specific fluorescence excitation wavelength results in only donor emission being detected. (**B**) Direct interaction between ECFP- and EYFP-tagged proteins results in a transfer of energy from the donor to acceptor, and acceptor emission is predominantly detected.

scanning microscopy [26,27], epifluorescence wide-field microscopy [28,29], scanning fluorimetry [30] and flow cytometry [31].

For ECFP and EYFP, the critical distance required for FRET is ~50 Å (5 nm) [32]. When the structure of GFP is considered [33,34], an inter-fluorophore distance of 50 Å corresponds to a separation between ECFP and EYFP of 25–35 Å [28]. Because this distance is relatively small, a positive FRET signal is generally interpreted as indicating a direct interaction between the two fluorophore-tagged proteins.

A number of studies have successfully used FRET to detect direct protein–protein interactions in living yeast cells. For example, FRET has been used extensively to map interactions between nuclear transport factors, such as those that occur between importins and exportins as they travel through the NPC (nuclear pore complex) [28], and to determine the overall structural organization of the NPC [29]. FRET has also been used to monitor protein oligomerization; for example, dimerization of the α-factor receptor protein Ste2 [27] and the mitochondrial outer membrane protein Tom70 [31].

FRET also has a broader applicability than simply detecting binary protein–protein interactions. For example, FRET has been used to monitor kinetic interactions between proteins, such as those observed during heterotrimeric G-protein activation [30]. More recently, FRET has also been used to quantify protein interactions; that is, to determine the fraction of a protein population that exists in a particular complex. Using spectrally resolved fluorescence microscopy, it has been shown that the majority of Ste2 in cells forms dimeric complexes [27].

In our laboratory, we have recently applied FRET as an experimental strategy to detect direct interactions between transcriptional activation domains and their targets *in vivo*. In particular, we used FRET to identify a direct target of Gal4 [26]. As mentioned above, one putative Gal4 target that had been identified is the SAGA complex. SAGA is a large multi-protein complex that in yeast is required for the normal transcription of approx. 10% of RNA polymerase II-dependent genes [35]. SAGA exhibits several properties expected of a direct target of Gal4. First, SAGA is recruited to the vicinity of Gal4 (i.e. the UAS) [22]. Secondly, recruitment of SAGA is dependent on Gal4, but not other factors, such as TBP and RNA polymerase II [22]. Thirdly, SAGA is required for recruitment of downstream factors: the UAS-bound SAGA facilitates the binding of TBP to the core promoter, thereby stimulating PIC assembly and transcription [22,23]. SAGA is essential for *GAL1* transcription: if SAGA is not recruited to the UAS, the PIC is not assembled and transcription does not occur [22,23].

Yeast SAGA is a 1.8 MDa complex that contains at least 14 subunits. The non-essential components of SAGA can be categorized into three groups on the basis of their distinct mutant phenotypes: (i) Gcn5, Ada2 and Ada3; (ii) Ada1, Spt7 and Spt20; and (iii) Spt3 and Spt8 [36–41]. The SAGA complex also includes a subset of TAFs that are also found in TFIID: TAF5, TAF6, TAF9, TAF10 and TAF12. Finally, SAGA contains the ATM (ataxia telangiectasia mutated)/phosphoinositide 3-kinase-related protein Tra1, which had been reported to be a direct target of certain activators, including Gal4 [24].

Our FRET analyses revealed that the SAGA subunit Tra1 is a direct *in vivo* target of Gal4 [26]. The Gal4–Tra1 interaction did not occur in the absence of Spt20, a SAGA subunit required for complex integrity, suggesting that Tra1 must be incorporated within an intact SAGA complex for interaction with Gal4 *in vivo*. Notably, four other putative targets of Gal4 – TBP, TFIIB, Srb4 and Gal11 – did not show a positive FRET signal, suggesting that these factors are not direct *in vivo* targets of Gal4. This study demonstrated the validity of the FRET assay for identifying proteins that interact directly with activators *in vivo*.

Advantages and potential limitations of FRET

FRET-based techniques offer several advantages over conventional methods for detecting interactions between proteins. The most notable advantage of FRET is that it can be used to detect protein–protein interactions in living cells. As such, FRET can be performed on intact, functional complexes rather than isolated subunits. The technique is also broadly applicable to studying a wide variety of proteins (i.e. cytoplasmic, nuclear or membrane-bound), and is highly sensitive to detecting an interaction at the level of a single cell. FRET is also a very versatile approach with respect to the equipment required to detect the FRET signal (i.e. scanning fluorimeter, laser scanning microscope or flow cytometer). Moreover, if the interactions are monitored using microscopy, it is possible to gain additional information regarding the subcellular localization of the protein and its interacting partner.

As with any technique for monitoring protein–protein interactions, FRET also has potential limitations. One potential caveat is that a false-positive FRET signal may arise if the two fluorophores are adventitiously positioned within ~30 Å of one another through an indirect interaction mediated by a third ('bridging') protein. There is also the possibility of obtaining a false-negative signal. One parameter thought to affect FRET efficiency is the relative orientation of the two fluorophores, raising the possibility that fluorophores may be positioned within the critical distance but may be not oriented properly to allow FRET to occur. However, a study has reported that fluorophore orientation is not an essential factor for intermolecular distance determination by FRET, implying that it may not be an important parameter governing energy transfer [42].

Another potential limitation of FRET is that the interpretation of a positive result becomes complicated for proteins that are present in multiple complexes. For example, detecting an interaction between an activator and a TAF protein does not distinguish whether the target is a TFIID subunit or a SAGA subunit. Moreover, the dynamic nature of some proteins/complexes may decrease the likelihood that a FRET signal will be observed. Finally, if the protein of interest is not abundant, it may be difficult to detect a FRET signal, as ECFP is not a particularly bright fluorophore.

Recent technical advances related to FRET

Recently, several improvements have been made to GFP-based fluorophores, with the goal of enhancing the detection limits of FRET. For example, yeast-specific versions of GFP, ECFP and EYFP have been constructed to reflect the yeast codon bias, and these are twice as detectable as unoptimized tags [43]. Other fluorophore improvements reflect the necessity of reducing the spectral overlap between the donor emission spectrum and the acceptor absorption spectrum, which is a parameter that affects FRET efficiency. In this regard, derivatives of ECFP and EYFP have been optimized for both fluorescence emission intensity and FRET ratio (i.e. EYFP peak/ECFP peak) by successive rounds of mutagenesis [44]. Similarly, site-directed mutagenesis has been used to enhance the spectroscopic properties of ECFP; the derivative, known as Cerulean, shows 2.5-fold higher intensity than the parent protein [45]. Finally,

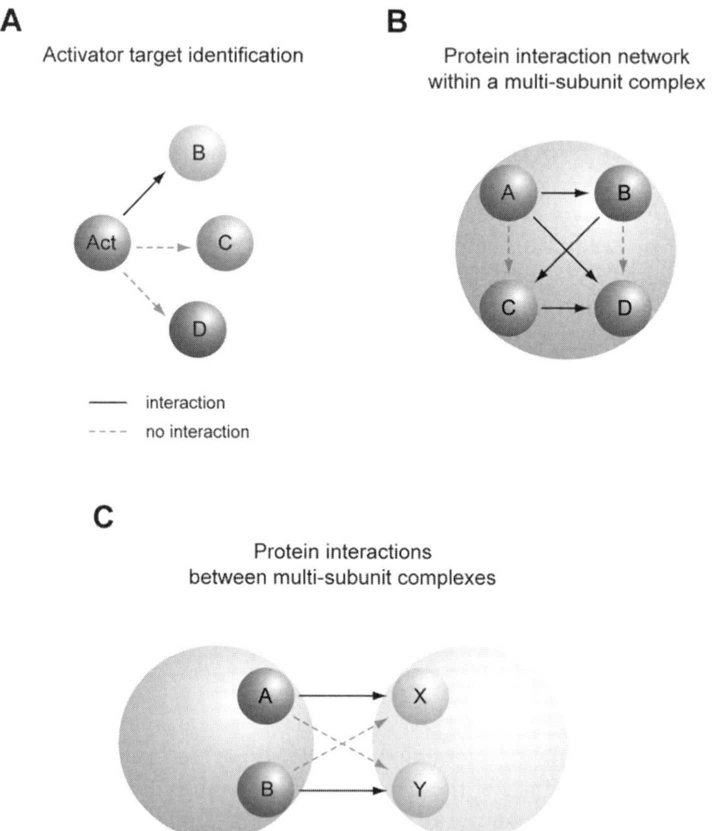

Figure 2 Applications of FRET in dissecting mechanisms of transcription activation. (**A**) Identifying the direct targets of activators. (**B**) Mapping subunit interactions within transcription complexes. (**C**) Delineating protein interactions between multi-subunit transcription complexes.

two studies have reported UV-excitable forms of GFP: Sapphire, which is twice as detectable as GFP [43], and GFP2, which shows increased FRET efficiency with EYFP compared with the traditional ECFP/EYFP fluorophore pair [46].

Applications of FRET in dissecting mechanisms of transcription activation

The success of FRET in detecting direct protein–protein interactions in living cells underscores its value in investigating mechanisms of transcription activation. The direct activator–target interactions detected by FRET will help to answer questions as to whether activators have multiple relevant targets, whether these targets are functionally redundant, and whether different activators function through common or distinct pathways and PICs. Moreover, FRET may be used to elucidate the mechanism by which activator–target interactions lead to transcription activation – whether this occurs through direct interaction with the transcription machinery, as appears to be the case for Gal4, or through interaction with other factors, such as chromatin remodelling proteins. In addition to identifying activator targets (Figure 2A), FRET can also be used, for example, to identify *in vivo* protein interaction partners of transcription factors. The use of FRET can also be extended to mapping interaction networks within large multi-subunit transcription complexes such as SAGA and Mediator (Figure 2B), as well as to delineating protein interactions between multi-subunit complexes (Figure 2C). The recent technical advances in improving the detection limits of FRET should facilitate such studies in yeast.

This work was supported by grants from the Leukemia and Lymphoma Society (D.P.A.) and the NIH (M.R.G.).

References

1. Burley, S.K. and Roeder, R.G. (1996) Annu. Rev. Biochem. **65**, 769–799
2. Albright, S.R. and Tjian, R. (2000) Gene **242**, 1–13
3. Green, M.R. (2000) Trends Biochem. Sci. **25**, 59–63
4. Lee, T.I. and Young, R.A. (2000) Annu. Rev. Genet. **34**, 77–137
5. Orphanides, G., Lagrange, T. and Reinberg, D. (1996) Genes Dev. **10**, 2657–2683
6. Ptashne, M. (1988) Nature (London) **335**, 683–689
7. Roeder, R.G. (1996) Trends Biochem. Sci. **21**, 327–335
8. Ptashne, M. and Gann, A. (1997) Nature (London) **386**, 569–577
9. Stringer, K.F., Ingles, C.J. and Greenblatt, J. (1990) Nature (London) **345**, 783–786
10. Lin, Y.S., Ha, I., Maldonado, E., Reinberg, D. and Green, M.R. (1991) Nature (London) **353**, 569–571
11. Xiao, H., Pearson, A., Coulombe, B., Truant, R., Zhang, S., Regier, J.L., Triezenberg, S.J., Reinberg, D., Flores, O., Ingles, C.J. et al. (1994) Mol. Cell. Biol. **14**, 7013–7024
12. Hengartner, C.J., Thompson, C.M., Zhang, J., Chao, D.M., Liao, S.M., Koleske, A.J., Okamura, S. and Young, R.A. (1995) Genes Dev. **9**, 897–910
13. Kobayashi, N., Boyer, T.G. and Berk, A.J. (1995) Mol. Cell. Biol. **15**, 6465–6473
14. Hall, D.B. and Struhl, K. (2002) J. Biol. Chem. **277**, 46043–46050
15. Wu, Y., Reece, R.J. and Ptashne, M. (1996) EMBO J. **15**, 3951–3963
16. Melcher, K. and Johnston, S.A. (1995) Mol. Cell. Biol. **15**, 2839–2848
16a. Klein, J., Nolden, M., Sanders, S.L., Kirchner, J., Weil, P.A. and Melcher, K. (2003) J. Biol. Chem. **278**, 6779–6786

17. Neely, K.E., Hassan, A.H., Brown, C.E., Howe, L. and Workman, J.L. (2002) Mol. Cell. Biol. **22**, 1615–1625
18. Koh, S.S., Ansari, A.Z., Ptashne, M. and Young, R.A. (1998) Mol. Cell **1**, 895–904
19. Park, J.M., Kim, H.S., Han, S.J., Hwang, M.S., Lee, Y.C. and Kim, Y.J. (2000) Mol. Cell. Biol. **20**, 8709–8719
20. Jeong, C.J., Yang, S.H., Xie, Y., Zhang, L., Johnston, S.A. and Kodadek, T. (2001) Biochemistry **40**, 9421–9427
21. Bryant, G.O. and Ptashne, M. (2003) Mol. Cell **11**, 1301–1309
22. Bhaumik, S.R. and Green, M.R. (2001) Genes Dev. **15**, 1935–1945
23. Larschan, E. and Winston, F. (2001) Genes Dev. **15**, 1946–1956
24. Brown, C.E., Howe, L., Sousa, K., Alley, S.C., Carrozza, M.J., Tan, S. and Workman, J.L. (2001) Science **292**, 2333–2337
25. Truong, K. and Ikura, M. (2001) Curr. Opin. Struct. Biol. **11**, 573–578
26. Bhaumik, S.R., Raha, T., Aiello, D.P. and Green, M.R. (2004) Genes Dev. **18**, 333–343
27. Raicu, V., Jansma, D.B., Miller, R.J. and Friesen, J.D. (2005) Biochem. J. **385**, 265–277
28. Damelin, M. and Silver, P.A. (2000) Mol. Cell **5**, 133–140
29. Damelin, M. and Silver, P.A. (2002) Biophys. J. **83**, 3626–3636
30. Yi, T.M., Kitano, H. and Simon, M.I. (2003) Proc. Natl. Acad. Sci. U.S.A. **100**, 10764–10769
31. Dye, B.T., Schell, K., Miller, D.J. and Ahlquist, P. (2005) Cytometry **63A**, 77–86
32. Heim, R. and Tsien, R.Y. (1996) Curr. Biol. **6**, 178–182
33. Ormo, M., Cubitt, A.B., Kallio, K., Gross, L.A., Tsien, R.Y. and Remington, S.J. (1996) Science **273**, 1392–1395
34. Yang, F., Moss, L.G. and Phillip s, Jr, G.N. (1996) Nat. Biotechnol. **14**, 1246–1251
35. Lee, T.I., Causton, H.C., Holstege, F.C., Shen, W.C., Hannett, N., Jennings, E.G., Winston, F., Green, M.R. and Young, R.A. (2000) Nature (London) **405**, 701–704
36. Horiuchi, J., Silverman, N., Pina, B., Marcus, G.A. and Guarente, L. (1997) Mol. Cell. Biol. **17**, 3220–3228
37. Roberts, S.M. and Winston, F. (1997) Genetics **147**, 451–465
38. Sterner, D.E., Grant, P.A., Roberts, S.M., Duggan, L.J., Belotserkovskaya, R., Pacella, L.A., Winston, F., Workman, J.L. and Berger, S.L. (1999) Mol. Cell. Biol. **19**, 86–98
39. Hampsey, M. (1997) Trends Genet. **13**, 427–429
40. Grant, P.A., Sterner, D.E., Duggan, L.J., Workman, J.L. and Berger, S.L. (1998) Trends Cell Biol. **8**, 193–197
41. Brown, C.E., Lechner, T., Howe, L. and Workman, J.L. (2000) Trends Biochem. Sci. **25**, 15–19
42. dos Remedios, C.G. and Moens, P.D. (1995) J. Struct. Biol. **115**, 175–185
43. Sheff, M.A. and Thorn, K.S. (2004) Yeast **21**, 661–670
44. Nguyen, A.W. and Daugherty, P.S. (2005) Nat. Biotechnol. **23**, 355–360
45. Rizzo, M.A., Springer, G.H., Granada, B. and Piston, D.W. (2004) Nat. Biotechnol. **22**, 445–449
46. Zimmermann, T., Rietdorf, J., Girod, A., Georget, V. and Pepperkok, R. (2002) FEBS Lett. **531**, 245–249

Biochem. Soc. Symp. **73**, 225–236
(Printed in Great Britain)

21

Core promoter-selective RNA polymerase II transcription

Petra Gross and Thomas Oelgeschläger[1]

Transcription Laboratory, Marie Curie Research Institute, The Chart, Oxted, Surrey RH8 0TL, U.K.

Abstract

The initiation of mRNA synthesis in eukaryotic cells is a complex and highly regulated process that requires the assembly of general transcription factors and RNAP II (RNA polymerase II; also abbreviated as Pol II) into a pre-initiation complex at the core promoter. The core promoter is defined as the minimal DNA region that is sufficient to direct low levels of activator-independent (basal) transcription by RNAP II *in vitro*. The core promoter typically extends approx. 40 bp up- and down-stream of the start site of transcription and can contain several distinct core promoter sequence elements. Core promoters in higher eukaryotes are highly diverse in structure, and each core promoter sequence element is only found in a subset of genes. So far, only TATA box and INR (initiator) element have been shown to be capable of directing accurate RNAP II transcription initiation independent of other core promoter elements. Computational analysis of metazoan genomes suggests that the prevalence of the TATA box has been overestimated in the past and that the majority of human genes are TATA-less. While TATA-mediated transcription initiation has been studied in great detail and is very well understood, very little is known about the factors and mechanisms involved in the function of the INR and other core promoter elements. Here we summarize our current understanding of the factors and mechanisms involved in core promoter-selective transcription and discuss possible pathways through which diversity in core promoter architecture might contribute to combinatorial gene regulation in metazoan cells.

Introduction

RNAP II (RNA polymerase II; also abbreviated as Pol II) is responsible for the transcription of protein-coding genes in eukaryotes. To ensure the correct

[1]To whom correspondence should be addressed (email t.oelgeschlager@mcri.ac.uk).

spatial and temporal expression of tens of thousands of genes, eukaryotic cells employ a large network of factors that regulate transcription in response to developmental or extracellular signals. In order to initiate transcription of specific genes, a number of different processes must occur in a concerted manner: de-condensation of packed chromatin, binding of a gene-specific set of transcription activators to regulatory sequences located either close to the transcription start site or many thousands of base pairs away (enhancers), and finally communication of these steps to the general RNAP II transcription machinery via co-activator complexes (reviewed in [1–3]).

The general RNAP II transcription machinery has been biochemically defined as the set of factors essential for accurate transcription initiation at TATA-containing promoters *in vitro*, and consists of the GTFs (general transcription factors) TFIIA, TFIIB, TFIID, TFIIE, TFIIF and TFIIH, and RNAP II [4–6]. RNAP II transcription initiation requires ordered assembly of RNAP II and GTFs into a PIC (pre-initiation complex) at the core promoter region.

The RNAP II core promoter is functionally defined as the minimal DNA region required to direct low levels of accurate RNAP II transcription initiation in the absence of activators *in vitro*. The core promoter typically extends approx. 40 bp upstream and downstream of the site of transcription initiation. The first RNAP II core promoters to be identified were found to be similar in structure, with a TATA box (consensus sequence TATAAA) located 25–30 bp upstream from the transcription initiation site [7]. For this reason, RNAP II core promoters were initially not considered to contribute actively to the regulation of RNAP II transcription. However, subsequent work demonstrated that metazoan RNAP II core promoters can be very complex in structure and that there is a large degree of diversity in RNAP II core promoter architecture in metazoan genomes [8–10].

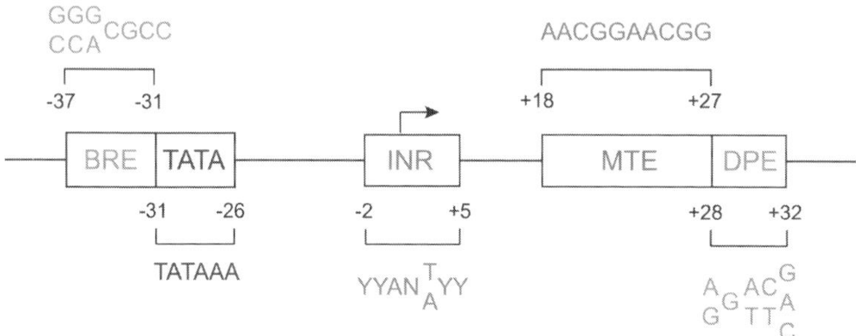

Figure 1 Known metazoan core promoter sequence elements. The consensus sequences shown for the BRE, TATA box and INR element are for mammals, and those for the MTE and DPE were determined for *Drosophila*. The position of each sequence element relative to the transcription start site (+1) is indicated. Note that each of these sequence motifs is only found in a subset of core promoters and that none of the elements shown is ubiquitously required for core promoter activity.

In addition to the TATA box, several other core promoter elements have been identified and functionally characterized [10] (Figure 1). The BRE (TFIIB recognition element) is found immediately upstream of the TATA box region of some core promoters and was originally identified as a core promoter element that enhanced transcription from the TATA box- and INR (initiator) element-containing Ad2ML (adenovirus-2 major late) promoter in a highly purified system reconstituted with human TBP (TATA-binding protein), TFIIB, TFIIE, TFIIF, TFIIH and RNAP II [11]. More recently, the BRE was shown to repress Ad2ML activity in a crude human nuclear extract, and this BRE-mediated repression can be reversed by a transcription activator [12].

The INR element was originally identified as a core promoter element essential for the activity of the TATA-less murine terminal deoxynucleotidyltransferase gene (*TdT*) [13]. The INR element encompasses the transcription start site, and in humans has the consensus sequence $YY^{+1}AN(T/A)YY$ (the superscript +1 denotes the nucleotide at which transcription starts). Like the TATA box, the INR element is sufficient to direct accurate RNAP II transcription initiation [13]. Furthermore, when spaced by ~25 bp, the TATA and INR elements can function in concert in a synergistic manner [14].

Core promoter elements downstream from the transcription start site have also been identified. A prevalent RNAP II promoter element in *Drosophila* is the DPE (downstream promoter element) [15,16]. The DPE is located 30 bp downstream from the transcription start and its function is dependent on the presence of an INR element. In humans, DPE motifs have been identified by sequence in some promoters, but these have not been extensively characterized so far. Finally, the MTE (motif ten element) was discovered recently based on its over-representation in a large database of *Drosophila* core promoters [17]. In *Drosophila*, the MTE is found precisely at positions +18 to +27 relative to the start site. Like the DPE, the MTE requires an INR element for function. However, the MTE appears to function independently of a TATA box or a DPE motif [17]. Interestingly, the same study also identified a functional MTE in the core promoter of the human sterol C5 desaturase-like gene (*SC5D*). Further studies are needed to establish a general role for DPE and MTE core promoter in mammalian cells. However, the growing list of functional core promoter elements underlines the notion that the complexity and diversity of core promoters, and their function in regulated RNAP II transcription, has yet to be fully explored.

Prevalence of core promoter elements in metazoan genomes

The availability of an increasing number of whole genome sequence databases has opened the door to extensive computational analysis of the prevalence of known core promoter elements in the genomes of different metazoan species. A computational analysis in *Drosophila* found TATA consensus sites in only 33% of 1941 promoters [18,19]. Similarly, an early database analysis found TATA boxes in only 32% of approx. 1000 potential core promoter regions in human genes [20]. The same study found an INR consensus sequence in 85% of the promoters analysed. A more recent bioinformatics study found a TATA box motif in only 11% of over 1770 human core promoter sequences [21]. The authors of that study attributed the significant under-representation of TATA motif elements compared with previous analyses to refinements in computational TATA box detection. In agreement with earlier studies, the INR element was found to be the most common RNAP II core promoter element. The most recent, and so far most comprehensive, bioinformatics study analysed two large sets of human RNAP II promoters [22]: 1871 promoters from the EPD (Eukaryotic Promoter Database) [23,24] and 8793 promoters from the DBTSS (Database of Transcriptional Start Sites) [25,26]. The study examined the statistical distribution of core promoter elements at their functional positions, as well as the occurrence of combinations of core promoter elements (i.e. TATA- and INR, BRE and TATA). The authors found that about half of the human RNAP II core promoters contain an INR consensus sequence at its functional position (49% in the EPD and 48.4% in the DBTSS). Only 21.8% of RNAP II core promoters in the EPD and 10.4% of RNAP II core promoters the DTSS were found to contain a TATA box motif. Interestingly, 60% of TATA-box-containing promoters were found to also contain an INR consensus site, compared with approx. 45% of TATA-less promoters. However, only 9.9% of promoters in the EPD and 4.3% of promoters in the DBTSS contain both TATA and INR elements separated by a distance permitting functional synergy [14]. Furthermore, the authors presented statistical evidence for novel functional combinations of core promoter elements. For example, an interesting result of the study is that the number of core promoter regions harbouring both a BRE and an INR element is comparable with the number of promoters containing a combination of TATA and INR elements. BRE functions have so far only been studied in the context of a functional TATA box [11,12,27]. However, only 11.8% (EPD) and 13.8% (DBTSS) of TATA-containing promoters were found to contain a BRE motif, compared with 28.1% (EPD) or 26.9% (DBTSS) of TATA-less promoters. This observation raises the interesting possibility that BRE functions, which are mediated through direct TFIIB–DNA interactions [11,12,27], may not be limited to promoters that contain a consensus TATA element and are therefore able to form a cognate TBP–TATA nucleoprotein complex [28–30].

Obviously, the functionality of core promoter sequence motifs within individual RNAP II promoters identified by computational analyses must be examined carefully on a case-by-case basis. However, the recent results of bioinformatics studies strongly suggest that the occurrence of the TATA box at RNAP II promoters in metazoan genomes has been overestimated in the past. Contrary to expectations, only a small fraction of RNAP II promoters appear to contain a TATA box. In contrast, a large proportion of RNAP II promoters in metazoan genomes appear to contain an INR element. Finally, approx. 25% of human promoters appear to lack known core promoter elements. This may point to the existence of additional core promoter sequence elements that remain to be identified and functionally characterized.

TATA compared with INR function: PIC assembly and structure

Of the known core promoter elements, only the TATA box and the INR element have been shown to direct accurate RNAP II transcription independently of other core promoter elements. The molecular events leading to PIC assembly and RNAP II transcription initiation on TATA-containing promoters have been studied in great detail and are well understood. On TATA box-containing promoters, PIC assembly is initiated through specific binding of the TFIID TBP subunit, the only sequence-specific DNA-binding polypeptide among the GTFs, to the TATA DNA sequence. TBP resembles a molecular 'saddle', and TBP binding results in bending of the TATA box DNA by 90° [28,29]. After formation of the TBP–TATA nucleoprotein complex, the remaining GTFs can be recruited in a stepwise manner: TFIIB and TFIIA are recruited first, followed by TFIIF/RNAP II, then TFIIE and finally TFIIH [4–6].

Importantly, the full set of GTFs and RNAP II is not sufficient to support INR-dependent transcription from TATA-less promoters or synergy between TATA and INR elements. So far, reconstitution of INR-specific transcription with purified factors has not been accomplished, and underlying molecular mechanisms remain poorly understood.

RNAP II transcription from TATA-containing promoters does not require the TFIID TAF (TBP-associated factor) subunits in purified systems and can be reconstituted with recombinant TBP instead of TFIID and in the absence of TFIIA. In contrast, in crude human nuclear extracts or in systems reconstituted with partially purified nuclear extract fractions, TFIID TAF subunits are absolutely required both for INR-dependent transcription from TATA-less promoters and for synergy between TATA and INR elements [31–33]. DNA-binding studies demonstrated that the TFIID complex can interact with promoter DNA downstream of the TATA region and up to ~35 bp downstream of the transcription start site [34–36]. These TAF–DNA interactions were shown subsequently to be sequence-dependent [36] and to be induced by the presence of a consensus INR element [32,37,38]. TFIID interactions with the INR element are likely to be mediated by TAF1 and TAF2. TAF1 within human TFIID can be cross-linked to the INR element of the Ad2ML promoter, and this interaction is greatly enhanced by the presence of TFIIA [39]. Moreover, *Drosophila* TAF2 has been shown to possess intrinsic DNA-binding activity [40]. Finally, partial TFIID complexes composed of recombinant TBP, TAF1 and TAF2 subunits have been shown to mediate INR functions in a partially purified *Drosophila* transcription system and to interact preferentially with a core promoter in the presence of a consensus INR sequence [41,42].

However, while TAFs within TFIID are necessary for INR function in systems reconstituted with highly purified GTFs and RNAP II, they are not sufficient [43]. This observation led to the discovery of TICs (TAF- and INR-dependent cofactors) in partially purified human nuclear extract fractions, which can restore INR function in purified systems [43]. A TIC-1 activity was shown to reconstitute TAF-dependent synergism between TATA and INR elements, whereas TIC-2 and TIC-3 activities were required to reconstitute INR-dependent transcription from the TATA-less *mTdT* promoter [43]. The polypeptides responsible for the different TIC activities remain to be identified.

The discovery that cofactors are required for INR-dependent transcription suggests that RNAP II PICs assembled on different core promoters can differ in composition. It further raises the question whether the order of events leading to PIC formation on INR-dependent TATA-less promoters is different from the pathway established for PIC assembly on TATA-dependent promoters. Studies with a mutant TFIID complex, harbouring a TBP mutant variant defective in TATA-specific DNA binding, suggest that the TATA box-binding activity of TBP is dispensable for INR-dependent transcription initiation at TATA-less promoters [44]. In agreement with this observation, isolated wild-type TFIID appears unable to bind specifically to TATA-less promoters in DNase I footprinting assays [44]. These observations may suggest that PIC formation at TATA-less promoters does not require formation of a cognate TBP nucleoprotein complex.

On TATA-containing promoters, TBP-induced bending of the TATA box DNA has been proposed to be a rate-limiting step in PIC formation [45–47] and essential for recruitment of TFIIB [30]. Intrinsic mechanical properties of the TATA box DNA contribute significantly to the specificity and stability of TBP interactions [48]. Consequently, the degree of TBP-induced bending is DNA sequence-dependent [49–51].

It is at present unclear if, and to what extent, non-cognate TBP–DNA interactions with the –30 promoter region contribute to PIC assembly and transcription initiation at TATA-less promoters [10]. One possibility might be that TBP forms an unstable 'unbent' protein–DNA complex in the absence of a TATA box sequence [47]. If this proves to be the case, the question arises as to how TFIIB is recruited under these conditions.

A recent computational analysis of flexibility profiles of a large number of human and mouse RNAP II promoter regions suggests that TATA boxes and INR elements possess very similar mechanical properties [52]. In addition, DNA directly upstream of TATA and INR elements tends to be more rigid than DNA downstream of these elements. The similarity in DNA flexibility profiles may indicate that recognition of the INR element, like recognition of the TATA box, involves protein-induced DNA bending. Consistent with this idea is the observation that the presence of a consensus INR sequence induces interactions of TFIID TAFs with DNA downstream of the TATA box region ([32,37,38]; P. Gross and T. Oelgeschläger, unpublished work), which can result in nucleosome-like DNA wrapping [39]. Although purified TFIID complex in isolation appears to be unable to form a stable complex with TATA-less promoters, it is generally thought that, similar to TATA-containing promoters, PIC assembly on TATA-less promoters is initiated by TFIID–promoter complex formation, perhaps in conjunction with TIC activities. Further work is necessary to clarify this important question.

Functional significance of core promoter diversity

What is the functional significance of the observed diversity in core promoter structure? In metazoan cells, the activity of core promoters is dependent on complex *cis*-acting regulatory DNA regions, which typically consist of multiple binding sites for transcription activators. *Cis*-acting regulatory sequences can be found promoter-proximal or in enhancers located many kilobase-pairs away from the transcription initiation site [8,9,53,54]. Activation of RNAP II transcription is typically achieved by the concerted action of multiple gene-specific transcription activators and in the context of a complex network of positive and negative cofactors [2,3].

Several studies have shown that the core promoter sequence context can significantly influence the 'responsiveness' of a given gene to gene-specific DNA-binding activators and repressors. The earliest investigations into the role of the TATA box element in gene activation revealed that different TATA box sequences respond differentially to activators [55–57]. For example, it was shown that the human Hsp70 (heat-shock protein 70) promoter becomes unresponsive to E1A when its natural TATA box is substituted by the SV40 (simian virus 40) TATA element [55]. Similarly, the response of the human myoglobin gene to its natural MSE (muscle-specific enhancer) is lost when its natural TATA element is replaced by the SV40 TATA box; conversely, the SV40 promoter can be made MSE-responsive by replacing the SV40 TATA box with the myoglobin TATA box [56]. Later studies investigated how the presence or absence of different core promoter elements (e.g. TATA box and INR element) affects activator responses. For example, the transcription activator c-Fos preferentially activates transcription from TATA-containing core promoters [58], while the Ets family member Elf-1 exhibits a preference for INR-containing core promoters [59]. Optimal transcription induction by the papillomavirus activator E2 [60] and by the strong artificial activator GAL4-VP16 [61] requires both TATA and INR elements in the target promoter. Interestingly, core promoter selectivity is not restricted to transcription activation, but is also observed in transcription repression. For example, p53 has been reported to repress transcription from promoters containing a consensus TATA motif, whereas promoters containing an INR element instead of a TATA box were resistant to repression by p53 [62]. Finally, recent studies in *Drosophila* have provided compelling evidence that core promoter structure plays an important role in the selectivity of enhancers for their target genes. These studies have identified enhancers that specifically stimulate transcription either from TATA-containing core promoters lacking a DPE or from TATA-less promoters containing a DPE [63,64].

What is the molecular basis for the core promoter-specific function of activators or repressors? A simple explanation is provided by the fact that transcription initiation nucleoprotein complexes that are assembled at different core promoters are not invariant, but differ in polypeptide composition and/or topology.

First, PIC formation at the core promoter region involves substantial changes in DNA conformation [65,66], which are core promoter sequence-dependent. Sequence-specific binding of TBP induces bending of the TATA DNA, and the magnitude of DNA bending within the TBP–TATA nucleoprotein

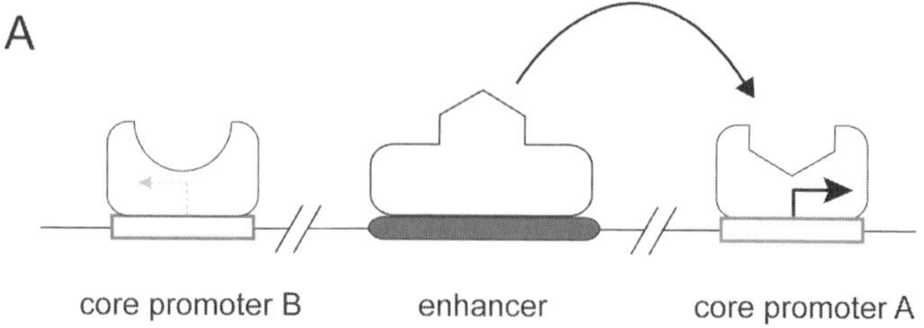

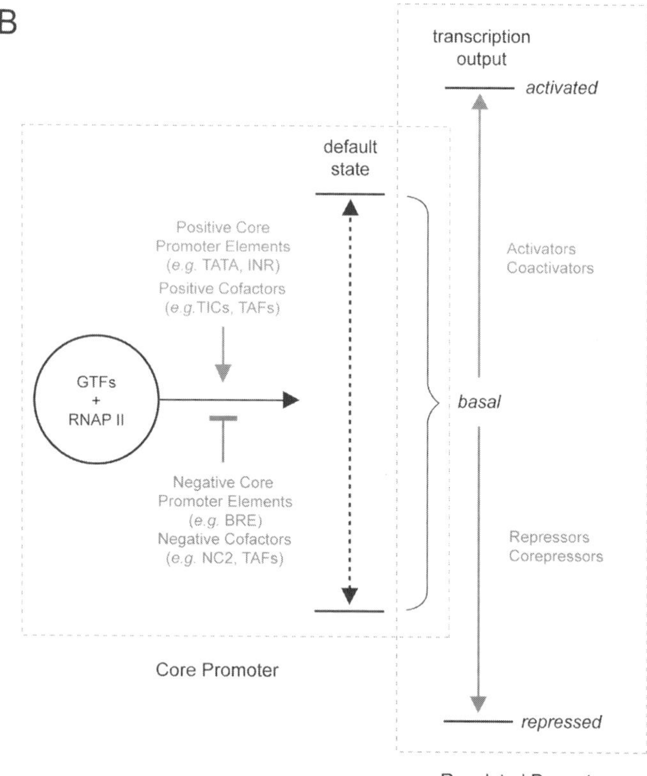

Figure 2 Models for core promoter-specific activator function. (**A**) Core promoter-specific differences in polypeptide composition and/or topology may affect the ability of transcription initiation complexes to interact with enhancer-bound activators. (**B**) A complex network of interactions between positive-acting (e.g. TATA box, INR element) and negative-acting (e.g. BRE) core promoter sequence elements, components of the core RNAP II transcription machinery (GTFs+RNAP II), and positive (e.g. TICs) and negative (e.g. NC2) cofactors affect transcription initiation complex structure and/or topology, and set a 'default state', which can be observed as basal core promoter activity *in vitro*. The default state determines the amplitude of transcription induction (or repression) and defines the requirement for specific activator functions.

complex differs significantly between different TATA box sequences [48–51]. Furthermore, TAF subunits of TFIID interact with core promoter regions downstream of the TATA box in a sequence-dependent manner and can induce nucleosome-like DNA wrapping [32,35–37,39]. Interestingly, TAF–DNA interactions downstream of the transcription start site can also influence the way TFIID engages in DNA interactions at the TATA box region [67]. Finally, TFIIB has been shown to engage in DNA-sequence-specific promoter contacts, both directly upstream of the TATA box region with the BRE [11,12] and downstream the TATA box region at or close to positions −17 and −20 [27].

Secondly, core promoter-selective transcription mediated by core promoter sequence elements other than the TATA box (e.g. INR element, DPE) cannot be recapitulated *in vitro* using the full set of basal GTFs (TFIIA, TFIIB, TFIID, TFIIE, TFIIF and TFIIH) and RNAP II. INR function requires TIC activity [43], and a recent report suggested that DPE function requires the co-activator PC4 and the protein kinase CK2 as cofactors [68]. In addition, metazoans express cell-type-specific paralogues of key GTFs. Paralogues identified to date include several TBP-related factors, a testis-specific TFIIAα/β-like factor, and tissue-specific TAFs [69–72]. It seems plausible that these specialized core components of the basal transcription apparatus contribute to core promoter-selective transcription mechanisms.

Differences in the composition and/or overall topology of transcription initiation nucleoprotein complexes could affect enhancer activity by: (i) affecting the ability to interact with enhancer-bound activators or their cognate co-activators [56] (Figure 2A), or (ii) differentially affecting activator function.

In human nuclear extracts, activator-independent (basal) transcription levels differ greatly between individual core promoters. For example, the TATA- and INR element-containing Ad2ML promoter supports up to two orders of magnitude higher basal transcription levels than the human Hsp70 promoter, which contains a consensus TATA element but lacks an INR. Consistent with this observation, introduction of a functional INR element into a TATA-containing model promoter or introduction of a consensus TATA box into a TATA-less INR-dependent promoter greatly increases basal promoter activity [13,33].

These core promoter-specific differences in basal transcription activity can be attributed to a complex functional interplay between core promoter sequence elements, positive- and negative-acting cofactors and the core RNAP II transcription machinery (Figure 2B). For example, TAFs in TFIID have been shown to either stimulate or repress the activity of TATA-containing promoters in a core promoter sequence-dependent manner [73,74]. Positive-acting TIC activities, in conjunction with TAFs, stimulate transcription from INR-containing promoters [43]. NC2, a ubiquitous repressor of RNAP II transcription, was shown previously to differentially repress transcription from different TATA box-containing core promoters [75]. Furthermore, we discovered recently that a functional INR element renders TATA-containing promoters resistant to repression by NC2. Importantly, INR-mediated resistance to NC2 requires TIC activity and TFIID TAFs (B. Malecová, P. Gross, M. Guittaut, S. Yavuz, and T. Oelgeschläger, unpublished work).

These observations support a model in which the core promoter sequence context sets a 'default state' of promoter activity by affecting the structure and function of the transcription initiation complex. The 'default state' determines the amplitude of transcription induction and defines the requirements for specific activator functions (Figure 2B). For example, an activator might strongly stimulate transcription from TATA-containing core promoters through reversal of repression by NC2. In the presence of an INR element and TAF and TIC cofactors, TATA-mediated transcription initiation becomes resistant to NC2 repression. This results in an elevated 'default state', which reduces the amplitude of induction. At the same time, the presence of an INR element eliminates the need for activator-mediated reversal of NC2 repression.

We speculate that core promoter regions co-evolved with their *cis*-acting regulatory sequences to establish specific regulatory requirements for the expression of different protein-encoding genes. It will be of interest to investigate whether classes of genes that are co-regulated by different activators or different combinations of activators share similarities in core promoter architecture and/or function.

Outlook

Recent observations corroborate the view that the core promoter architecture plays an important role in the combinatorial regulation of RNAP II transcription in metazoan cells. However, the complexity and diversity in core promoter structure remains to be fully appreciated. Novel core promoter elements have been identified recently, and it is anticipated that additional elements will be discovered in the future. TATA box-mediated transcription initiation has been investigated in great detail, but the function of other core promoter sequence elements remains poorly understood. Reconstitution of core promoter-specific transcription directed by elements other than the TATA box with purified factors has not yet been achieved. A major focus of current and future investigations must be the identification of required core promoter sequence-specific cofactors, which is prerequisite for investigations into the precise molecular mechanisms underlying core-promoter selective transcription.

References

1. Lemon, B. and Tjian, R. (2000) Genes Dev. **14**, 2551–2569
2. Roeder, R.G. (2005) FEBS Lett. **579**, 909–915
3. Roeder, R.G. (1998) Cold Spring Harbor Symp. Quant. Biol. **63**, 201–218
4. Hahn, S. (2004) Nat. Struct. Mol. Biol. **11**, 394–403
5. Orphanides, G., Lagrange, T. and Reinberg, D. (1996) Genes Dev. **10**, 2657–2683
6. Roeder, R.G. (1996) Trends Biochem. Sci. **21**, 327–335
7. Breathnach, R. and Chambon, P. (1981) Annu. Rev. Biochem. **50**, 349–383
8. Butler, J.E.F. and Kadonaga, J.T. (2002) Genes Dev. **16**, 2583–2592
9. Smale, S.T. (2001) Genes Dev. **15**, 2503–2508
10. Smale, S.T. and Kadonaga, J.T. (2003) Annu. Rev. Biochem. **72**, 449–479
11. Lagrange, T., Kapanidis, A.N., Tang, H., Reinberg, D. and Ebright, R.H. (1998) Genes Dev. **12**, 34–44
12. Evans, R., Fairley, J.A. and Roberts, S.G.E. (2001) Genes Dev. **15**, 2945–2949

13. Smale, S.T. and Baltimore, D. (1989) Cell **57**, 103–113
14. O'Shea-Greenfield, A. and Smale, S.T. (1992) J. Biol. Chem. **267**, 6450
15. Burke, T.W., Willy, P.J., Kutach, A.K., Butler, J.E. and Kadonaga, J.T. (1998) Cold Spring Harbor Symp. Quant. Biol. **63**, 75–82
16. Kadonaga, J.T. (2002) Exp. Mol. Med. **34**, 259–264
17. Lim, C.Y., Santoso, B., Boulay, T., Dong, E., Ohler, U. and Kadonaga, J.T. (2004) Genes Dev. **18**, 1606–1617
18. Kutach, A.K. and Kadonaga, J.T. (2000) Mol. Cell. Biol. **20**, 4754–4764
19. Ohler, U., Liao, G.C., Niemann, H. and Rubin, G.M. (2002) Genome Biol. **3**, research0087.1–0087.12
20. Suzuki, Y., Tsunoda, T., Sese, J., Taira, H., Mizushima-Sugano, J., Hata, H., Ota, T., Isogai, T., Tanaka, T., Nakamura, Y. et al. (2001) Genome Res. **11**, 677–684
21. Bajic, V.B., Choudhary, V. and Hock, C.K. (2004) In Silico Biol. **4**, 109–125
22. Gershenzon, N.I. and Ioshikhes, I.P. (2005) Bioinformatics **21**, 1295–1300
23. Praz, V., Perier, R., Bonnard, C. and Bucher, P. (2002) Nucleic Acids Res. **30**, 322–324
24. Schmid, C.D., Praz, V., Delorenzi, M., Perier, R. and Bucher, P. (2004) Nucleic Acids Res. **32** (Database issue), D82–85
25. Suzuki, Y., Yamashita, R., Nakai, K. and Sugano, S. (2002) Nucleic Acids Res. **30**, 328–331
26. Suzuki, Y., Yamashita, R., Sugano, S. and Nakai, K. (2004) Nucleic Acids Res. **32** (Database issue), D78–D81
27. Fairley, J.A., Evans, R., Hawkes, N.A. and Roberts, S.G.E. (2002) Mol. Cell. Biol. **22**, 6697–6705
28. Kim, J.L., Nikolov, D.B. and Burley, S.K. (1993) Nature (London) **365**, 520–527
29. Kim, Y., Geiger, J.H., Hahn, S. and Sigler, P.B. (1993) Nature (London) **365**, 512–520
30. Nikolov, D.B., Chen, H., Halay, E.D., Usheva, A.A., Hisatake, K., Lee, D.K., Roeder, R.G. and Burley, S.K. (1995) Nature (London) **377**, 119–128
31. Smale, S.T., Schmidt, M.C., Berk, A.J. and Baltimore, D. (1990) Proc. Natl. Acad. Sci. U.S.A. **87**, 4509–4513
32. Kaufmann, J. and Smale, S.T. (1994) Genes Dev. **8**, 821–829
33. Martinez, E., Chiang, C.-M., Ge, H. and Roeder, R.G. (1994) EMBO J. **13**, 3115–3126
34. Parker, C.S. and Topol, J. (1984) Cell **36**, 357–369
35. Sawadogo, M. and Roeder, R.G. (1985) Cell **43**, 165–175
36. Nakajima, N., Horikoshi, M. and Roeder, R.G. (1988) Mol. Cell. Biol. **8**, 4028–4040
37. Wang, J.C. and Van Dyke, M.W. (1993) Biochim. Biophys. Acta **1216**, 73–80
38. Purnell, B.A., Emanuel, P.A. and Gilmour, D.S. (1994) Genes Dev. **8**, 830–842
39. Oelgeschläger, T., Chiang, C.-M. and Roeder, R.G. (1996) Nature (London) **382**, 735–738
40. Verrijzer, C.P., Yokomori, K., Chen, J.-L. and Tjian, R. (1994) Science **264**, 933–941
41. Verrijzer, C.P., Chen, J.-L., Yokomori, K. and Tjian, R. (1995) Cell **81**, 1115–1125
42. Chalkley, G.E. and Verrijzer, C.P. (1999) EMBO J. **18**, 4835–4845
43. Martinez, E., Ge, H., Tao, Y., Yuan, C.-X., Palhan, V. and Roeder, R.G. (1998) Mol. Cell. Biol. **18**, 6571–6583
44. Martinez, E., Zhou, Q., L'Etoile, N.D., Oelgeschläger, T., Berk, A.J. and Roeder, R.G. (1995) Proc. Natl. Acad. Sci. U.S.A. **92**, 11864–11868
45. Hoopes, B.C., LeBlanc, J.F. and Hawley, D.K. (1992) J. Biol. Chem. **267**, 11539–11547
46. Parvin, J.D., McCormick, R.J., Sharp, P.A. and Fisher, D.E. (1995) Nature (London) **373**, 724–727
47. Zhao, X. and Herr, W. (2002) Cell **108**, 615–627
48. Juo, Z.S., Chiu, T.K., Lieberman, P.M., Baikalov, I., Berk, A.J. and Dickerson, R.E. (1996) J. Mol. Biol. **261**, 239–254
49. Starr, D.B., Hoopes, B.C. and Hawley, D.K. (1995) J. Mol. Biol. **250**, 434–446
50. Wu, J., Parkhurst, K.M., Powell, R.M., Brenowitz, M. and Parkhurst, L.J. (2001) J. Biol. Chem. **276**, 14614–14622
51. Powell, R.M., Parkhurst, K.M. and Parkhurst, L.J. (2002) J. Biol. Chem. **277**, 7776–7784
52. Fukue, Y., Sumida, N., Nishikawa, J. and Ohyama, T. (2004) Nucleic Acids Res. **32**, 5834–5840
53. Blackwood, E.M. and Kadonaga, J.T. (1998) Science **281**, 60–63
54. Merika, M. and Thanos, D. (2001) Curr. Opin. Genet. Dev. **11**, 205–208
55. Simon, M.C., Fisch, T.M., Benecke, B.J., Nevins, J.R. and Heintz, N. (1988) Cell **52**, 723–729

56. Wefald, F.C., Devlin, B.H. and Williams, R.S. (1990) Nature (London) **344**, 260–262
57. Taylor, I.C.A. and Kingston, R.E. (1990) Mol. Cell. Biol. **10**, 165–175
58. Metz, R., Bannister, A.J., Sutherland, J.A., Hagemeier, C., O'Rourke, E.C., Cook, A., Bravo, R. and Kouzarides, T. (1994) Mol. Cell. Biol. **14**, 6021–6029
59. Ernst, P., Hahm, K., Trinh, L., Davis, J.N., Roussel, M.F., Turck, C.W. and Smale, S.T. (1996) Mol. Cell. Biol. **16**, 6121–6131
60. Ham, J., Steger, G. and Yaniv, M. (1994) EMBO J. **13**, 147–157
61. Emami, K.H., Navarre, W.W. and Smale, S.T. (1995) Mol. Cell. Biol. **15**, 5906–5916
62. Mack, D.H., Vartikar, J., Pipas, J.M. and Laimins, L.A. (1993) Nature (London) **363**, 281–283
63. Ohtsuki, S., Levine, M. and Cai, H.N. (1998) Genes Dev. **12**, 547–556
64. Butler, J.E.F. and Kadonaga, J.T. (2001) Genes Dev. **15**, 2515–2519
65. Nikolov, D.B. and Burley, S.K. (1997) Proc. Natl. Acad. Sci. U.S.A. **94**, 15–22
66. Coulombe, B. and Burton, Z.F. (1999) Microbiol. Mol. Biol. Rev. **63**, 457–478
67. Knutson, A., Castaño, E., Oelgeschläger, T., Roeder, R.G. and Westin, G. (2000) J. Biol. Chem. **275**, 14190–14197
68. Lewis, B.A., Sims, III, R.J., Lane, W.S. and Reinberg, D. (2005) Mol. Cell **18**, 471–481
69. Müller, F. and Tora, L. (2004) EMBO J. **23**, 2–8
70. Hochheimer, A. and Tjian, R. (2003) Genes Dev. **17**, 1309–1320
71. Veenstra, G.J. and Wolffe, A.P. (2001) Trends Biochem. Sci. **26**, 665–671
72. Kimmins, S., Kotaja, N., Davidson, I. and Sassone-Corsi, P. (2004) Reproduction **128**, 5–12
73. Oelgeschläger, T., Tao, Y., Kang, Y.K. and Roeder, R.G. (1998) Mol. Cell **1**, 925–931
74. Guermah, M., Tao, Y. and Roeder, R.G. (2001) Mol. Cell. Biol. **21**, 6882–6894
75. Kim, J., Parvin, J.D., Shykind, B.M. and Sharp, P.A. (1996) J. Biol. Chem. **271**, 18405–18412

Subject index